FILTER BANK TRANSCEIVERS FOR OFDM AND DMT SYSTEMS

Providing key background material together with advanced topics, this self-contained book is written in an easy-to-read style and is ideal for newcomers to multicarrier systems.

Early chapters provide a review of basic digital communication, starting from the equivalent discrete-time channel and including a detailed review of the MMSE receiver. Later chapters then provide extensive performance analysis of OFDM and DMT systems, with discussions of many practical issues such as implementation and power spectrum considerations. Throughout, theoretical analysis is presented alongside practical design considerations, whilst the filter bank transceiver representation of OFDM and DMT systems opens up possibilities for further optimization such as minimum bit error rate, minimum transmission power, and higher spectral efficiency.

With plenty of insightful real-world examples and carefully designed end-of-chapter problems, this is an ideal single-semester textbook for senior undergraduate and graduate students, as well as a self-study guide for researchers and professional engineers.

YUAN-PEI LIN is a Professor in Electrical Engineering at the National Chiao Tung University, Hsinchu, Taiwan. She is a recipient of the Ta-You Wu Memorial Award, the Chinese Institute of Electrical Engineering's Outstanding Youth Electrical Engineer Award, and of the Chinese Automatic Control Society's Young Engineer in Automatic Control Award.

SEE-MAY PHOONG is a Professor in the Graduate Institute of Communication Engineering and the Department of Electrical Engineering at the National Taiwan University (NTU). He is a recipient of the Charles H. Wilts Prize for outstanding independent doctoral research in electrical engineering at the California Institute of Technology, and the Chinese Institute of Electrical Engineering's Outstanding Youth Electrical Engineer Award.

P. P. VAIDYANATHAN is a Professor in Electrical Engineering at the California Institute of Technology, where he has been a faculty member since 1983. He is an IEEE Fellow and has authored over 400 technical papers, four books, and many invited chapters in leading journals, conferences, and handbooks. He was a recipient of the Award for Excellence in Teaching at the California Institute of Technology three times, and he has received numerous other awards including the F. E. Terman Award of the American Society for Engineering Education and the Technical Achievement Award of the IEEE Signal Processing Society.

FILTER BANK TRANSCEIVERS FOR OFDM AND DMT SYSTEMS

YUAN-PEI LIN
National Chiao Tung University, Taiwan

SEE-MAY PHOONG
National Taiwan University

P. P. VAIDYANATHAN
California Institute of Technology

CAMBRIDGE
UNIVERSITY PRESS

CAMBRIDGE
UNIVERSITY PRESS

University Printing House, Cambridge CB2 8BS, United Kingdom

One Liberty Plaza, 20th Floor, New York, NY 10006, USA

477 Williamstown Road, Port Melbourne, VIC 3207, Australia

314-321, 3rd Floor, Plot 3, Splendor Forum, Jasola District Centre, New Delhi - 110025, India

79 Anson Road, #06-04/06, Singapore 079906

Cambridge University Press is part of the University of Cambridge.

It furthers the University's mission by disseminating knowledge in the pursuit of education, learning and research at the highest international levels of excellence.

www.cambridge.org
Information on this title: www.cambridge.org/9781107002739

© Cambridge University Press 2011

First published 2011

A catalogue record for this publication is available from the British Library

ISBN 978-1-107-00273-9 Hardback

To our families
— Yuan-Pei Lin and See-May Phoong

To Usha, Vikram, Sagar, and my parents
— P. P. Vaidyanathan

Contents

Preface

Recent years have seen the great success of OFDM (orthogonal frequency division multiplexing) and DMT (discrete multitone) transceivers in many applications. The OFDM system has found many applications in wireless communications. It has been adopted in IEEE 802.11 for wireless local area networks, DAB for digital audio broadcasting, and DVB for digital video broadcasting. The DMT system is the enabling technology for high-speed transmission over digital subscriber lines. It is used in ADSL (asymmetric digital subscriber lines) and VDSL (very-high-speed digital subscriber lines). The OFDM and DMT systems are both examples of DFT transceivers that employ redundant guard intervals for equalization. Having a guard interval can greatly simplify the task of equalization at the receiver and it is now one of the most effective approaches for channel equalization. In this book we will study the OFDM and DMT under the framework of filter bank transceivers. Under such a framework, there are numerous possible extensions. The freedom in the filter bank transceivers can be exploited to better the systems for various design criteria. For example, transceivers can be optimized for minimum bit error rate, for minimum transmission power, or for higher spectral efficiency. We will explore all these possible optimization problems in this book.

The first three chapters describe the major building blocks relevant for the discussion of signal processing for communication and give the tools useful for solving problems in this area. Chapters 4–5 introduce the multirate building blocks and filter bank transceivers, and the basic idea of guard intervals for channel equalization. Chapter 6 gives a detailed discussion of OFDM and DMT systems. Chapters 7–10 consider the design of filter bank transceivers for different criteria and channel environments. A detailed outline is given at the end of Chapter 1. This book has been used as a textbook for a first-year graduate course at National Chiao Tung University, Taiwan, and at National Taiwan University. Most of the chapters can be covered in 16–18 weeks. Homework problems are given for Chapters 2–10.

It is our pleasure to thank our families for the patience and support during all phases of this time-consuming project. We would like to thank our universities, National Chiao Tung University and National Taiwan University, and the National Science Council of Taiwan for their generous support during the writing of this book. We would also like to thank our students Chien-Chang Li, Chun-Lin Yang, Chen-Chi Lo, and Kuo-Tai Chiu for generating some of the plots. PPV wishes to acknowledge the California Institute of Technology, the National Science Foundation (USA), and the Office of Naval Research (USA), for all the support and encouragement.

1

Introduction

The goal of a communication system is to transmit information efficiently and accurately to another location. In the case of digital communications, the information is a sequence of "ones" and "zeros" called the bit stream. The transmitter takes in the bit stream and generates an information-bearing continuous-time signal $x_a(t)$, as in Fig. 1.1. When the signal propagates through the channel, such as wirelines, atmosphere, etc., distortion is inevitably introduced into the transmitted signal $x_a(t)$. As a result, the received signal $r_a(t)$ at the receiver is in general different from the transmitted signal $x_a(t)$. The task of the receiver is to mitigate the distortion and reproduce a bit stream with as few errors as possible.

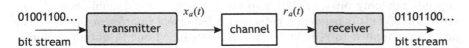

Figure 1.1. Digital communication system.

A digital communication system in general consists of many building blocks. Figure 1.2 shows a block diagram consisting of the major building blocks that are relevant to the topic of signal processing for communications. At the transmitter, we have a sequence of bits to be sent to the receiver. The *bits-to-symbol mapping* block takes several bits of input and maps the bits to a real or complex *modulation symbol* $s(n)$. Some processing may be applied to these symbols and the discrete-time output $x(n)$ is then converted to a continuous-time signal $x_a(t)$. The transmitted signal $x_a(t)$ propagates through the channel. At the receiver, the received signal $r_a(t)$ is converted to a discrete-time signal $r(n)$. Usually some signal processing is applied to $r(n)$ before the receiver makes a decision on the transmitted symbols and obtains $\widehat{s}(n)$ (*symbol detection*). The *symbol-to-bits mapping* block maps the symbols $\widehat{s}(n)$ back to bit stream. The reader can find relevant background material in [50, 67, 120, 137].

When a signal propagates through the channel, distortion is invariably

1

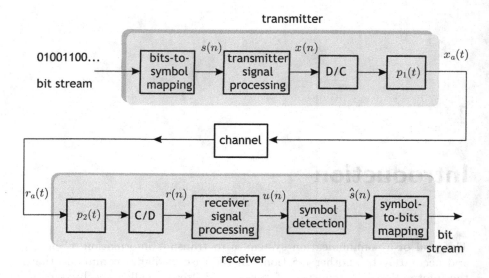

Figure 1.2. Simple block diagram for a digital communication system.

introduced to the transmitted signal. In addition to channel noise, there is also interference from other symbols. At time n the received signal $r(n)$ depends not only on $s(n)$, but also on past transmitted symbols $s(n-1)$, $s(n-2), \ldots$ This dependency is termed *intersymbol interference* (ISI). The processing applied to $r(n)$ at the receiver is carried out to obtain estimates of the transmitted symbols before symbol detection. The process is generally known as equalization and the signal processing block is called an equalizer. When the receiver can perfectly regenerate the transmitted symbols $s(n)$ in the absence of channel noise, we say the equalization is *zero-forcing*. In many applications, the transmitter also helps with equalization. In this case some signal processing is applied to the symbols $s(n)$, and the resulting output $x(n)$ is transmitted as shown in Fig. 1.2.

One way that the transmitter can greatly ease the task of equalization at the receiver is to divide the transmitted signal into blocks and add redundant samples, also called a guard interval, to each block. Figure 1.3 shows two examples of guard intervals called *zero padding* and *cyclic prefix*. In the zero-padding scheme, the guard intervals consist of "zeros." With cyclic prefix, the last few samples of each block are copied and inserted at the beginning of the block as shown in the figure. The guard interval acts as a buffer between consecutive blocks. If the guard interval is sufficiently long, the interblock interference (IBI) can be avoided or can be later removed at the receiver by discarding the redundant samples. When there is no IBI, interference comes only from the same block. In this case, intrablock interference can be canceled easily using matrix operations.

The most notable example of systems that use cyclic prefix as a guard interval is the DFT (Discrete Fourier Transform)-based transceiver shown in

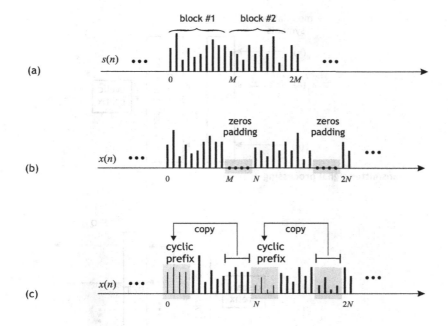

Figure 1.3. Two examples of guard intervals. (a) A signal $s(n)$ with samples divided into blocks; (b) the sequence $x(n)$ after zeros are padded to each block of $s(n)$; (c) the sequence $x(n)$ after a cyclic prefix is inserted in each block of $s(n)$.

Fig. 1.4. The signal processing at the transmitter applies IDFT (Inverse Discrete Fourier Transform) to the input block of modulation symbols and adds a cyclic prefix to the IDFT outputs. The receiver discards the prefix and performs a DFT of each block. Due to the combination of cyclic prefix and IDFT/DFT operations, zero-forcing equalization can be achieved by only a set of simple scalars called frequency domain equalizers (FEQs). When ISI is canceled, the overall system from the transmitter inputs to the receiver outputs is equivalent to a set of parallel subchannels as shown in Fig. 1.5. In general the subchannels have different noise variances. If the information of the subchannel noise variances is available to the transmitter, the symbols $s_i(n)$ can be further designed to better the performance. For example, the symbols of different subchannels can carry different numbers of bits (bit loading) [23], and the power of the symbols can also be different (power loading). The transmitter can optimize bit loading and power loading to maximize the transmission rate [24].

The cyclic-prefixed DFT-based system is widely used in both wired and wireless communication systems. It is generally called an OFDM (orthogonal frequency division multiplexing) system [27] in wireless transmission and a DMT (discrete multitone) system [24] in wired DSL (digital subscriber lines) transmission. For wireless applications, the channel state information is usually not available to the transmitter. The transmitter is typically in-

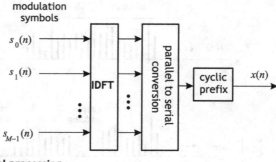

transmitter signal processing

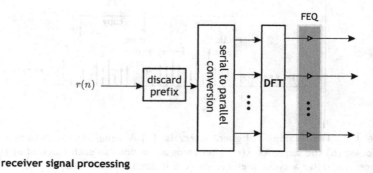

receiver signal processing

Figure 1.4. DFT-based transceiver with cyclic prefix added as a guard interval.

dependent of the channel and there is no bit or power loading. Having a channel-independent transmitter is also a very useful feature for broadcasting systems. For broadcast applications, there is only one transmitter and there are many receivers, each with a different transmission path. It is impossible for the transmitter to optimize for different channels simultaneously. In OFDM systems for wireless applications, usually without bit and power allocation, the transmitters have the desirable channel-independence property. The channel-dependent part of the transceiver is the set of FEQ coefficients at the receiver. In DMT systems for wired DSL applications, signals are transmitted over copper lines. The channel does not vary rapidly. This gives the receiver time to send back the channel state information to the transmitter. The transmitter can then allocate bits and power to the subchannels to maximize the transmission rate. More details on DSL transmission can be found in [14, 122, 144, 145].

Both the OFDM and DMT systems have been shown to be very useful transmission systems. The DMT system was adopted in standards for ADSL (asymmetric digital subscriber lines) [7] and VDSL (very-high-speed digital subscriber lines) [8] transmission. The OFDM systems have been adopted in standards for digital audio broadcasting [39], digital video broadcasting

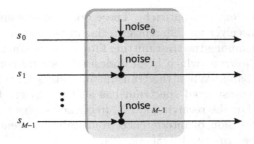

Figure 1.5. Equivalent parallel subchannels.

[40], wireless local area networks [54], and broadband wireless access [55]. A variation of the cyclic-prefixed DFT-based transceiver is the so-called cyclic-prefixed single-carrier (SC-CP) system [129]. The modulation symbols are directly sent out after a cyclic prefix is added. As in the OFDM system, the redundant cyclic prefix greatly facilitates equalization at the receiver. The SC-CP system is part of the broadband wireless access standard [55].

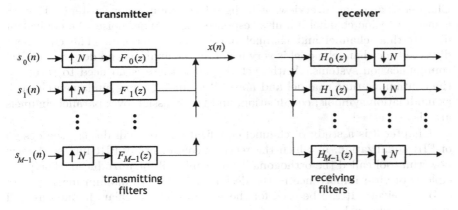

Figure 1.6. Filter bank transceiver, in which only the transmitter signal processing and receiver signal processing parts are shown.

The insertion and removal of redundant samples can be represented using multirate building blocks. (Definitions of multirate building blocks will be given in Chapter 4.) Based on the multirate formulation the DFT-based system can also be viewed as a discrete-time *filter bank transceiver* (Fig. 1.6), or a *transmultiplexer*. The transmitter and receiver each consists of a bank of discrete-time filters. Such a formulation lends itself to the frequency domain analysis of the transceiver. For example, for the transmitter side it offers additional insight on the effect of individual transmitting filters on the transmitted spectrum. For the receiver side, we can analyze the subchannel noise using a frequency domain approach. These observations are very useful for designing the transceiver for different criteria. In DSL applications, good frequency separation among the transmitting filters is important for reducing the so-called *spectral leakage*, which is an undesired spectral component outside the transmission band. When the transmitting filters have higher stopband attenuation, the transmitted spectrum has a faster spectral rolloff and less spectral leakage. For the receiving filters, frequency separation is also important for the suppression of interference from radio frequency signals which share the same spectrum with DSL signals.

The filter bank framework is also useful for designing transceivers with better spectral efficiency. In the DFT-based transceiver, a long guard interval is required if the channel impulse response is long. The use of a long redundant guard interval decreases the spectral efficiency, so we would like the guard interval to be as short as possible. On the other hand, it is desirable that the guard interval be long enough so that FIR equalization is possible. The filter bank transceiver can be used to introduce guard intervals of a very general form. In most cases, zero-forcing equalization can be achieved using a guard interval much shorter than what is needed in the DFT-based transceiver.

Outline

Chapter 2 gives an overview of a digital communication system. From a continuous-time channel impulse response and channel noise, the equivalent discrete-time channel and channel noise will be derived. The equivalent discrete-time channel model is very useful in the analysis and design of digital communication systems. With such a model, there is no need to revert to the continuous-time channel and noise. Terminology and fundamentals such as modulation symbols, equalization, and transmission over parallel channels are also reviewed.

Chapter 3 is a study of channel equalization. We will discuss the design of FIR equalization, in which the receiver contains only FIR filters. A very powerful tool called the orthogonality principle will be introduced. The principle is of vital importance in the design of MMSE (minimum mean square error) receivers. It can be used for the equalization of scalar channels as well as parallel channels.

Chapter 4 gives the basics of multirate signal processing. Multirate building blocks are introduced. The operations of blocking and unblocking that appear frequently in digital transmission are described using multirate building blocks. In addition, polyphase decomposition of filters is reviewed. Based on the decomposition, the polyphase representation of filter banks can be derived and efficient polyphase implementation can be obtained. Readers who are familiar with multirate systems and filter banks can skip this chapter.

Chapter 5 formulates some modern digital communication systems using multirate building blocks. The filter bank transceiver is introduced and conditions on the transmitter and receiver for zero ISI are derived. Using the multirate formulation, redundant samples can be inserted in the transmitted signal. Two types of redundant samples are discussed in detail: cyclic prefix and zero padding. The matrix form representations of these systems are used frequently in the discussions of applications in later chapters.

Chapter 6 is devoted to the study of some useful DFT-based transceivers. The OFDM, DMT, and SC-CP systems will be presented and the performance will be analyzed. The corresponding filter bank representation will also be derived. These transceivers have found many practical applications due to the fact that they can be implemented efficiently using fast algorithms.

Chapter 7 deals with the design of optimal transceivers when the transmitter does not have the channel state information, which is usually the case for wireless applications. As the transmitter does not have the channel knowledge, there is no bit/power allocation. We consider the design of minimum bit error rate (BER) transceivers by adding a unitary precoder at the transmitter and a post-coder at the receiver of the OFDM system. We will see that the derivation of the minimum BER transceiver nicely ties the OFDM and the SC-CP systems together.

Chapter 8 deals with the design of optimal transceivers when the channel state information is available to the transmitter. In addition to bit and power allocation, the transmitter and receiver can be jointly optimized. For a given error rate and target transmission rate, the transceiver will be designed to minimize the transmission power.

Chapter 9 describes a method to improve the frequency separation among the subchannels for the DFT-based transceivers. Some short FIR filters called subfilters are introduced in the subchannels to enhance the stopband attenuation of the transmitting and receiving filters. By using a slightly longer guard interval, we can include the subfilters without changing the ISI-free property. When subfilters are added to the receiver, the transmission rate can be increased considerably in the presence of narrowband RFI (radio frequency interference). For the transmitter side, the subfilters can improve the spectral rolloff of the transmitted spectrum while having little effect on the transmission rate.

Chapter 10 is a study of minimum redundancy for FIR equalization. For a given channel, we consider the minimum redundancy that is required to ensure the existence of FIR equalizers. We will see that the answer is directly tied to what we call the congruous zeros of the channel. The minimum redundancy can be determined by inspection once the zeros of the channel are known.

1.1 Notations

- Boldfaced lower case letters represent vectors and boldfaced upper case letters are reserved for matrices. The notation $\mathbf{A}^T$ denotes the transpose of $\mathbf{A}$, and $\mathbf{A}^\dagger$ denotes the transpose-conjugate of $\mathbf{A}$.

- The function $E[y]$ denotes the expected value of a random variable y.

- The notation $\mathbf{I}_M$ is used to represent the $M \times M$ identity matrix and $\mathbf{0}_{mn}$ denotes an $m \times n$ matrix whose entries are all equal to zero. The subscript is omitted when the size of the matrix is clear from the context.

- The determinant of a square matrix $\mathbf{A}$ is denoted as $\det(\mathbf{A})$. The notation $\text{diag}[\lambda_0 \quad \lambda_1 \quad \ldots \quad \lambda_{M-1}]$ denotes an $M \times M$ diagonal matrix with the kth diagonal element equal to λ_k.

- The notation $\mathbf{W}$ is used to represent the $M \times M$ unitary DFT matrix, given by

$$[\mathbf{W}]_{kn} = \frac{1}{\sqrt{M}} e^{-j\frac{2\pi}{M}kn} \quad \text{for} \ \ 0 \le k, n \le M - 1.$$

- For a discrete-time sequence $c(n)$, the z-transform is denoted as $C(z)$ and the Fourier transform as $C(e^{j\omega})$. For a continuous-time function $x_a(t)$, the Fourier transform is denoted as $X_a(j\Omega)$.

2

Preliminaries of digital communications

In this chapter, we shall review some introductory materials that are useful for our discussion in subsequent chapters. For convenience, we reproduce in Fig. 2.1 the block diagram for digital communication systems introduced in Chapter 1.

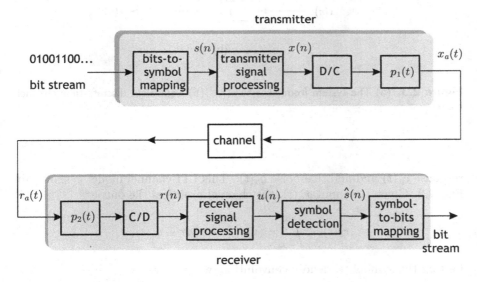

Figure 2.1. Simple block diagram for a digital communication system.

2.1 Discrete-time channel models

In the study of communication systems, the transmission channel is often modeled as a continuous-time linear time invariant (LTI) system with impulse

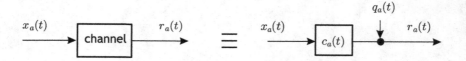

Figure 2.2. LTI channel model.

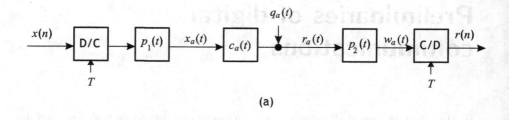

(a)

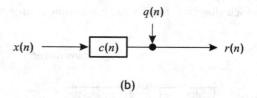

(b)

Figure 2.3. (a) The system from $x(n)$ to $r(n)$. (b) Equivalent discrete-time channel model.

response $c_a(t)$ and additive noise $q_a(t)$. This LTI channel model is shown in Fig. 2.2. Given the input $x_a(t)$, the channel produces the output

$$r_a(t) = \int_{-\infty}^{\infty} c_a(\tau)x_a(t-\tau)d\tau + q_a(t).$$

Letting the symbol '∗' denote convolution, we can write

$$r_a(t) = (x_a * c_a)(t) + q_a(t).$$

Though the channel is a continuous-time system, it is often more convenient to work directly on an equivalent discrete-time system. In many aspects of digital communications, a discrete-time channel model is often adequate and much easier to work with. In this section, we shall show that the system from $x(n)$ at the transmitter to $r(n)$ at the receiver (Fig. 2.1) is equivalent to a discrete-time LTI system.

The system from $x(n)$ to $r(n)$ is shown separately in Fig. 2.3(a). Suppose that the samples $x(n)$ are spaced apart by T seconds. The D/C converter takes the discrete-time sequence $x(n)$ and produces a continuous-time impulse train spaced apart by T:

$$\sum_n x(n)\delta_a(t - nT),$$

where $\delta_a(t)$ is the continuous-time Dirac delta function. After the impulse train passes through the transmitting pulse $p_1(t)$, we get a continuous-time signal $x_a(t)$:

$$x_a(t) = \sum_{k=-\infty}^{\infty} x(k)p_1(t - kT).$$

The signal $x_a(t)$ is transmitted through the channel. At the receiving end, the received signal is

$$r_a(t) = (x_a * c_a)(t) + q_a(t) = \sum_{k=-\infty}^{\infty} x(k)(p_1 * c_a)(t - kT) + q_a(t).$$

The received signal $r_a(t)$ is first passed through a receiving pulse $p_2(t)$, which produces

$$w_a(t) = (r_a * p_2)(t) = \sum_{k=-\infty}^{\infty} x(k)(p_1 * c_a * p_2)(t - kT) + (q_a * p_2)(t). \quad (2.1)$$

Then $w_a(t)$ is uniformly sampled every T seconds to produce the discrete-time output $r(n) = w_a(nT)$. This uniform sampling operation is denoted by the box labeled C/D. Define the effective continuous-time channel and effective noise, respectively, as follows:

$$c_e(t) = (p_1 * c_a * p_2)(t) \quad \text{and} \quad q_e(t) = (q_a * p_2)(t).$$

Then the received discrete-time signal is

$$r(n) = \sum_{k=-\infty}^{\infty} x(k)c_e(nT - kT) + q_e(nT).$$

The above expression can be rewritten as

$$r(n) = \sum_{k=-\infty}^{\infty} x(k)c(n - k) + q(n),$$

where $c(n)$ and $q(n)$ are, respectively, the **discrete-time equivalent channel and noise** given by

$$\begin{aligned} c(n) &= (p_1 * c_a * p_2)(t)\Big|_{t=nT}, \\ q(n) &= (q_a * p_2)(t)\Big|_{t=nT}. \end{aligned} \quad (2.2)$$

Thus, the system shown in Fig. 2.3(a) can be represented as in Fig. 2.3(b), which contains only discrete-time signals and systems. The transfer function of the discrete-time channel is given by

$$C(z) = \sum_{n=-\infty}^{\infty} c(n)z^{-n}.$$

Observe that $c(n)$ is the sampled version of the cascade of the transmitting pulse $p_1(t)$, the channel $c_a(t)$, and the receiving pulse $p_2(t)$. Choosing different transmitting and receiving pulses will affect the discrete-time channel. In practice, the channel is often modeled as a finite impulse response (FIR) filter. From (2.2), we can see that the channel length is inversely proportional to the sampling period T. Reducing T by one-half will double the length of $c(n)$. When the channel $C(z)$ has more than one nonzero tap, say $c(0)$ and $c(1)$, it will introduce *interference* between the received symbols. To see this, suppose that there are no "signal processing" blocks at the transmitter and receiver in Fig. 2.1, then the transmitted signal $x(n) = s(n)$. The received signal will be

$$r(n) = c(0)s(n) + c(1)s(n-1) + q(n);$$

the current symbol $s(n)$ is contaminated by the past symbol $s(n-1)$. This phenomenon is known as **intersymbol interference** (ISI). The task of symbol recovery is complicated by both the additive noise $q(n)$ and ISI.

Example 2.1 Consider a transmission system with effective continuous-time channel $c_e(t) = (p_1 * c_a * p_2)(t)$ given in the Fig. 2.4.

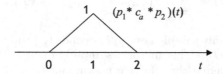

Figure 2.4. An example of $(p_1 * c_a * p_2)(t)$.

Suppose we send one sample of $x(n)$ per second, i.e. the sampling period $T = 1$. Then the discrete-time equivalent channel is $c(n) = \delta(n-1)$, a delay. The channel does not introduce ISI. When we increase our transmission rate to two samples per second, then the sampling period becomes $T = 0.5$ and the discrete-time equivalent channel is $C(z) = 0.5z^{-1} + z^{-2} + 0.5z^{-3}$. The channel becomes a three-tap FIR channel. We see that the faster we send the samples $x(n)$, the longer the FIR channel $c(n)$.   ■

Note that there is no carrier modulation in the system shown in Fig. 2.1. This is known as **baseband communication**. In wireless communications, the signal $x_a(t)$ is modulated to a carrier frequency f_c for transmission, as shown in Fig. 2.5. This is known as **passband transmission**. For passband communications, after carrier modulation the signal that is transmitted through the channel $c_a(t)$ is given by

$$v_a(t) = 2Re\{x_a(t)e^{j2\pi f_c t}\},$$

where $Re\{\bullet\}$ denotes the real part. At the receiver, the received signal in this case becomes $y_a(t) = (v_a * c_a)(t) + q_a(t)$. Carrier demodulation is performed to obtain the baseband signal[1]

$$r_a(t) = y_a(t)e^{-j2\pi f_c t}.$$

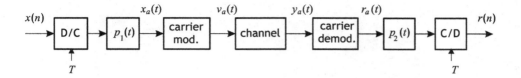

Figure 2.5. Passband communication channel.

By following a similar procedure, one can show (Problem 2.3) that the system sandwiched between the C/D and D/C converters in a passband communication system is also equivalent to a discrete-time LTI system. In this case, the equivalent discrete-time channel and noise are, respectively, given by

$$c(n) = (p_1 * \tilde{c}_a * p_2)(t)\Big|_{t=nT},$$
$$q(n) = (\tilde{q}_a * p_2)(t)\Big|_{t=nT}, \tag{2.3}$$

where $\tilde{c}_a(t) = c_a(t)e^{-j2\pi f_c t}$ and $\tilde{q}_a(t) = q_a(t)e^{-j2\pi f_c t}$. From this relation, one can clearly see the effect of carrier modulation: what the transmitted signal $x_a(t)$ sees is a frequency-shifted version of the original channel $c_a(t)$ and noise $q_a(t)$. Note that both the channel impulse response $c(n)$ and the channel noise $q(n)$ are complex for passband transmission due to the term $e^{-j2\pi f_c t}$. For baseband transmission, these quantities are real.

Channel noise Throughout this book, we will assume that the channel noise $q(n)$ is a *zero-mean wide sense stationary (WSS) Gaussian* random process. For baseband transmission, $q(n)$ is real and the probability density function (pdf) of a zero-mean Gaussian noise $q(n)$ is given by

$$f_q(q) = \frac{1}{\sqrt{2\pi \mathcal{N}_0}} e^{-q^2/2\mathcal{N}_0}, \tag{2.4}$$

where $\mathcal{N}_0$ is the noise variance. Figure 2.6 shows the well-known bell-shaped Gaussian pdf for different values of $\mathcal{N}_0$, more widespread for a larger variance. For passband transmission, the channel noise $q(n)$ is in general modeled as a zero-mean *circularly symmetric complex Gaussian* random variable whose pdf is given by

$$f_q(q) = \frac{1}{\pi \mathcal{N}_0} e^{-(q_0^2 + q_1^2)/\mathcal{N}_0}, \tag{2.5}$$

where $q = q_0 + jq_1$ and $\mathcal{N}_0$ is the variance of q. The real and imaginary parts are both zero-mean Gaussian and they have equal variance: $E[q_0^2] = E[q_1^2] = \mathcal{N}_0/2$. This pdf is shown in Fig. 2.7 for $\mathcal{N}_0 = 1$.

In the following, we will describe some commonly used models of equivalent discrete-time channels. These models, though simplified, are useful for the analysis of digital communication systems. They are also frequently employed in numerical simulations to evaluate the system performance.

---

[1]The high-frequency component centered around $2f_c$ is in general eliminated by a low-pass filter in the process of carrier demodulation.

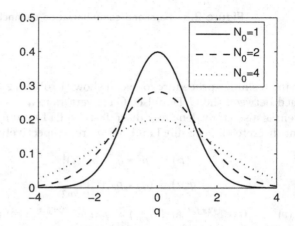

Figure 2.6. The pdf of a zero-mean real Gaussian random variable.

AGN and AWGN channels   A channel is called an AGN (additive Gaussian noise) channel when the channel satisfies the following two properties.

(1) Its channel impulse response is

$$c(n) = \delta(n) = \begin{cases} 1, & n = 0; \\ 0, & \text{otherwise.} \end{cases}$$

(2) The channel noise $q(n)$ is a Gaussian random process.

If in addition to being a Gaussian random process, $q(n)$ is also white, that is, $E\{q(n)q^*(m)\} = 0$ whenever $m \neq n$, then the channel is an AWGN (additive white Gaussian noise) channel.   When the channel is an AGN or AWGN channel, there is no ISI and the transmission error comes from the channel noise only.

FIR channels   In many applications, the channel not only introduces additive noise $q(n)$, but also distorts the transmitted signal. The channel $c(n)$ is no longer an impulse, and in general it has a causal infinite impulse response (IIR). For the purpose of analysis, the channel is often modeled as a **finite impulse response** (FIR) filter, that is

$$C(z) = \sum_{n=0}^{L} c(n)z^{-n}. \tag{2.6}$$

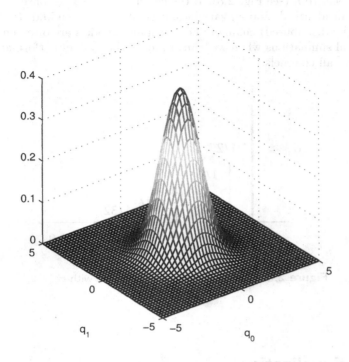

Figure 2.7. The pdf of a circularly symmetric complex Gaussian random variable.

The impulse response is nonzero only for a finite number of coefficients (or taps). The integers L and $L+1$ are, respectively, the channel order and channel length. By making L large enough, a causal IIR filter can be well approximated by an FIR filter. In the frequency domain, the magnitude response of the FIR channel $|C(e^{j\omega})|$ is not flat unless $c(n)$ has only one nonzero tap. The channel has different *gains* for different frequency components. Thus such channels are also known as **frequency-selective channels**. When $c(n)$ has only one nonzero tap, it is called **frequency-nonselective**.

Random channels with uncorrelated taps   In many situations, the exact channel impulse response may not be available, and only the statistics of the channel is known. One of the widely adopted channel models assumes that the taps are zero-mean uncorrelated random variables with known variances. In this case, $c(n)$ satisfies the following conditions:

$$
\begin{aligned}
&(1) \quad E[c(n)] = 0, \\
&(2) \quad E[c(n)c^*(n-k)] = \sigma_n^2 \delta(k).
\end{aligned}
\tag{2.7}
$$

The set of quantities $\{\sigma_n^2\}$ is called the power delay profile. We say that the channel has an exponential **power delay profile** when σ_n^2 decays exponentially with respect to n (see Fig. 2.8). If the channel impulse responses $c(n)$ are independent identical random variables, it is often called an i.i.d. (independent identically distributed) channel. These channel models are often employed in numerical simulations when we want to learn the system performance "averaged over all channels."

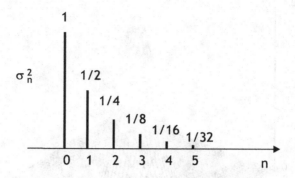

Figure 2.8. Exponential power delay profile with $\sigma_n^2 = 2^{-n}$.

2.2   Equalization

In wideband communication systems, the transmission channel often introduces some degree of intersymbol interference (ISI) to the received signal. Suppose a signal $x(n)$ is transmitted over an FIR channel with impulse response $c(n)$, for $0 \le n \le L$. Then the received signal is

$$r(n) = c(0)x(n) + c(1)x(n-1) + \cdots + c(L)x(n-L) + q(n),$$

where $q(n)$ is the channel noise. At the receiver, we would like to process $r(n)$ so that its output $\hat{x}(n)$ is "close" to $x(n-n_0)$, a delayed version of the transmitted signal $x(n)$. The integer n_0 is called the system delay. Such processing is generally known as **equalization**. There are many equalization techniques. One approach is to use an LTI filter $a(n)$ as shown in Fig. 2.9. This is also known as linear equalization and the filter $a(n)$ is called an **equalizer**. Below we look at some simple equalization techniques. In later chapters, more advanced methods for channel equalization will be explored.

Consider the FIR channel $C(z) = \sum_{n=0}^{L} c(n)z^{-n}$. When $C(z)$ is known, there are many methods to eliminate or mitigate ISI. One simple way is to use an IIR filter $A(z) = 1/C(z)$. When $r(n)$ passes through the equalizer $a(n)$, the output will be

$$\hat{x}(n) = x(n) + (q * a)(n);$$

there is no ISI. Such an equalizer is said to be **zero-forcing (ZF)** and the system is called an **ISI-free** system. Define the output error $e(n)$ as

$$e(n) = \hat{x}(n) - x(n-n_0),$$

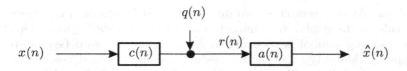

Figure 2.9. LTI equalizer.

where the system delay $n_0 = 0$ in this case. Then the error $e(n) = (q * a)(n)$ consists only of noise. In practice, the IIR zero-forcing equalizer $1/C(z)$ is not frequently used because the equalizer $1/C(z)$ will be unstable when $C(z)$ has zeros outside the unit circle. To avoid this problem, we can use an FIR equalizer $a(n)$. The output of the equalizer is

$$\widehat{x}(n) = (a * c * x)(n) + (a * q)(n).$$

There are two common ways of designing $a(n)$. One is to choose $a(n)$ such that the ISI is small in some sense. That is, we would like the convolution $(c * a)(n)$ to be as close to a delay $\delta(n - n_0)$ as possible. Another way of designing $a(n)$ is to include the effect of both channel noise and ISI. Note that when $C(z)$ has a zero $z_0 \neq 0$, the product $A(z)C(z)$ cannot be a delay z^{-n_0} for any FIR equalizer $A(z)$ because $A(z)C(z)$ will have a zero at z_0. In particular, when an FIR channel has more than one nonzero tap, an FIR equalizer can never be zero-forcing; the output error will contain both the channel noise and ISI. We will study these solutions of FIR equalizers in more detail in Chapter 3.

Signal to noise ratio (SNR)   In a digital communication system, we often measure the performance by evaluating the ratio of signal power $\mathcal{E}_x$ to the mean squared error $\sigma_e^2 = E[|\widehat{x}(n) - x(n - n_0)|^2]$. This ratio is known as the **signal to noise ratio (SNR)** and it is given by

$$\beta = \frac{\mathcal{E}_x}{\sigma_e^2}.$$

When the equalizer is not zero-forcing, the error $e(n) = \widehat{x}(n) - x(n - n_0)$ contains not only noise, but also ISI terms. In this case, β is also known as the signal to noise interference ratio (SINR), but we shall refer to it simply as SNR.

2.3   Digital modulation

In digital communication systems, the transmitted bit stream consisting of "zeros" and "ones," is often partitioned into segments of length, say, b. Each segment (codeword) is then mapped to one member in a set of 2^b real or complex numbers. This process is known as digital modulation. The real or complex numbers representing the codewords are known as **modulation**

symbols. At the receiver, based on the received information a decision will be made on the symbol transmitted. This process is called symbol detection. The resulting symbol is then decoded to a b-bit codeword (symbol-to-bits mapping). Many types of digital modulations have been developed. In the following, we will describe two widely used digital modulation schemes known as the pulse amplitude modulation (PAM) and quadrature amplitude modulation (QAM). We will analyze their performance for transmission over an AWGN channel. Note that in a baseband communication system, where the channel is real, PAM is often employed, whereas in a passband system the channel is complex, and QAM is usually employed.

2.3.1   Pulse amplitude modulation (PAM)

In b-bit pulse amplitude modulation (PAM), a codeword of b bits is mapped to a real number

$$s = \pm(2k+1)\Delta, \quad \text{where} \quad k \in \{0, 1, \ldots, 2^{b-1} - 1\}. \quad (2.8)$$

Figure 2.10 shows the possible values of a PAM symbol for $b = 2$ and $b = 3$, respectively [120]. The Gray code indicated in the figure will be explained later. Such figures are called **signal constellations**. The minimum distance between two constellation points is 2Δ. Assume that all constellation points are equiprobable. Then the signal power of a b-bit PAM constellation is given by

$$\mathcal{E}_s = E[s^2] = \frac{\Delta^2}{3}(2^{2b} - 1). \quad (2.9)$$

The signal power is proportional to the square of the minimum distance 2Δ. When the minimum distance is fixed, the signal power will increase by roughly 6 dB for every additional bit.

(a)

$-3\Delta \quad -\Delta \quad \Delta \quad 3\Delta$

(b)
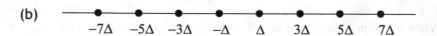

$-7\Delta \quad -5\Delta \quad -3\Delta \quad -\Delta \quad \Delta \quad 3\Delta \quad 5\Delta \quad 7\Delta$

Figure 2.10. PAM constellations: (a) 2-bit PAM; (b) 3-bit PAM.

Suppose that a b-bit PAM symbol s of the form in (2.8) is transmitted through a real zero-mean AWGN channel with noise variance $\mathcal{N}_0$. The received signal is $r = s + q$. Assume that s and q are independent. The conditional pdf of the received signal r given that s is transmitted is

$$f_{r|s}(r|s) = f_{r|s}(s + q|s) = f_q(q),$$

where $f_q(q)$ is the Gaussian pdf given in (2.4). From the received signal r, we make a decision on the transmitted symbol. The commonly adopted decision rule is the **nearest neighbor decision rule (NNDR)**. In an NNDR, we make a decision

$$\hat{s} = (2i + 1)\Delta, \quad \text{if } |r - (2i + 1)\Delta| \leq |r - (2j + 1)\Delta| \text{ for all } j. \quad (2.10)$$

The decision $\hat{s}$ is the constellation point that is closest to the received signal r. For all the interior constellation points, the symbol will be detected erroneously if the channel noise has $|q| > \Delta$. For the two exterior constellation points $s = (2^b - 1)\Delta$ and $s = (-2^b + 1)\Delta$, we will make an error when $q < -\Delta$ and $q > \Delta$, respectively. Therefore the probability that the detection is erroneous is given by

$$P(\hat{s} \neq s|s) = \begin{cases} P(q > \Delta), & \text{for } s = (-2^b + 1)\Delta; \\ P(q < -\Delta), & \text{for } s = (2^b - 1)\Delta; \\ P(|q| > \Delta), & \text{otherwise.} \end{cases}$$

Using the formula for the Gaussian pdf given in (2.4), the conditional probability of symbol error is given by

$$P(\hat{s} \neq s|s) = \begin{cases} Q\left(\sqrt{\dfrac{3\mathcal{E}_s}{(2^{2b} - 1)\mathcal{N}_0}}\right), & \text{for } s = \pm(2^b - 1)\Delta; \\ 2Q\left(\sqrt{\dfrac{3\mathcal{E}_s}{(2^{2b} - 1)\mathcal{N}_0}}\right), & \text{otherwise,} \end{cases}$$

where the function $Q(x)$ is the area under a Gaussian tail, defined as

$$Q(x) = \frac{1}{\sqrt{2\pi}} \int_x^\infty e^{-\tau^2/2} \, d\tau. \quad (2.11)$$

For equiprobable PAM symbols, the **symbol error rate (SER)** is given by

$$SER_{pam}(b) = \frac{1}{2^b} \sum_s P(\hat{s} \neq s|s) = 2(1 - 2^{-b})Q\left(\sqrt{\frac{3\mathcal{E}_s}{(2^{2b} - 1)\mathcal{N}_0}}\right). \quad (2.12)$$

As the function $Q(x)$ decays rapidly with respect to x, for a moderate SNR value most symbol errors happen between adjacent constellation points. We can use a mapping in which the codewords of adjacent constellation points differ by only one bit. An error between adjacent constellation points results only in one bit error. The widely used **Gray code** is a mapping scheme that possesses this property. Figures 2.11(a) and (b) show a Gray code mapping for 2-bit PAM and 3-bit PAM, respectively. When the Gray code is employed in a b-bit PAM modulation, the **bit error rate (BER)** and SER are related approximately by [120]

$$BER_{pam}(b) \approx \frac{1}{b} SER_{pam}(b). \quad (2.13)$$

From the above formulas, we see that the bit error rate depends on the signal power $\mathcal{E}_s$ and the noise power $\mathcal{N}_0$. The BER curves are often plotted against the SNR. For AWGN channels, the SNR is simply the ratio $\mathcal{E}_s/\mathcal{N}_0$. In order to evaluate the accuracy of the BER formula derived above, we compute the BER curves through **Monte Carlo simulation** in Fig. 2.12. In the Monte Carlo simulation, sufficiently many rounds of simulation are carried out and the results are averaged to give an accurate estimate of the actual BER. As a rule of thumb, for a BER of 10^{-l}, we need to generate at least $100 * 10^l$ bits in the simulations to obtain an accurate estimate. In Fig. 2.12, the BER approximations obtained from (2.13) are plotted in the dotted curves and the BER curves obtained from the Monte Carlo simulation are plotted in the solid curves (as the dotted curves almost overlap with the solid curves, we see only the solid curves in the figure). Since the two curves match almost perfectly, the formula in (2.13) gives a very good approximation of the true BER values. Comparing the PAM of different constellation sizes, we see that for a BER of 10^{-4}, we need an SNR of around 11.7 dB, 18.2 dB, and 24.2 dB for 1-bit, 2-bit, and 3-bit PAM, respectively. To achieve the same BER, the required SNR increases roughly by 6 dB for every additional bit.

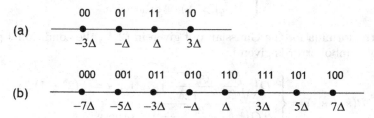

Figure 2.11. Gray code mapping: (a) 2-bit PAM; (b) 3-bit PAM.

Equation (2.12) relates the error rate to the SNR $\mathcal{E}_s/\mathcal{N}_0$. It can be used to obtain the number of bits that can be transmitted for a given SNR and target error rate. By rearranging the terms in (2.12), we get

$$b = \frac{1}{2} \log_2 \left(1 + \frac{\mathcal{E}_s/\mathcal{N}_0}{\Gamma_{pam}} \right), \qquad (2.14)$$

where

$$\Gamma_{pam} = \frac{1}{3} \left[Q^{-1} \left(\frac{SER_{pam}}{2(1 - 2^{-b})} \right) \right]^2.$$

If one compares the formula for b with the channel capacity, which is given by

$$0.5 \log_2 \left(1 + \mathcal{E}_s/\mathcal{N}_0 \right) \qquad \text{(bits per use)},$$

the quantity Γ_{pam} represents the difference in the required SNR between the PAM scheme and the channel capacity. Therefore Γ_{pam} is also known as the **SNR gap**. For a moderate error rate, the inverse Q function is relatively flat. Therefore the SNR gap is well approximated by

$$\Gamma_{pam} \approx \frac{1}{3} \left[Q^{-1} \left(SER_{pam}/2 \right) \right]^2. \qquad (2.15)$$

The SNR gap is a quantity that depends only on the error rate. In Table 2.1, we list the values of Γ_{pam} for some typical SER_{pam}.

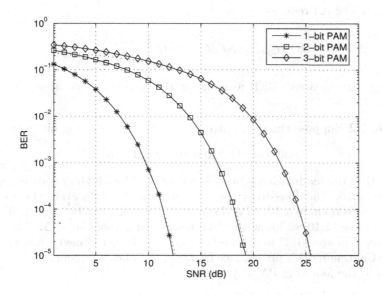

Figure 2.12. BER performance of PAM in a real zero-mean AWGN channel. The solid curves are the experimental values obtained from the Monte Carlo simulation and the dotted curves (almost indistinguishable from the solid curves) are the theoretical values obtained from the formula in (2.13).

SER_{pam}	Γ_{pam}	Γ_{pam} in dB
10^{-2}	2.21	3.44
10^{-3}	3.61	5.57
10^{-4}	5.05	7.03
10^{-5}	6.50	8.13
10^{-6}	7.98	9.02
10^{-7}	9.46	9.76

Table 2.1. The SNR gap Γ_{pam} in (2.15)

Binary phase shift keying (BPSK) modulation For the special case of PAM with $b = 1$, there are only two constellation points, $+\Delta$ and $-\Delta$, and this is more commonly known as binary phase shift keying (BPSK) modulation. For BPSK symbols, the bit error rate and symbol error rate are the same.

The formula (2.12) reduces to

$$BER_{bpsk} = SER_{bpsk} = Q\left(\sqrt{\frac{\mathcal{E}_s}{\mathcal{N}_0}}\right).$$

Unlike (2.13), the above BER formula for BPSK is exact; no approximation is made.

Example 2.2 Suppose the transmitter is to send the following bit stream:

$$000\ 010\ 111\ 110\ 010\ 100\ 101\ 110.$$

Assume that the modulation scheme is a 3-bit PAM with Gray code mapping as in Fig. 2.10. The "bits-to-symbol mapping" block takes every three input bits and maps them to a 3-bit PAM symbol. The first three bits are "000" and thus from Fig. 2.10 we know that the first PAM symbol is -7Δ. Then the next three bits are "010" and from the figure we have the next PAM symbol as $-\Delta$. Continuing this process, we find that the above sequence of 24 bits is mapped to the following PAM symbols:

$$s(n):\ -7\Delta,\ -\Delta,\ 3\Delta,\ \Delta,\ -\Delta,\ 7\Delta,\ 5\Delta,\ \Delta.$$

Now suppose that the above PAM symbols are transmitted over an AWGN channel and due to channel noise the received sequence is

$$r(n):\ -9\Delta,\ -0.7\Delta,\ 1.9\Delta,\ 1.1\Delta,\ -1.3\Delta,\ 6.6\Delta,\ 4.8\Delta,\ 2.1\Delta.$$

Assume that at the receiver there is no additional signal processing and that NNDR is applied directly to $r(n)$. The output of the symbol detector will be

$$\widehat{s}(n):\ -7\Delta,\ -\Delta,\ \boldsymbol{\Delta},\ \Delta,\ -\Delta,\ 7\Delta,\ 5\Delta,\ \boldsymbol{3\Delta}.$$

Comparing $\widehat{s}(n)$ with $s(n)$, we find that we have made two symbol errors out of eight transmitted symbols (indicated by boldfaced symbols). In this experiment, the symbol error rate is given by $SER = 0.25$. After the "symbol-to-bits mapping" block using Gray code, we obtain the following sequence:

$$000\ 010\ \mathbf{110}\ 110\ 010\ 100\ 101\ \mathbf{111}.$$

Comparing the decoded sequence with the transmitted sequence, two bits (indicated by boldfaced numbers) are received erroneously. The bit error rate is $BER = 1/12$, which is equal to $SER/3$ in this example because the two erroneously detected symbols are adjacent to the actual transmitted symbols. ∎

2.3.2 Quadrature amplitude modulation (QAM)

Unlike PAM symbols, QAM symbols are complex numbers. For $2b$-bit QAM, a codeword of $2b$ bits is mapped to a symbol of the form[2]

$$s = \pm(2k+1)\Delta \pm j(2l+1)\Delta, \quad \text{where} \quad k, l \in \{0, 1, \ldots, 2^{b-1} - 1\}. \quad (2.16)$$

---

[2]Unless mentioned otherwise, the QAM symbols in this book have a square constellation. Hence each QAM symbol carries an even number of bits.

Figure 2.13(a) and (b) show, respectively, the signal constellations for 2-bit and 4-bit QAM with the corresponding Gray codes. The special case of 2-bit QAM is also known as quadrature phase shift keying (QPSK). All the four constellation points in QPSK have the same magnitude. From (2.8) and (2.16), we see that the real and imaginary parts of a $2b$-bit QAM symbol can be viewed as two b-bit PAM symbols. Using this relation, many results for the QAM symbol can be obtained by modifying those of the PAM symbols. For example, the signal power of the $2b$-bit QAM symbol is

$$\mathcal{E}_s = E[|s|^2] = \frac{2\Delta^2}{3}(2^{2b} - 1).$$

It is twice that of a b-bit PAM symbol.

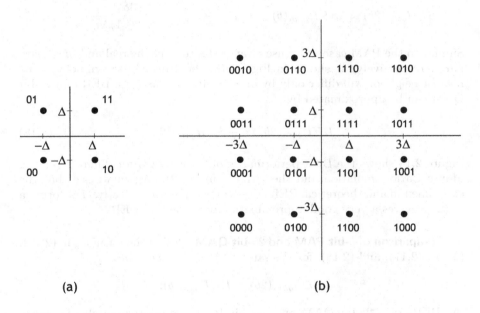

(a)   (b)

Figure 2.13. QAM constellation and its Gray code mapping: (a) 2-bit QAM (also known as QPSK); (b) 4-bit QAM.

Suppose that a $2b$-bit QAM symbol with power $\mathcal{E}_s$ is transmitted through a zero-mean complex AWGN channel with noise variance $\mathcal{N}_0$. Suppose that the noise is circularly symmetric so that its pdf is as given in (2.5). Then the real part and the imaginary part are both Gaussian with variance $\mathcal{N}_0/2$. Therefore the transmission of a $2b$-bit QAM symbol through a complex AWGN channel with noise variance $\mathcal{N}_0$ can be viewed as the transmission of two b-bit PAM symbols, each with power $\mathcal{E}_s/2$, through two real AWGN channels, each with noise variance $\mathcal{N}_0/2$. From earlier discussions, we know that when a b-bit PAM symbol with power $\mathcal{E}_s/2$ is transmitted through a real AWGN channel with noise variance $\mathcal{N}_0/2$, the SER is given by

$$SER_{pam}(b) = 2\Big(1 - \frac{1}{2^b}\Big)Q\Big(\sqrt{\frac{3\mathcal{E}_s/2}{(2^{2b}-1)\mathcal{N}_0/2}}\Big).$$

A QAM symbol is correctly decoded when both the real and imaginary parts are correctly decoded. The probability for this is $(1 - SER_{pam}(b))^2$. Thus the SER of a $2b$-bit QAM symbol is given by

$$SER_{qam}(2b) = 2SER_{pam}(b) - SER_{pam}^2(b).$$

When the error rate is small, we can ignore the second-order term and the SER is well approximated by

$$SER_{qam}(2b) \approx 2SER_{pam}(b) = 4\Big(1 - \frac{1}{2^b}\Big)Q\Big(\sqrt{\frac{3\mathcal{E}_s}{(2^{2b}-1)\mathcal{N}_0}}\Big). \qquad (2.17)$$

Similar to the PAM case, if we use Gray codes to map the real and imaginary parts, respectively, as shown in Fig. 2.13, then any QAM symbol and its nearest neighbor will differ only by one bit. In this case, the BER of a $2b$-bit QAM can be approximated by

$$BER_{qam}(2b) \approx \frac{1}{2b}SER_{qam}(2b). \qquad (2.18)$$

Figure 2.14 shows the BER performance of $2b$-bit QAM for different b using Monte Carlo simulation and the formula in (2.18). Again we see that the experimental and theoretical BER curves match almost perfectly. The formula in (2.18) gives a very good approximation of the actual BER.

Comparison of b-bit PAM and $2b$-bit QAM Using the formulas in (2.12), (2.13), (2.17), and (2.18), for the same SNR $\mathcal{E}_s/\mathcal{N}_0$ we have

$$BER_{qam}(2b) \approx BER_{pam}(b);$$

the BERs of a $2b$-bit QAM and a b-bit PAM are approximately the same, but the bit rate of QAM is twice that of PAM. However, we should note that the comparison is based on different channel settings. For PAM, the symbols are real and the channel noise is also real with variance $\mathcal{N}_0$. For QAM, the symbols are complex and the channel noise is complex with the variances of both the real and imaginary parts equal to $\mathcal{N}_0/2$. In other words, in passband communication if we choose a QAM over a PAM of the same bit rate, we will have a gain in SNR. For example, for a BER of 10^{-4}, we see from Fig. 2.12 and Fig. 2.14 that a 2-bit QAM needs an SNR of around 11.7 dB whereas a 2-bit PAM needs an SNR of around 18.2 dB; we have a saving of 6.5 dB by using QAM.

For QAM symbols we can also express $2b$ in terms of the SNR, $\mathcal{E}_s/\mathcal{N}_0$, as we did for PAM symbols in (2.14). By rearranging (2.17), we obtain

$$2b = \log_2\Big(1 + \frac{\mathcal{E}_s/\mathcal{N}_0}{\Gamma_{qam}}\Big), \qquad (2.19)$$

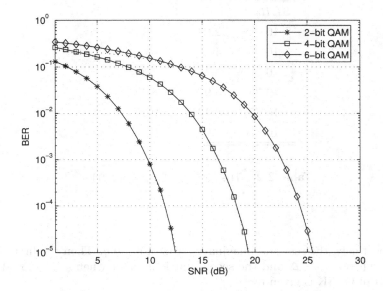

Figure 2.14. BER performance of QAM in zero-mean AWGN channels. The solid curves are the experimental values obtained from the Monte Carlo simulation and the dotted curves (almost indistinguishable from the solid curves) are the theoretical values obtained from the formula in (2.18).

where Γ_{qam} is the SNR gap given by

$$\Gamma_{qam} = \frac{1}{3}\left[Q^{-1}\left(\frac{SER_{qam}}{4(1-2^{-b})}\right)\right]^2. \qquad (2.20)$$

Again for moderate error rates, the following expression gives a very accurate approximation of the SNR gap:

$$\Gamma_{qam} \approx \frac{1}{3}\left[Q^{-1}\left(SER_{qam}/4\right)\right]^2. \qquad (2.21)$$

In Table 2.2, we list the values of Γ_{qam} for some typical SER_{qam}. Although the formula in (2.19) is derived for even-bit QAM symbols, the right-hand side is also used for estimating the number of bits that can be transmitted when there is no even-bit constraint [38].

Quadrature phase shift keying (QPSK) modulation When a QAM symbol has only two bits, it is also commonly known as a QPSK symbol. The constellation and a Gray code mapping for QPSK are shown in Fig. 2.13(a). Suppose a QPSK symbol with signal power $\mathcal{E}_s$ is transmitted through an AWGN channel with noise variance $\mathcal{N}_0$. For equiprobable QPSK symbols, we can compute the BER by computing the BER for any constellation point because of the symmetry. Suppose that "11" is transmitted. Let q_r and q_i

SER_{qam}	Γ_{qam}	Γ_{qam} in dB
10^{-2}	2.63	4.19
10^{-3}	4.04	6.06
10^{-4}	5.48	7.39
10^{-5}	6.95	8.42
10^{-6}	8.42	9.25
10^{-7}	9.91	9.96

Table 2.2. The SNR gap Γ_{qam} in (2.21)

be, respectively, the real and imaginary parts of noise q. Then the first bit is in error when $q_r < -\Delta$ and the second bit is in error when $q_i < -\Delta$. Thus the BER of QPSK is given by

$$BER_{qpsk} = 0.5P(q_r < -\Delta) + 0.5P(q_i < -\Delta) = Q\left(\sqrt{\frac{\mathcal{E}_s}{\mathcal{N}_0}}\right).$$

The above BER formula for QPSK is exact and it is identical to that of BPSK.

Other modulation schemes   PAM and QAM are the most commonly used modulation schemes due to their simplicity. There are many other modulation schemes, for example phase shift keying (PSK), frequency shift keying (FSK), differential phase shift keying (DPSK), minimum shift keying (MSK), and so forth. In practice, we may choose one modulation scheme over others, depending on the application. For example, in some applications it might be desirable to have modulation symbols with a constant magnitude, that is $|s_k| = \mathcal{E}_s$ for all k. In this case, we can use PSK modulation (shown in Fig. 2.15). In a PSK modulation scheme, all the constellation points are uniformly distributed on a circle and the radius of the circle determines the symbol power. For a more detailed and complete coverage of various digital modulations, the readers are referred to [120] and [137].

Example 2.3 Suppose we want to send the 24-bit sequence in Example 2.2 using 4-bit QAM symbols. Let the constellation and Gray code mapping be as shown in Fig. 2.13(b). As each symbol carries four bits, we group the 24 bits into codewords of four bits:

$$0000 \; 1011 \; 1110 \; 0101 \; 0010 \; 1110.$$

From Fig. 2.13(b), we find the corresponding six QAM symbols $s(n)$. These QAM symbols are sent over an AWGN channel. Suppose the received signals $r(n)$ are as given in Table 2.3. The error probabilities in this example are for the sake of demonstration. The actual errors are usually much smaller, such as 10^{-2}, 10^{-4}, etc. Assume that at the receiver there is no additional signal

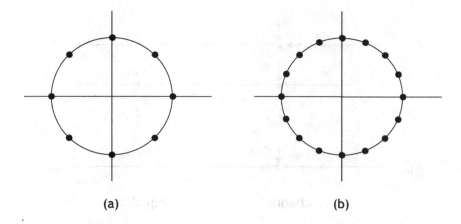

Figure 2.15. Phase shift keying modulation: (a) 8-PSK; (b) 16-PSK.

n	$s(n)$	$r(n)$	$\widehat{s}(n)$
0	$-3\Delta - 3\Delta j$	$-4.1\Delta - 2.6\Delta j$	$-3\Delta - 3\Delta j$
1	$3\Delta + \Delta j$	$3.7\Delta + 2.1\Delta j$	$3\Delta + 3\Delta j$
2	$\Delta + 3\Delta j$	$0.9\Delta + 2.8\Delta j$	$\Delta + 3\Delta j$
3	$-\Delta - \Delta j$	$-1.1\Delta - \Delta j$	$-\Delta - \Delta j$
4	$-3\Delta + 3\Delta j$	$-4\Delta + 1.7\Delta j$	$-3\Delta + \Delta j$
5	$\Delta + 3\Delta j$	$2.1\Delta + 1.1\Delta j$	$3\Delta + \Delta j$

Table 2.3. Transmitted QAM symbols $s(n)$, received signals $r(n)$, and detected symbols $\widehat{s}(n)$.

processing and the NNDR is applied directly to $r(n)$. After symbol detection we get $\widehat{s}(n)$. Comparing $\widehat{s}(n)$ with $s(n)$, we find that we have made three symbol errors ($\widehat{s}(1)$, $\widehat{s}(4)$, $\widehat{s}(5)$) out of six transmitted symbols. The symbol error rate is $SER = 1/2$. After symbol-to-bits mapping using the Gray code provided in Fig. 2.13(b), we obtain the following sequence:

$$0000 \ 101\mathbf{0} \ 1110 \ 0101 \ 001\mathbf{1} \ \mathbf{1011}.$$

Comparing the decoded sequence with the transmitted sequence, four bits (indicated by boldfaced numbers) are received erroneously. The bit error rate is $BER = 4/24 = 1/6$, which is larger than $SER/4 = 1/8$. This is because $\widehat{s}(5)$ is not adjacent to $s(5)$, which causes an error of two bits rather than one bit. ∎

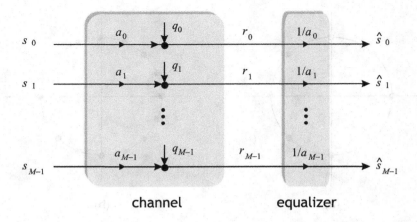

Figure 2.16. A set of M parallel channels and the corresponding zero-forcing equalizer.

2.4 Parallel subchannels

In many wideband communication systems, a wideband channel is divided into a set of subchannels, each with a smaller bandwidth. Examples include the widely used OFDM and DMT systems, which will be studied in detail in Chapter 6. In these systems, an FIR channel is converted to a set of parallel ISI-free channels as shown in Fig. 2.16. The received signal of the ith subchannel is

$$r_i = a_i s_i + q_i,$$

where s_i is the symbol transmitted over the ith subchannel. The quantities a_i and q_i are, respectively, the ith subchannel gain and noise. Because each subchannel has only a single tap, zero-forcing equalization can be done by using simple scalar multipliers $1/a_i$ (if $a_i \neq 0$) as indicated in the figure. Let $\mathbf{s}$ and $\widehat{\mathbf{s}}$ be, respectively, the input and output vectors. Define the output error vector as

$$\mathbf{e} = \widehat{\mathbf{s}} - \mathbf{s}.$$

Then we can redraw the parallel channels as in Fig. 2.17. It is clear that the output error of the ith subchannel is $e_i = q_i/a_i$. The error variance for the ith subchannel is

$$\sigma_{e_i}^2 = \frac{\mathcal{N}_i}{|a_i|^2},$$

where $\mathcal{N}_i$ is the variance of q_i. For many applications, the subchannels have the same noise variances $\mathcal{N}_i = \mathcal{N}_0$, but the subchannel gains can be very different. Thus the error variances $\sigma_{e_i}^2$ can be very different for different subchannels. Signals transmitted over different subchannels encounter different levels of distortion. For subchannels with large error variances, the bit error rate will be high and the overall performance of the parallel channels will be limited by these *bad* subchannels. To see this effect, let us consider the following example with only two subchannels.

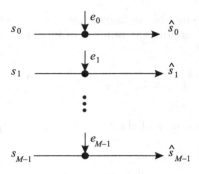

Figure 2.17. Equivalent parallel channels of Fig. 2.16.

Example 2.4 Suppose that there are only two subchannels and the gains are $a_0 = 1$ and $a_1 = 0.1$, respectively. The transmission power is fixed at 10. The subchannel noises q_i are AWGN with variance $\mathcal{N}_i = 1$. A quick calculation shows that the output error variances are

$$\sigma_{e_0}^2 = 1, \quad \sigma_{e_1}^2 = 100.$$

Suppose that the maximum signal power allowed on each subchannel is $\mathcal{E}_{max} = 10$. Let the transmitted signals s_i be BPSK symbols with $s_i = \pm\sqrt{10}$ so that the power is $\mathcal{E}_i = 10$. For BPSK modulation, BER is equal to SER. From (2.12), the BER of the zeroth subchannel is $Q(\sqrt{10}) = 7.83 \times 10^{-4}$, whereas the BER of the first subchannel is $Q(\sqrt{0.1}) = 0.38$. The average BER of the parallel channels is approximately

$$0.5 * Q(\sqrt{0.1}) = 0.19.$$

The performance of the system is severely limited by the first subchannel. Instead of transmitting BPSK symbols on both the subchannels, suppose that we now transmit a 2-bit PAM symbol on the zeroth subchannel and the first subchannel is not utilized for transmission; the zeroth and first subchannels are allocated two bits and zero bits, respectively, so that the total number of bits transmitted is still two. Using the BER formula for 2-bit PAM, we find that $BER \approx 3/4 * Q(\sqrt{10/5}) = 0.059$, which is much smaller than 0.19. ∎

From the above example, we see that by properly loading the bits among the subchannels, we can significantly improve the BER performance for the same transmission rate. In what follows, we will show how to achieve this.

Bit loading When the powers on the subchannels are fixed at $\mathcal{E}_{max}$ and the subchannel error variances $\sigma_{e_i}^2$ are different for different i, the bits assigned to the subchannels b_i can be adjusted to improve the error rate. This is called **bit loading** (also known as bit allocation). Below, we will first consider bit loading for the PAM case under the peak power constraint. Suppose that the error rates of all the subchannels are the same and their SERs are equal to SER_0. For PAM symbols, we know that the number of bits is related to the SNR by (2.14). Therefore the number of bits that can be transmitted on the ith subchannel is

$$b_i = \frac{1}{2} \log_2 \left(1 + \frac{\mathcal{E}_{max}/\sigma_{e_i}^2}{\Gamma_{pam}} \right), \tag{2.22}$$

where $\sigma_{e_i}^2$ is the noise power of the ith subchannel output. The average bit rate is given by

$$b = \frac{1}{M} \sum_{i=0}^{M-1} b_i.$$

The bits computed in (2.22) are not integer in general. When the constraint of integer bit is applied, the average bit rate is $b = (1/M) \sum_{i=0}^{M-1} \lfloor b_i \rfloor$, where $\lfloor x \rfloor$ denotes the largest integer less than or equal to x.

For QAM symbols, suppose the ith subchannel carries $2b_i$ bits. Then, from (2.19), we get the number of bits that can be sent through the ith subchannel as

$$2b_i = \log_2 \left(1 + \frac{\mathcal{E}_{max}/\sigma_{e_i}^2}{\Gamma_{qam}} \right).$$

The average bit rate is given by

$$2b = \frac{1}{M} \sum_{i=0}^{M-1} 2b_i.$$

For the case of integer bit loading, the average bit rate is $1/M \sum_{i=0}^{M-1} 2\lfloor b_i \rfloor$.

Example 2.5 Consider the parallel channels in Example 2.4. Suppose that PAM symbols are sent, the desired SER is $SER_0 = 10^{-7}$, and the maximum transmission power allowed is $\mathcal{E}_{max} = 1000$. Using Table 2.1, we have $\Gamma_{pam} = 9.46$. The subchannel SNRs are

$$SNR_0 = 1000, \quad SNR_1 = 10.$$

The maximum achievable bits for the two subchannels are

$$b_0 = 3.37, \quad b_1 = 0.52.$$

Thus the overall maximum achievable bit rate is $b = 1.95$. If an integer bit allocation is desired, this value becomes $b = 1.5$. Note that although $b_1 = 0.52$, we round it down to zero so that the desired quality-of-service of $SER_0 = 10^{-7}$ is not violated. ∎

In the above discussions, we assume that the peak signal power of each subchannel is limited by $\mathcal{E}_{max}$. In some applications, we may be concerned about the average signal power rather than the peak signal power. The problem of bit loading for this case will be studied in Chapter 8.

2.5 Further reading

Some basic concepts for digital communication systems were briefly reviewed in this chapter. There are many textbooks that provide a more detailed and comprehensive treatment of these topics. The interested readers are referred to [50, 67, 120], to name just a few.

2.6 Problems

2.1 Suppose we have a communication system with the waveform $(p_1 * c_a * p_2)(t)$ given by

$$(p_1 * c_a * p_2)(t) = \begin{cases} -0.5|t - 2| + 1, & \text{for } 0 \leq t \leq 4; \\ 0, & \text{otherwise.} \end{cases}$$

Find the discrete-time equivalent channel $c(n)$ when the sampling period is $T = 1$. What would $c(n)$ be if we increase the sampling period to $T = 2$?

2.2 *Multipath channels.* In many applications, the transmission channels have the form

$$c_a(t) = \sum_{k=1}^{L} a_k \delta_a(t - \tau_k),$$

where $\delta_a(t)$ is a continuous-time impulse (which is also known as a Dirac function). These channels are known as multipath channels. The parameters a_k and τ_k are, respectively, called the gain and delay of the kth path. The expression above shows that $c_a(t)$ has L paths. Suppose that we have a two-path channel with $a_1 = 1$, $a_2 = -0.5$, $\tau_1 = 0$, and $\tau_2 = 3$. Assume that the convolution of the transmitting and receiving pulses yields

$$(p_1 * p_2)(t) = \begin{cases} t, & \text{if } 0 \leq t \leq 1; \\ 2 - t & \text{if } 1 < t \leq 2; \\ 0, & \text{otherwise.} \end{cases}$$

Let the sampling period be $T = 1$. Find the discrete-time equivalent channel $c(n)$. If the delay of the second path is changed to $\tau_2 = 1$, what is $c(n)$?

2.3 Derive the discrete-time equivalent channel model for the passband communication channel shown in Fig. 2.5 by showing that the channel and noise are as given in (2.3).

2.4 Consider the system in Fig. 2.3(a). Let the channel be $c_a(t) = \delta_a(t)$ and let the transmitting and receiving pulses be

$$p_1(t) = p_2(t) = \begin{cases} 1, & \text{if } 0 \leq t \leq 1; \\ 0, & \text{otherwise.} \end{cases}$$

Suppose that $T = 1$ and the transmitted symbols are $x(n) = (-1)^n$ for $0 \leq n \leq 5$ and zero otherwise. In the absence of noise: (a) plot $x_a(t)$ and $w_a(t)$; (b) find $r(n) = w_a(nT)$; (c) verify the relation $r(n) = (c * x)(n)$, where $c(n)$ is the discrete-time equivalent channel.

2.5 Repeat Problem 2.4 for $T = 0.5$.

2.6 Let the continuous-time channel noise $q_a(t)$ be a white real WSS process
with power spectral density (psd) $S_{q_a}(j\Omega) = \mathcal{N}_a$ for all Ω.

(a) Suppose that the receiving pulse $p_2(t)$ is an ideal lowpass filter with
the passband region $[-0.5/T,\ 0.5/T]$, where T is the underlying
symbol spacing. Is the equivalent discrete-time channel noise $q(n)$
also a white WSS process? What is the noise power?

(b) Now let the passband of $p_2(t)$ be $[-1/T,\ 1/T]$. Repeat part (a).

2.7 Consider a system where the 2-bit PAM symbols $s(n)$ are transmitted over a hypothetical noiseless channel with impulse response $c(n) =
\alpha\delta(n)$, where $0 < \alpha \leq 1$. The received signal $r(n) = \alpha s(n)$ is an attenu-
ated version of the transmitted symbol $s(n)$. Suppose the receiver does
not know about the attenuation factor α and it makes its decisions $\widehat{s}(n)$
using the NNDR based on the original constellation of $s(n)$. Compute
the symbol error rate (SER). Express your answer in terms of α.

2.8 Consider a set of parallel channels with four subchannels as in Fig. 2.16.
Let the noise be zero-mean AWGN with variance $\mathcal{N}_k = 10$ and the gains
a_k be given by the following table:

k	0	1	2	3
a_k	5	1	10	$\sqrt{40}$

Suppose that the zero-forcing equalizer is employed.

(a) When BPSK symbols with signal power $\mathcal{E}_i = 10$ are transmitted,
compute the subchannel SNRs and the average BER.

(b) Suppose that we do not send any bits over the first two subchannels and transmit 2-bit PAM over the last two subchannels. The
maximum signal power allowed on each subchannel is $\mathcal{E}_{max} = 10$.
Compute the average BER assuming that the Gray code is used
for the mapping.

2.9 Suppose PAM symbols are transmitted over the set of parallel channels
in Problem 2.8. Let the desired SER be $SER_0 = 10^{-6}$, and the maximum transmission power be $\mathcal{E}_{max} = 1000$. Compute the number of bits
b_k (integer) that can be sent through the kth subchannel for $0 \leq k \leq 3$.
What is the average bit rate b?

3

FIR equalizers

In wideband communication systems, the transmission channels are usually frequency-selective and thus will introduce some degree of intersymbol interference (ISI) in addition to the noise. At the receiver, some signal processing is often carried out to alleviate the effect of these distortions. Such processing is in general known as **channel equalization**. Figure 3.1 shows the block diagram of a transmission system with a linear equalizer $a(n)$. In many applications, the channel can be modeled as an FIR LTI filter by making the order L_c sufficiently large:

$$C(z) = \sum_{n=0}^{L_c} c(n)z^{-n}.$$

The channel noise $q(n)$ is a zero-mean wide sense stationary (WSS) process. In this chapter, we shall consider FIR equalization. The equalizer $A(z)$ has transfer function

$$A(z) = \sum_{n=0}^{L_a} a(n)z^{-n}.$$

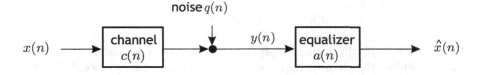

Figure 3.1. Transmission system with an LTI equalizer $a(n)$.

Letting $x(n)$ be the transmitted signal, the equalizer output is given by

$$\widehat{x}(n) = (a * c * x)(n) + (a * q)(n), \tag{3.1}$$

where "$*$" denotes convolution. The goal of channel equalization is to produce an output signal $\widehat{x}(n)$ that is "close" to the transmitted signal $x(n)$ in some sense. Some equalization techniques will be introduced below.

The zero-mean assumption Unless mentioned otherwise, in this book we assume that all the random variables have zero-mean. One consequence of this is that the mean squared value is equal to the variance. That is,

$$E[|v|^2] = \sigma_v^2,$$

for the zero-mean random variable v. The mean squared value $E[|\bullet|^2]$ and the variance σ^2 are used interchangeably in the book.

3.1 Zero-forcing equalizers

Consider Fig. 3.1. The transfer function from the transmitted signal $x(n)$ to the equalizer output $\widehat{x}(n)$ is given by

$$T(z) = A(z)C(z).$$

The equalizer $A(z)$ is **zero-forcing** if $T(z) = z^{-n_0}$. The output of a zero-forcing equalizer can be expressed as

$$\widehat{x}(n) = x(n - n_0) + q_0(n),$$

where $q_0(n) = (q * a)(n)$ is the equalizer output noise. The integer n_0 is the system delay. As the channel noise $q(n)$ does not affect the design of a zero-forcing equalizer (though it affects the system performance), we will assume $q(n) = 0$ in this section.

For an FIR channel $C(z)$ and an FIR equalizer $A(z)$, their product $A(z)C(z)$ is also FIR. The transfer function $T(z)$ can be a delay if and only if both $C(z)$ and $A(z)$ are delays. When the channel is frequency-selective, $C(z)$ has more than one tap. Therefore there does not exist an FIR zero-forcing equalizer for frequency-selective channels. An alternative solution would be to find $A(z)$ so that $A(z)C(z)$ is "close" to a delay z^{-n_0}. To do this, let us define

$$d(n) = (a * c)(n) - \delta(n - n_0).$$

When $d(n) = 0$, the equalizer $A(z)$ is zero-forcing. However, this is not possible unless the channel $c(n)$ has only one nonzero tap. So we design the equalizer $a(n)$ to minimize $\sum_n |d(n)|^2$. This problem can be formulated using matrix notation as follows. Let the vectors $\mathbf{t}$ and $\mathbf{a}$ be given by

$$\mathbf{t} = \begin{bmatrix} t(0) \\ t(1) \\ \vdots \\ t(L) \end{bmatrix} \quad \text{and} \quad \mathbf{a} = \begin{bmatrix} a(0) \\ a(1) \\ \vdots \\ a(L_a) \end{bmatrix}, \tag{3.2}$$

where $L = L_c + L_a$ is the order of $T(z)$. Then they are related by

$$\mathbf{t} = \mathbf{C}_{low}\mathbf{a}, \tag{3.3}$$

where the $(L+1) \times (L_a+1)$ matrix $\mathbf{C}_{low}$ is a lower triangular Toeplitz matrix given by

$$
\mathbf{C}_{low} = \begin{bmatrix}
c(0) & 0 & \cdots & 0 & \cdots & 0 \\
c(1) & c(0) & & 0 & & 0 \\
\vdots & & \ddots & & & \vdots \\
c(L_c) & & & c(0) & & \\
0 & \ddots & & & \ddots & \\
\vdots & & c(L_c) & & \cdots & c(0) \\
0 & \cdots & 0 & c(L_c) & & c(1) \\
\vdots & & \vdots & & \ddots & \vdots \\
0 & \cdots & 0 & & & c(L_c)
\end{bmatrix}.
\tag{3.4}
$$

The subscript *"low"* serves as a reminder that the matrix is a "lower triangular" matrix. Define an $(L+1) \times 1$ vector

$$
\mathbf{1}_{n_0} = [\ 0 \quad \cdots \quad 0 \quad \overset{n_0}{1} \quad 0 \quad \cdots \quad 0\ ]^T,
\tag{3.5}
$$

which has only one nonzero entry in the n_0th location. Then we can write

$$
\mathcal{E}_d(n_0) = \sum_{n=0}^{L} |d(n)|^2 = \|\mathbf{t} - \mathbf{1}_{n_0}\|^2 = \|\mathbf{C}_{low}\mathbf{a} - \mathbf{1}_{n_0}\|^2,
\tag{3.6}
$$

where $\| \bullet \|$ denotes the Euclidean norm of the vector. The quantity $\mathcal{E}_d(n_0)$ is a measure of the closeness of the transfer function $T(z)$ to the delay z^{-n_0}. Thus the problem of finding the optimal $a(n)$ becomes a least-squares problem of finding $\mathbf{a}$ to minimize $\|\mathbf{C}_{low}\mathbf{a} - \mathbf{1}_{n_0}\|^2$. The product $\mathbf{C}_{low}\mathbf{a}$ is a linear combination of the columns of $\mathbf{C}_{low}$. We would like to find the vector in the column space[1] of $\mathbf{C}_{low}$ that is the closest to $\mathbf{1}_{n_0}$. From linear algebra theory, we know that the closest vector is the orthogonal projection of $\mathbf{1}_{n_0}$ onto the column space of $\mathbf{C}_{low}$. As the matrix $\mathbf{C}_{low}$ has full column rank unless $C(z) = 0$ (Problem 3.1), the least-squares solution is unique and it is given by [52]

$$
\mathbf{a}_{ls} = (\mathbf{C}_{low}^{\dagger} \mathbf{C}_{low})^{-1} \mathbf{C}_{low}^{\dagger} \mathbf{1}_{n_0},
\tag{3.7}
$$

where † denotes transpose conjugation. The corresponding equalizer $a_{ls}(n) = [\mathbf{a}_{ls}]_n$ will be called the **least-squares equalizer**. The inverse $(\mathbf{C}_{low}^{\dagger} \mathbf{C}_{low})^{-1}$ always exists as $\mathbf{C}_{low}$ has full column rank. Substituting the expression in (3.7) into (3.6), we obtain

$$
\mathcal{E}_{d,ls}(n_0) = \|\mathbf{B}_{ls} \mathbf{1}_{n_0} - \mathbf{1}_{n_0}\|^2,
\tag{3.8}
$$

where $\mathbf{B}_{ls}$ is the positive semidefinite matrix given by

$$
\mathbf{B}_{ls} = \mathbf{C}_{low} (\mathbf{C}_{low}^{\dagger} \mathbf{C}_{low})^{-1} \mathbf{C}_{low}^{\dagger}.
\tag{3.9}
$$

---

[1] The column space of a matrix is the subspace spanned by its column vectors.

Expanding the right-hand side of (3.8) we get

$$\mathcal{E}_{d,ls}(n_0) = \mathbf{1}_{n_0}^\dagger \mathbf{B}_{ls}^\dagger \mathbf{B}_{ls} \mathbf{1}_{n_0} - \mathbf{1}_{n_0}^\dagger \mathbf{B}_{ls} \mathbf{1}_{n_0} - \mathbf{1}_{n_0}^\dagger \mathbf{B}_{ls}^\dagger \mathbf{1}_{n_0} + 1.$$

Using the facts that $\mathbf{B}_{ls}^\dagger = \mathbf{B}_{ls}$ and $\mathbf{B}_{ls}^\dagger \mathbf{B}_{ls} = \mathbf{B}_{ls}$, we can simplify the above expression to

$$\mathcal{E}_{d,ls}(n_0) = 1 - [\mathbf{B}_{ls}]_{n_0,n_0}, \tag{3.10}$$

where $[\mathbf{B}_{ls}]_{n_0,n_0}$ is the n_0th diagonal entry of $\mathbf{B}_{ls}$. One can choose the system delay n_0 to minimize $\mathcal{E}_{d,ls}(n_0)$. It follows from (3.10) that the smallest $\mathcal{E}_{d,ls}(n_0)$ is achieved when n_0 is chosen such that the n_0th diagonal entry of $\mathbf{B}_{ls}$ is the largest. When the system delay n_0 is chosen optimally, the equalizer is called a *delay-optimized* least-squares equalizer. It is left as an exercise to show that all the diagonal entries of the matrix $\mathbf{B}_{ls}$ satisfy $0 \le [\mathbf{B}_{ls}]_{kk} \le 1$. The maximum of $[\mathbf{B}_{ls}]_{kk}$ can be 0 or 1 only for some very special cases (Problems 3.4 and 3.5). We summarize the design procedure as follows:

- compute the matrix $\mathbf{A} = (\mathbf{C}_{low}^\dagger \mathbf{C}_{low})^{-1} \mathbf{C}_{low}^\dagger$;

- find n_0 so that the n_0th diagonal entry of the matrix $\mathbf{B}_{ls} = \mathbf{C}_{low} \mathbf{A}$ is the largest.

From (3.7), the delay-optimized least-squares equalizer is given by the n_0th column of $\mathbf{A}$.

Example 3.1 Let us consider the following two FIR channels:

$$C_0(z) = 1 + 2z^{-1},$$
$$C_1(z) = 1 + 0.95z^{-1}.$$

For the channel $C_0(z)$, a causal stable zero-forcing equalizer $a_{zf}(n)$ does not exist because $1/C_0(z)$ has a pole at $z = -2$, which is outside the unit circle. On the other hand, the inverse $1/C_1(z)$ is causal and stable. Now we consider FIR least-squares equalizers with different orders. The system delay n_0 is chosen optimally. Figure 3.2 shows the plot of $\mathcal{E}_{d,ls}$ versus L_a, and Table 3.1 shows the smallest L_a needed to achieve the listed target values of $\mathcal{E}_{d,ls}$ for $C_0(z)$ and $C_1(z)$, respectively. Observe that for the channel $C_0(z)$, with a relatively small order L_a, we are able to make the overall impulse response very close to an impulse $\delta(n - n_0)$. On the other hand, it is not as easy to equalize the channel $C_1(z)$. To achieve the same target $\mathcal{E}_{d,ls}$, we need a much longer equalizer for $C_1(z)$. For example, when the target $\mathcal{E}_{d,ls} = 10^{-3}$, the smallest equalizer order needed is $L_a = 4$ for $C_0(z)$ and $L_a = 44$ for $C_1(z)$. This can be explained as follows. To get a small $\mathcal{E}_d$, the equalizer $a(n)$ should approximate the impulse response of $z^{-n_0}/C_1(z)$, which has the form $(-0.95)^{n-n_0}$, a quantity decaying slowly with n. Thus we need a long equalizer $a(n)$. On the other hand, for the channel $C_0(z)$, if we choose the equalizer as

$$A(z) = 0.5z^{-L_a} \left(1 - 0.5z + 0.5^2 z^2 - \cdots + (-0.5)^{L_a} z^{L_a}\right),$$

then the transfer function will be

$$T(z) = A(z)C_0(z) = 0.5(-0.5)^{L_a} + z^{-L_a-1}.$$

If the system delay is chosen as $n_0 = L_a + 1$, then the quantity $\mathcal{E}_d(n_0)$ is equal to 0.5^{2L_a+2}, which decays at a much faster rate. This explains why we are able to achieve the target values of $\mathcal{E}_{d,ls}$ listed in Table 3.1 with a relatively small L_a. Note that the fact that $1/C_0(z)$ is unstable does not matter when we design the FIR least-squares equalizers. ∎

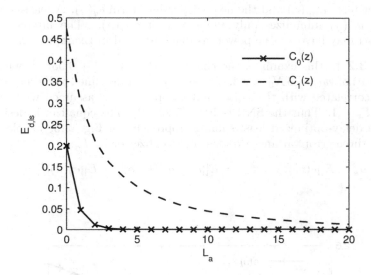

Figure 3.2. See Example 3.1. A plot of $\mathcal{E}_{d,ls}$.

Target $\mathcal{E}_{d,ls}$	L_a for $C_0(z)$	L_a for $C_1(z)$
0.1	1	6
0.05	1	10
0.01	3	23
0.005	3	29
0.001	4	44

Table 3.1. See Example 3.1. Smallest equalizer order L_a needed for $C_0(z)$ and $C_1(z)$ to achieve the target $\mathcal{E}_{d,ls}$.

Effect of channel noise The channel noise has been ignored in the above analysis. Suppose now there is channel noise $q(n)$. The noisy output $\widehat{x}(n)$ is given by (3.1). Define the output error $e(n) = \widehat{x}(n) - x(n - n_0)$. In this case, the output error is given by

$$e(n) = \underbrace{(a * c * x)(n) - x(n - n_0)}_{e_{sig}(n)} + \underbrace{(a * q)(n)}_{e_q(n)}. \qquad (3.11)$$

When the convolution $(a * c)(n)$ is not an impulse $\delta(n - n_0)$, the quantity $e_{sig}(n)$ is not zero. The output error consists of two terms: the signal-dependent term $e_{sig}(n)$ and the noise-dependent term $e_q(n)$. A least-squares equalizer $a_{ls}(n)$ minimizes only the power of $e_{sig}(n)$. The filtered noise $(a_{ls} * q)(n)$ may have a large power, as demonstrated in Example 3.2.

Example 3.2 In this example, the channel is $C_1(z) = 1 + 0.95z^{-1}$, with an AWGN $q(n)$ of variance $\mathcal{N}_0 = 0.3$. The signal $x(n)$ is assumed to be zero-mean WSS, uncorrelated with the noise, and its spectrum is assumed white with variance $\mathcal{E}_x = 1$. Thus the SNR is $10/3$ (5.23 dB). The equalizer is designed using the delay-optimized least-squares approach. In Fig. 3.3, we plot the following three error variances versus the equalizer order L_a:

$$\sigma_e^2 = E[|e(n)|^2], \quad \sigma_{e_{sig}}^2 = E[|e_{sig}(n)|^2], \quad \sigma_{e_q}^2 = E[|e_q(n)|^2].$$

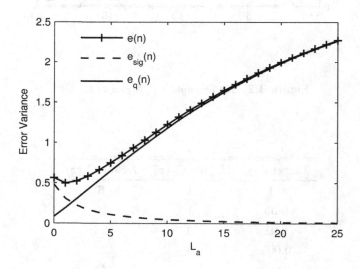

Figure 3.3. See Example 3.2. The variances of $e(n)$, $e_{isi}(n)$, and $e_q(n)$ defined in (3.11) versus the equalizer orders L_a.

From the plot, we see that the variance of the signal-dependent term $e_{sig}(n)$ decreases monotonically with respect to L_a, but the variance of the noise-dependent term $e_q(n)$ increases as L_a increases. As a result, when L_a is large the output error variance σ_e^2 is dominated by the noise power $\sigma_{e_q}^2$, which increases as L_a increases. Increasing the length of the least-squares equalizer leads to a larger output error!

The noise amplification can also be understood from a frequency domain viewpoint. When L_a increases, the approximation $c_1(n) * a_{ls}(n) \approx \delta(n - n_0)$ becomes more accurate. In the frequency domain, $A_{ls}(e^{j\omega})$ approaches $1/C_1(e^{j\omega})$ when L_a is large. As $C_1(e^{j\omega})$ is a lowpass filter with $C_1(e^{j\pi}) = 0.05$, the equalizer $A_{ls}(e^{j\omega})$ is a highpass filter with $A_{ls}(e^{j\pi}) \approx 20$; the high-frequency component of the channel noise $q(n)$ is severely amplified by the least-squares equalizer. ∎

As we can see from the above example, though a least-squares equalizer $a_{ls}(n)$ can effectively reduce the signal-dependent term $e_{sig}(n)$, its ability to combat the channel noise is limited. A better design of the equalizer should take both $e_{sig}(n)$ and $e_q(n)$ into consideration. By designing an equalizer that minimizes the output error variance $E[|e(n)|^2]$, we are able to minimize the combined effect of the signal-dependent term and the noise term. Before we proceed to study such equalizers, we introduce in Section 3.2 a useful tool called the orthogonality principle. This principle plays an important role in many areas of signal processing and communications.

3.2 Orthogonality principle and linear estimation

In digital communications, we frequently face the problem of estimating a random variable x from some noisy observations $y_0, y_1, \ldots, y_{K-1}$. The estimate $\widehat{x}$ is often obtained by taking a linear combination of the observed samples; that is,

$$\widehat{x} = a_0 y_0 + a_1 y_1 + \cdots + a_{K-1} y_{K-1}.$$

This is known as a linear estimator. We would like to find a_i so that $\widehat{x}$ is close to x in some sense. Define the estimation error e as

$$e = \widehat{x} - x.$$

One commonly used criterion is the mean squared error (MSE) $E[|e|^2]$. The optimal estimate $\widehat{x}$ that minimizes the MSE is known as the **minimum mean squared error (MMSE)** solution. A powerful tool for finding the MMSE solution is the **orthogonality principle**, outlined in Theorem 3.1.

Theorem 3.1 The linear estimate $\widehat{x}$ minimizes the MSE $E[|e|^2]$ if and only if

$$E[ey_i^*] = 0, \quad \text{for } i = 0, 1, \ldots, K - 1; \tag{3.12}$$

that is, if and only if the estimation error e is orthogonal to every observed sample y_i. ∎

Proof. Suppose $\widehat{x}_\perp$ is the estimate that satisfies (3.12) and $e_\perp = \widehat{x}_\perp - x$. Let $\widehat{x}$ be another linear estimate of x. Let us rewrite the MSE for x as

$$E[|\widehat{x} - x|^2] = E[|\widehat{x}_\perp - x + \widehat{x} - \widehat{x}_\perp|^2].$$

Expanding the right-hand side of the above equation, we have

$$E[|e_\perp|^2] + E[|\widehat{x} - \widehat{x}_\perp|^2] + E[e_\perp(\widehat{x} - \widehat{x}_\perp)^*] + E[e_\perp^*(\widehat{x} - \widehat{x}_\perp)].$$

Note that $\widehat{x} - \widehat{x}_\perp$ is a linear combination of y_i. Because $e_\perp$ is orthogonal to every y_i, it is orthogonal to $\widehat{x} - \widehat{x}_\perp$. Thus we have

$$E[|\widehat{x} - x|^2] = E[|e_\perp|^2] + E[|\widehat{x} - \widehat{x}_\perp|^2] \geq E[|e_\perp|^2].$$

The inequality becomes an equality if and only if $\widehat{x} = \widehat{x}_\perp$. ∎

The orthogonality principle finds many applications. For example, it has been successfully applied to the topic of linear prediction theory [164]. We will show how to apply the principle to the design of MMSE equalizers in the next section.

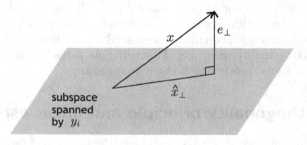

Figure 3.4. Geometrical interpretation of the MMSE estimate.

The orthogonality principle also has a nice geometrical interpretation. Note that the estimate $\widehat{x}$ is a linear combination of y_i and thus belongs to the subspace spanned by y_i. We wish to find the element in this subspace that is "closest" to x. From the theory of linear algebra, this closest element $\widehat{x}_\perp$ is the orthogonal projection of x onto the subspace spanned by y_i. This is precisely what is stated in the orthogonality principle: the MMSE estimate $\widehat{x}_\perp$ is such that the error $e_\perp$ is orthogonal to every y_i and hence all of their linear combinations. Therefore the three quantities x, $\widehat{x}_\perp$, and $e_\perp$ form a right triangle with x, $\widehat{x}_\perp$, and $e_\perp$ being the hypotenuse, the base, and the height, respectively. Figure 3.4 illustrates this geometrical relationship. In particular, it implies that

$$E[\widehat{x}_\perp e_\perp^*] = 0;$$

that is, the MMSE estimate and its estimation error are orthogonal. From the Pythagorean theorem, we know that they satisfy

$$E[|x|^2] = E[|e_\perp|^2] + E[|\widehat{x}_\perp|^2],$$

a relation that can also be verified directly by using the orthogonality principle. It follows from the above equation that the power of the MMSE estimate $\widehat{x}_\perp$ is always less than that of x, unless the estimation error is zero.

3.2.1 Biased and unbiased linear estimates

A linear estimate $\widehat{x}$ of x is said to be an **unbiased estimate** if

$$E[\widehat{x}|x] = x;$$

otherwise it is said to be **biased**. In many digital communication systems, the linear estimate $\widehat{x}$ is usually the output of an equalizer. It is often possible to write the estimate in the form

$$\widehat{x} = \alpha x + \tau, \tag{3.13}$$

where τ is a random variable related to the channel noise and interfering symbols. Usually we can make the following assumptions[2] about τ:

(i) τ has zero-mean,

$$E[\tau] = 0;$$

(ii) τ is statistically independent of x.

From (3.13), we can write

$$E[\widehat{x}|x] = \alpha x + E[\tau|x].$$

Using the above two assumptions, $E[\tau|x] = E[\tau] = 0$ and we have

$$E[\widehat{x}|x] = \alpha x.$$

Therefore we conclude that the estimate $\widehat{x}$ expressed in the form in (3.13) is biased if $\alpha \neq 1$. Note that the fact that τ is statistically independent of x implies that x and τ are uncorrelated, but the converse may not be true.

As we will see in Section 3.3, the output of an MMSE equalizer is a biased estimate of the desired signal. If the bias is not removed before applying symbol detection at the receiver, the bit error rate (BER) will increase (Section 3.4). The bias can be removed easily by dividing $\widehat{x}$ by α; the result will be an unbiased estimate

$$\widehat{x}_{unb} = \frac{\widehat{x}}{\alpha} = x + \frac{\tau}{\alpha}. \tag{3.14}$$

Biased and unbiased SNRs Let $\widehat{x}$ be an estimate of x and assume that they are related by (3.13). Define the error as $e = \widehat{x} - x$. We say the SNR,

$$\beta_{biased} = \frac{\sigma_x^2}{\sigma_e^2}, \tag{3.15}$$

is a biased SNR if $\widehat{x}$ is biased. When the bias is removed as in (3.14), we obtain an unbiased estimate $\widehat{x}_{unb}$. The unbiased SNR is given by

$$\beta = \frac{|\alpha|^2 \sigma_x^2}{\sigma_\tau^2}. \tag{3.16}$$

---

[2]In most practical communication systems, these assumptions are often valid because (i) the channel noise is usually zero-mean and independent of the desired signal x, and (ii) the transmitted symbols are usually iid with zero mean, which implies that the interfering symbols are also independent of x and of zero-mean.

Note that we have used variances in the above definitions of the biased and
unbiased SNRs. Because all random variables are assumed to have zero-mean,
their variances are the same as their mean squared errors.

As we will see below, the MMSE estimate $\widehat{x}_\perp$ is always biased unless its
estimation error $e_\perp = 0$, which is impossible when the observed samples y_i
contain noise. There is a simple relation between the biased and unbiased
SNRs of an MMSE estimate, as stated in the following lemma [28].

Lemma 3.1 Suppose that the MMSE estimate $\widehat{x}_\perp$ can be expressed as $\widehat{x}_\perp =$
$\alpha x + \tau$, where τ satisfies Assumptions (i) and (ii). Then the biased SNR
β_{biased} and the unbiased SNR β are related by

$$\beta_{biased} = \beta + 1.$$

Moreover the constant α is real, satisfying $0 \leq \alpha \leq 1$, and the biased and
unbiased SNRs can be respectively expressed as

$$\beta_{biased} = \frac{1}{1 - \alpha}, \tag{3.17}$$

$$\beta = \frac{\alpha}{1 - \alpha}. \tag{3.18}$$

∎

Proof. Let $\widehat{x}_\perp$ be the MMSE estimate of x. Using (3.13), we can write
the estimation error as

$$e_\perp = \widehat{x}_\perp - x = (\alpha - 1)x + \tau.$$

Note that the unconditional mean $E[e_\perp]$ is zero because the expectation
operation is over all random variables. The MSE and variance of $e_\perp$ are
the same. As x and τ are uncorrelated, the variance (or MSE) of $e_\perp$
can be written as

$$\sigma_{e_\perp}^2 = |\alpha - 1|^2 \sigma_x^2 + \sigma_\tau^2. \tag{3.19}$$

On the other hand, we have $E[e_\perp e_\perp^*] = E[e_\perp(\widehat{x}_\perp^* - x^*)]$, which is equal
to $E[-e_\perp x^*]$ because $e_\perp$ is orthogonal to $\widehat{x}_\perp$. So we can write

$$\sigma_{e_\perp}^2 = E[-e_\perp x^*] = E[-((\alpha - 1)x + \tau)x^*] = (1 - \alpha)\sigma_x^2. \tag{3.20}$$

One direct implication of the above expression is that, when we write
an MMSE estimate as $\widehat{x}_\perp = \alpha x + \tau$, the constant α is a real number $\leq$
1. Using the fact that $\sigma_{e_\perp}^2 \leq \sigma_x^2$, Eq. (3.20) implies that $\alpha \geq 0$.
Substituting (3.20) into (3.15), we immediately get the expression of
biased SNR given in (3.17). Using (3.19) and (3.20), we find that

$$\sigma_\tau^2 = \alpha(1 - \alpha)\sigma_x^2.$$

Substituting the above expression into (3.16), we get

$$\beta = \frac{\alpha^2 \sigma_x^2}{\alpha(1 - \alpha)\sigma_x^2},$$

which simplifies to (3.18). It follows from the two expressions (3.18) and
(3.17) that $\beta_{biased} = \beta + 1$.   ∎

Note that the expression (3.20) implies that $\alpha = 1$ if and only if the estimation error $e_{\perp} = 0$. In other words, *an MMSE estimate is always biased unless* $\widehat{x}_{\perp} = x$. Because an MMSE estimator minimizes the mean squared error $E[|e|^2]$, it maximizes the biased SNR $\beta_{biased} = \sigma_x^2/E[|e|^2]$. Our next lemma shows that if we remove the bias of the MMSE estimate as in (3.14), this **bias-removed MMSE estimate** is also the best linear unbiased estimate. Therefore to get an optimal unbiased estimate, we can first find an MMSE estimate and then remove the bias of the MMSE estimate before symbol detection. As bias is invariably removed before symbol detection, the bias removal operation is implicit in all MMSE receivers and it is usually not shown in the system block diagram.

Lemma 3.2 [28] Suppose that the MMSE estimate can be expressed in the form of (3.13). Define the unbiased estimate

$$\widehat{x}_{unb,\perp} = \widehat{x}_{\perp}/\alpha.$$

Then any other linear unbiased estimate of x cannot have a larger SNR than $\widehat{x}_{unb,\perp}$. ∎

Proof. Let us assume the contrary. Suppose that there is an unbiased linear estimate with a higher SNR. Let

$$\widehat{x}' = x + \theta$$

be such an unbiased linear estimate. Its unbiased SNR is given by

$$\beta' = \frac{\sigma_x^2}{\sigma_\theta^2},$$

which is assumed to be larger than β, the SNR of $\widehat{x}_{unb,\perp}$. Using the fact that $\beta_{biased} = \beta + 1$, we have $\beta_{biased} < \beta' + 1$, which can be rewritten as

$$\beta_{biased} = \frac{\sigma_x^2}{\sigma_{e_{\perp}}^2} < \frac{\sigma_x^2}{\sigma_\theta^2} + 1,$$

where $\sigma_{e_{\perp}}^2$ is the MSE of $\widehat{x}_{\perp}$. Rearranging the terms, one gets

$$\sigma_{e_{\perp}}^2 > \frac{\sigma_x^2 \sigma_\theta^2}{\sigma_x^2 + \sigma_\theta^2}. \qquad (3.21)$$

Now construct a new estimate as

$$\widehat{x}'' = \frac{\sigma_x^2}{\sigma_x^2 + \sigma_\theta^2}\widehat{x}' = \frac{\sigma_x^2}{\sigma_x^2 + \sigma_\theta^2}(x + \theta).$$

The new estimation error $e'' = \widehat{x}'' - x$ is

$$e'' = \frac{-\sigma_\theta^2}{\sigma_x^2 + \sigma_\theta^2}x + \frac{\sigma_x^2}{\sigma_x^2 + \sigma_\theta^2}\theta.$$

As the estimate $\widehat{x}'$ is unbiased, we have $E[\theta|x] = 0$, which implies that the zero-mean random variables x and θ are uncorrelated. So the new MSE is

$$\sigma_{e''}^2 = \left(\frac{\sigma_\theta^2}{\sigma_x^2 + \sigma_\theta^2}\right)^2 \sigma_x^2 + \left(\frac{\sigma_x^2}{\sigma_x^2 + \sigma_\theta^2}\right)^2 \sigma_\theta^2.$$

Simplifying the above expression, we get

$$\sigma_{e''}^2 = \frac{\sigma_x^2 \sigma_\theta^2}{\sigma_x^2 + \sigma_\theta^2}.$$

Using (3.21), we will have $\sigma_{e''}^2 < \sigma_{e_\perp}^2$, a contradiction! Thus we conclude that there does not exist any unbiased linear estimate with a higher SNR than $x_{unb,\perp}$. ∎

3.2.2   Estimation of multiple random variables

The orthogonality principle can also be applied to solve the problem of estimating more than one random variable. Suppose we are to estimate M random variables x_i, $0 \le i \le M-1$, from K observed samples y_j, $0 \le j \le K-1$. A linear estimate of x_i has the form

$$\widehat{x}_i = \sum_{j=0}^{K-1} a_{ij} y_j.$$

Defining the vectors $\mathbf{x} = [x_0 \ \ldots \ x_{M-1}]^T$, $\mathbf{y} = [y_0 \ \ldots \ y_{K-1}]^T$, and $\widehat{\mathbf{x}} = [\widehat{x}_0 \ \ldots \ \widehat{x}_{M-1}]^T$, the problem becomes one of estimating the random vector $\mathbf{x}$ from the observed vector $\mathbf{y}$:

$$\widehat{\mathbf{x}} = \mathbf{A}\mathbf{y},$$

where $\mathbf{A}$ is the $M \times K$ matrix given by $[\mathbf{A}]_{ij} = a_{ij}$. Define the error vector $\mathbf{e} = \widehat{\mathbf{x}} - \mathbf{x}$, where the kth entry is the kth estimation error $e_k = \widehat{x}_k - x_k$. Our aim is to find $\mathbf{A}$ such that $E[\mathbf{e}^\dagger \mathbf{e}]$ is minimized. Note that minimizing $E[\mathbf{e}^\dagger \mathbf{e}]$ is equivalent to minimizing $E[|e_k|^2]$ for every k. By the orthogonality principle, the estimate $\widehat{x}_i$ is optimal in the MMSE sense if and only if every estimation error $e_i = \widehat{x}_i - x_i$ is orthogonal to every y_j. That is,

$$E[e_i y_j^*] = 0, \quad \text{for } 0 \le i \le M-1, \ 0 \le j \le K-1,$$

or equivalently in matrix form

$$E[\mathbf{e}\mathbf{y}^\dagger] = \mathbf{0}. \tag{3.22}$$

The MMSE estimate $\widehat{\mathbf{x}}_\perp$ is such that the error vector $\mathbf{e}_\perp$ is orthogonal to the observed vector $\mathbf{y}$. As the MMSE estimate has the form $\widehat{\mathbf{x}}_\perp = \mathbf{A}\mathbf{y}$, the error vector is orthogonal to $\widehat{\mathbf{x}}_\perp$ as well:

$$E[\mathbf{e}_\perp \widehat{\mathbf{x}}^\dagger] = \mathbf{0}.$$

In other words, every $e_{i,\perp}$ is orthogonal to every $\widehat{x}_{j,\perp}$. Using the fact that $E[\mathbf{e}_\perp^\dagger \widehat{\mathbf{x}}_\perp] = 0$ (which follows directly from the above expression), it can be shown that the lengths of the three vectors $\widehat{\mathbf{x}}_\perp$, $\mathbf{e}_\perp$, and $\mathbf{x}$ satisfy the Pythagorean theorem:

$$E[\mathbf{x}^\dagger \mathbf{x}] = E[\widehat{\mathbf{x}}_\perp^\dagger \widehat{\mathbf{x}}_\perp] + E[\mathbf{e}_\perp^\dagger \mathbf{e}_\perp];$$

the same right triangle relation continues to hold. As in the case of one random variable, the MMSE estimate $\widehat{\mathbf{x}}_\perp$ is a biased estimate unless the estimation error $\mathbf{e}_\perp = \mathbf{0}$. Thus the following SNR for the kth estimate is biased:

$$\beta_{k,biased} = \frac{\sigma^2_{x_{k,\perp}}}{\sigma^2_{e_{k,\perp}}}.$$

When the kth estimate can be expressed as $\widehat{x}_{k,\perp} = \alpha_k x_k + \tau_k$ for some zero-mean random variable τ_k that is statistically independent of x_k, then it can also be shown that the biased and unbiased SNRs are, respectively, given by

$$\beta_{k,biased} = \frac{1}{1-\alpha_k} \quad \text{and} \quad \beta_k = \frac{\alpha_k}{1-\alpha_k}.$$

The relation $\beta_{k,biased} = \beta_k + 1$ continues to hold for all k. Moreover the bias-removed MMSE estimate $\widehat{x}_{k,\perp}/\alpha_k$ also maximizes the unbiased SNR β_k among all unbiased estimate of x_k.

3.3 MMSE equalizers

We are now ready to derive **minimum mean squared error (MMSE)** equalizers. Consider the transmission system in Fig. 3.1. An equalizer is called an MMSE equalizer if it minimizes the quantity

$$E[|e(n)|^2] = E[|\widehat{x}(n) - x(n-n_0)|^2],$$

where the integer n_0 is the system delay. In this section, we will illustrate how the orthogonality principle is used to derive MMSE equalizers for FIR channels and MIMO (multi-input multi-output) frequency nonselective channels. We consider only FIR equalizers. Readers can find more details in [64].

3.3.1 FIR channels

For an L_cth order channel $c(n)$ with noise $q(n)$, the received signal in Fig. 3.1 is

$$y(n) = \sum_{k=0}^{L_c} c(k)x(n-k) + q(n).$$

It is assumed that the transmitted signal $x(n)$ is a zero-mean wide sense stationary (WSS) random process and the noise $q(n)$ is a zero-mean WSS process uncorrelated with $x(n)$. The output of an L_ath order equalizer $a(n)$ is given by

$$\widehat{x}(n) = \sum_{k=0}^{L_a} a(k)y(n-k). \tag{3.23}$$

Define the output error $e(n) = \hat{x}(n) - x(n - n_0)$, where n_0 is the system delay. Then our goal is to find $a(n)$ so that $E[|e(n)|^2]$ is minimized. That is, we wish to find the MMSE equalizer. From (3.23), we see that finding the MMSE equalizer is equivalent to the problem of finding the MMSE estimate of $x(n - n_0)$ from the $(L_a + 1)$ received samples $y(n)$, $y(n - 1)$, ..., $y(n - L_a)$. The orthogonality principle says that the estimate is optimal if and only if the output error $e(n)$ satisfies

$$E[e(n)y^*(n - k)] = 0, \quad \text{for } 0 \leq k \leq L_a. \tag{3.24}$$

We have $(L_a + 1)$ equations and we can solve for the $(L_a + 1)$ unknowns $a(n)$, $0 \leq n \leq L_a$. A matrix approach to solving such a problem is given below.

Matrix formulation

Define the following vectors:

$$\mathbf{a} = \begin{bmatrix} a(0) \\ a(1) \\ \vdots \\ a(L_a) \end{bmatrix}, \ \mathbf{y} = \begin{bmatrix} y(n) \\ y(n-1) \\ \vdots \\ y(n-L_a) \end{bmatrix}, \ \mathbf{q} = \begin{bmatrix} q(n) \\ q(n-1) \\ \vdots \\ q(n-L_a) \end{bmatrix}, \ \mathbf{x} = \begin{bmatrix} x(n) \\ x(n-1) \\ \vdots \\ x(n-L) \end{bmatrix} \tag{3.25}$$

where $L = L_a + L_c$. Then we can write the error as $e(n) = \mathbf{y}^T\mathbf{a} - x(n - n_0)$. The orthogonality condition in (3.24) becomes

$$E\left[\mathbf{y}^*\left(\mathbf{y}^T\mathbf{a} - x(n - n_0)\right)\right] = \mathbf{0},$$

where the superscript * indicates complex conjugation. Let us denote the autocorrelation matrices of $\mathbf{y}$, $\mathbf{q}$, and $\mathbf{x}$ as $\mathbf{R}_y$, $\mathbf{R}_q$, and $\mathbf{R}_x$, respectively, and define the cross correlation vector $\mathbf{r}_{xy}(n_0) = E[x(n-n_0)\mathbf{y}^*]$. Then the MMSE equalizer is given by[3]

$$\mathbf{a}_\perp = [\mathbf{R}_y^*]^{-1}\mathbf{r}_{xy}(n_0).$$

Next we will express $\mathbf{a}_\perp$ in terms of the channel taps $c(n)$. Using the relation that $y(n) = (c * x)(n) + q(n)$, we can write the vector $\mathbf{y}$ as

$$\mathbf{y} = \mathbf{C}_{low}^T\mathbf{x} + \mathbf{q},$$

where $\mathbf{C}_{low}$ is the $(L + 1) \times (L_a + 1)$ Toeplitz matrix in (3.4). Under the assumption that the noise $q(n)$ and transmitted signal $x(n)$ are uncorrelated, we have

$$\mathbf{R}_y = \mathbf{C}_{low}^T\mathbf{R}_x\mathbf{C}_{low}^* + \mathbf{R}_q.$$

Similarly we can express the cross correlation vector $\mathbf{r}_{xy}(n_0)$ as

$$\mathbf{r}_{xy}(n_0) = \mathbf{C}_{low}^\dagger\mathbf{R}_x^*\mathbf{1}_{n_0},$$

where $\mathbf{1}_{n_0}$ is the unit vector defined in (3.5). The product $\mathbf{R}_x^*\mathbf{1}_{n_0}$ is simply the n_0th column of $\mathbf{R}_x^*$. Substituting the expressions of $\mathbf{R}_y$ and $\mathbf{r}_{xy}(n_0)$ into the expression of $\mathbf{a}_\perp$, we obtain

$$\mathbf{a}_\perp = \left[\mathbf{C}_{low}^\dagger\mathbf{R}_x^*\mathbf{C}_{low} + \mathbf{R}_q^*\right]^{-1}\mathbf{C}_{low}^\dagger\mathbf{R}_x^*\mathbf{1}_{n_0}. \tag{3.26}$$

---

[3]Because each entry of $\mathbf{y}$ is a mixture of channel noise and transmitted signal, the autocorrelation matrix $\mathbf{R}_y$ is in general invertible, except for some very special cases.

The output signal of the MMSE equalizer is

$$\widehat{x}_\perp(n) = \mathbf{y}^T \mathbf{a}_\perp = \mathbf{x}^T \mathbf{C}_{low} \mathbf{a}_\perp + \mathbf{q}^T \mathbf{a}_\perp. \tag{3.27}$$

Next we proceed to derive the minimized mean squared error. Because the error $e_\perp(n) = \widehat{x}_\perp(n) - x(n - n_0)$ is orthogonal to $\mathbf{y}$, it is also orthogonal to $\widehat{x}_\perp(n)$. Therefore we can write $\mathcal{E}_\perp(n_0) = E[|e_\perp(n)|^2]$ as

$$\mathcal{E}_\perp(n_0) = E\left[-\left(\widehat{x}_\perp(n) - x(n - n_0)\right) x^*(n - n_0)\right].$$

Substituting (3.27) into the above expression and simplifying the result, we have

$$\mathcal{E}_\perp(n_0) = \mathcal{E}_x - \mathbf{1}_{n_0}^T \mathbf{R}_x^T \mathbf{C}_{low} \left[\mathbf{C}_{low}^\dagger \mathbf{R}_x^* \mathbf{C}_{low} + \mathbf{R}_q^*\right]^{-1} \mathbf{C}_{low}^\dagger \mathbf{R}_x^* \mathbf{1}_{n_0}, \tag{3.28}$$

where $\mathcal{E}_x = r_x(0) = E[|x(n)|^2]$ is the signal power. Let us define the matrix

$$\mathbf{B}_\perp = \frac{1}{\mathcal{E}_x} \mathbf{R}_x^T \mathbf{C}_{low} \left(\mathbf{C}_{low}^\dagger \mathbf{R}_x^* \mathbf{C}_{low} + \mathbf{R}_q^*\right)^{-1} \mathbf{C}_{low}^\dagger \mathbf{R}_x^*.$$

Then the minimized mean squared error can be rewritten as

$$\mathcal{E}_\perp(n_0) = \mathcal{E}_x(1 - [\mathbf{B}_\perp]_{n_0,n_0}). \tag{3.29}$$

One direct consequence of the above expression is that all the diagonal entries of the matrix $\mathbf{B}_\perp$ satisfy $0 \leq [\mathbf{B}_\perp]_{kk} \leq 1$ (the non-negativity of $[\mathbf{B}_\perp]_{kk}$ follows from the fact that $\mathbf{B}_\perp$ is positive semidefinite). Therefore the optimal system delay n_0 is such that the n_0th diagonal entry of the matrix $\mathbf{B}_\perp$ is the largest. One can follow a procedure similar to that of the least-squares equalizer to obtain a delay-optimized MMSE equalizer.

Zero-mean iid inputs

In many systems, the transmitted samples $x(n)$ are zero-mean and iid (independent and identically distributed). This, in particular, implies that the autocorrelation matrix $\mathbf{R}_x = \mathcal{E}_x \mathbf{I}$. In this special case, the MMSE equalizer and the matrix $\mathbf{B}_\perp$ become, respectively,

$$\mathbf{a}_\perp = \left(\mathbf{C}_{low}^\dagger \mathbf{C}_{low} + \frac{1}{\mathcal{E}_x} \mathbf{R}_q^*\right)^{-1} \mathbf{C}_{low}^\dagger \mathbf{1}_{n_0}, \tag{3.30}$$

$$\mathbf{B}_\perp = \mathbf{C}_{low} \left(\mathbf{C}_{low}^\dagger \mathbf{C}_{low} + \frac{1}{\mathcal{E}_x} \mathbf{R}_q^*\right)^{-1} \mathbf{C}_{low}^\dagger. \tag{3.31}$$

It follows from the above expressions that $\mathbf{a}_\perp$ and $\mathbf{B}_\perp$ are related by $\mathbf{C}_{low} \mathbf{a}_\perp = \mathbf{B}_\perp \mathbf{1}_{n_0}$. Substituting this result into the expression of $\widehat{x}_\perp(n)$ in (3.27), we can find that the MMSE estimate can be expressed as

$$\widehat{x}_\perp(n) = \alpha x(n - n_0) + \tau(n),$$

where

$$\alpha = [\mathbf{B}_\perp]_{n_0,n_0},$$
$$\tau(n) = \sum_{k \neq n_0} [\mathbf{B}_\perp]_{k,n_0} x(n - k) + \mathbf{q}^T \mathbf{a}_\perp.$$

Note that the quantity $\tau(n)$ has zero-mean and it is statistically independent of the desired signal $x(n - n_0)$. Because $[\mathbf{B}_\perp]_{n_0,n_0} < 1$ (except for the unrealistic case that $q(n) = 0$ and the channel $c(n)$ has only one nonzero tap, see Problem 3.15), the MMSE estimate is a biased estimate. We can conclude from Lemma 3.1 that the unbiased SNR β and the biased SNR β_{biased} are, respectively, given by

$$\beta = \frac{[\mathbf{B}_\perp]_{n_0,n_0}}{1 - [\mathbf{B}_\perp]_{n_0,n_0}} \quad \text{and} \quad \beta_{biased} = \frac{1}{1 - [\mathbf{B}_\perp]_{n_0,n_0}}.$$

The relation $\beta_{biased} = \beta + 1$ continues to hold.

Uncorrelated noise   When the noise $q(n)$ is also uncorrelated, i.e. $\mathbf{R}_q = \mathcal{N}_0\mathbf{I}$, the expressions of $\mathbf{a}_\perp$ and $\mathbf{B}_\perp$ can, respectively, be simplified as

$$\mathbf{a}_\perp = \left(\mathbf{C}_{low}^\dagger \mathbf{C}_{low} + \frac{1}{\gamma}\mathbf{I}\right)^{-1} \mathbf{C}_{low}^\dagger \mathbf{1}_{n_0},$$

$$\mathbf{B}_\perp = \mathbf{C}_{low} \left(\mathbf{C}_{low}^\dagger \mathbf{C}_{low} + \frac{1}{\gamma}\mathbf{I}\right)^{-1} \mathbf{C}_{low}^\dagger,$$

where γ is the SNR $\mathcal{E}_x/\mathcal{N}_0$. It is well known that when the SNR approaches infinity, an MMSE equalizer reduces to a least-squares equalizer. One can see that as the quantity $1/\gamma$ goes to zero, $\mathbf{a}_\perp$ and $\mathbf{B}_\perp$ in the above formulas reduce, respectively, to $\mathbf{a}_{ls}$ and $\mathbf{B}_{ls}$ in (3.7) and (3.9). Using the matrix identity given in Problem 3.18(b), the equalizer and the matrix can also be expressed as

$$\mathbf{a}_\perp = \mathbf{C}_{low}^\dagger \left(\mathbf{C}_{low}\mathbf{C}_{low}^\dagger + \frac{1}{\gamma}\mathbf{I}\right)^{-1} \mathbf{1}_{n_0}, \tag{3.32}$$

$$\mathbf{B}_\perp = \mathbf{C}_{low}\mathbf{C}_{low}^\dagger \left(\mathbf{C}_{low}\mathbf{C}_{low}^\dagger + \frac{1}{\gamma}\mathbf{I}\right)^{-1}. \tag{3.33}$$

These alternative formulas are sometimes useful for simplifying expressions.

3.3.2   MIMO frequency-nonselective channels

As we will see in later chapters, MIMO frequency-nonselective channels arise in many wideband communication systems. Consider a transmission scheme where the transmitted signal is an $M \times 1$ vector $\mathbf{x}$ and the received signal is a $K \times 1$ vector $\mathbf{y}$. For frequency-nonselective channels, these vectors are related by

$$\mathbf{y} = \mathbf{Cx} + \mathbf{q}, \tag{3.34}$$

where $\mathbf{C}$ is a $K \times M$ channel matrix and $\mathbf{q}$ is a $K \times 1$ noise vector, which is independent of $\mathbf{x}$. Let the equalizer be an $M \times K$ matrix $\mathbf{A}$. Then the equalizer output is

$$\hat{\mathbf{x}} = \mathbf{Ay}.$$

The output error vector is given by $\mathbf{e} = \hat{\mathbf{x}} - \mathbf{x}$. By the orthogonality principle, the MMSE equalizer satisfies

$$E[\mathbf{ey}^\dagger] = \mathbf{0}.$$

Substituting the expression $\mathbf{e} = \mathbf{Ay} - \mathbf{x}$ into the above equation, one can show that the MMSE equalizer is given by

$$\mathbf{A}_\perp = \mathbf{R}_{xy}\mathbf{R}_y^{-1},$$

where the auto- and cross correlation matrices are, respectively, $\mathbf{R}_y = E[\mathbf{yy}^\dagger]$ and $\mathbf{R}_{xy} = E[\mathbf{xy}^\dagger]$. Using the fact that $\mathbf{x}$ and $\mathbf{q}$ are uncorrelated, one can express these correlation matrices in terms of the channel matrix $\mathbf{C}$ and arrive at

$$\mathbf{A}_\perp = \mathbf{R}_x\mathbf{C}^\dagger\left[\mathbf{CR}_x\mathbf{C}^\dagger + \mathbf{R}_q\right]^{-1}. \tag{3.35}$$

It can be verified that the autocorrelation matrix of the corresponding output error vector $\mathbf{e}_\perp$ is given by

$$\mathbf{R}_{e_\perp} = E[\mathbf{e}_\perp\mathbf{e}_\perp^\dagger] = \mathbf{R}_x - \mathbf{R}_x\mathbf{C}^\dagger\left[\mathbf{CR}_x\mathbf{C}^\dagger + \mathbf{R}_q\right]^{-1}\mathbf{CR}_x.$$

If both $\mathbf{R}_x$ and $\mathbf{R}_q$ are invertible, we can apply the matrix inversion lemma in Appendix A to rewrite the above expression more compactly as

$$\mathbf{R}_{e_\perp} = \left(\mathbf{R}_x^{-1} + \mathbf{C}^\dagger\mathbf{R}_q^{-1}\mathbf{C}\right)^{-1}.$$

The minimized mean squared error for the kth input signal x_k is given by $E[|e_{k,\perp}|^2] = [\mathbf{R}_{e_\perp}]_{kk}$. The minimized total mean squared error is trace $(\mathbf{R}_{e_\perp})$.

Special case of scalar frequency-selective channels   Observe that when $\mathbf{C} = [c(0)\ c(1)\ \ldots\ c(L_c)]$, $\mathbf{x} = [x(n)\ x(n-1)\ \ldots\ x(N-L_c)]^T$, $\mathbf{y} = y(n)$, and $\mathbf{q} = q(n)$, the MIMO channel model in (3.34) reduces to the special case of scalar channels discussed in the previous subsection. Therefore all the results of the scalar case can be obtained from the formulas for the MIMO case. For example, the MMSE equalizer in (3.26) can be derived from the formula in (3.35) by making some appropriate substitutions (see Problem 3.16).

Zero-mean iid inputs

When the input signals x_i are zero-mean and iid, the autocorrelation matrix reduces to $\mathbf{R}_x = \mathcal{E}_x\mathbf{I}$. The MMSE equalizer $\mathbf{A}_\perp$ and the autocorrelation matrix $\mathbf{R}_{e_\perp}$ become

$$\mathbf{A}_\perp = \mathbf{C}^\dagger\left[\mathbf{CC}^\dagger + \frac{1}{\mathcal{E}_x}\mathbf{R}_q\right]^{-1},$$

$$\mathbf{R}_{e_\perp} = \mathcal{E}_x\left[\mathbf{I} - \mathbf{B}_\perp\right], \tag{3.36}$$

where

$$\mathbf{B}_\perp = \mathbf{C}^\dagger\left(\mathbf{CC}^\dagger + \frac{1}{\mathcal{E}_x}\mathbf{R}_q\right)^{-1}\mathbf{C}.$$

Comparing the matrix $\mathbf{B}_\perp$ with that of the FIR channel case in (3.31), one finds that they are identical if we set $\mathbf{C} = \mathbf{C}_{low}^T$. Therefore many results are similar to those of the FIR channel case. For example, all the diagonal entries of $\mathbf{B}_\perp$ are smaller than one and the MMSE equalizer is always biased. The biased SNR of the kth estimate $\widehat{x}_{k,\perp}$ is given by

$$\beta_{k,biased} = \frac{1}{1 - [\mathbf{B}_\perp]_{kk}}.$$

The unbiased SNR is given by $\beta_k = \beta_{k,biased} - 1$. If the noise vector $\mathbf{q}$ is also uncorrelated, one can further simplify the results by substituting $\mathbf{R}_q = \mathcal{N}_0 \mathbf{I}$ into the expressions of $\mathbf{A}_\perp$ and $\mathbf{B}_\perp$.

3.3.3 Examples

Example 3.3 *Frequency-nonselective scalar channel.* Let us consider the transmission of the signal x over a frequency-nonselective channel. The received signal is

$$y = cx + q,$$

where the channel noise q is assumed to be zero-mean and independent of x. Suppose we are to use the linear estimate $\widehat{x} = ay$. By the orthogonality principle, the MMSE solution is such that the estimation error $e = \widehat{x} - x$ is orthogonal to the observation y, i.e. $E[(\widehat{x} - x)y^*] = 0$. This means $E[xy^*] = E[\widehat{x}y^*]$. Observe that because x and q are orthogonal, $E[xq^*] = 0$; we have $E[xy^*] = c^*\sigma_x^2$ and $E[\widehat{x}y^*] = a(|c|^2\sigma_x^2 + \sigma_q^2)$. We arrive at

$$a_\perp = \frac{c^*\sigma_x^2}{|c|^2\sigma_x^2 + \sigma_q^2} = \frac{\gamma c^*}{\gamma|c|^2 + 1}, \tag{3.37}$$

where $\gamma = \sigma_x^2/\sigma_q^2$. The MMSE estimate is

$$\widehat{x}_\perp = \underbrace{\frac{\gamma|c|^2}{\gamma|c|^2 + 1}}_{\alpha} x + \underbrace{\frac{\gamma c^*}{\gamma|c|^2 + 1}}_{\tau} q.$$

The output error $e_\perp = \widehat{x}_\perp - x$ is

$$e_\perp = \frac{-1}{\gamma|c|^2 + 1} x + \frac{\gamma c^*}{\gamma|c|^2 + 1} q,$$

which contains contributions from both q and x. The minimized MSE is given by

$$E[|e_\perp|^2] = \frac{\sigma_x^2}{\gamma|c|^2 + 1}. \tag{3.38}$$

The biased SNR is therefore given by

$$\beta_{biased} = \frac{\sigma_x^2}{E[|e_\perp|^2]} = \gamma|c|^2 + 1.$$

To remove the bias, we can multiply $\widehat{x}$ by $1/\alpha$ to obtain the bias-removed MMSE output as $\widehat{x}_{unb} = x + q/c$. It follows that the unbiased SNR is

$$\beta = \gamma|c|^2,$$

which is equal to $\beta_{biased} - 1$, as expected from the results of Lemma 3.1. If one is to use a zero-forcing equalizer, the output will be

$$\widehat{x}_{zf} = y/c = x + q/c\,;$$

it is identical to the unbiased estimate $\widehat{x}_{unb}$ of the MMSE equalizer. In this case, the MMSE equalizer has the same unbiased SNR as the zero-forcing equalizer. ■

In the above example, we see that when the channel has only one nonzero tap, the unbiased SNR of an MMSE equalizer is the same as that of a zero-forcing receiver. When the bias of an MMSE equalizer is removed, the resulting output is the same as a zero-forcing equalizer and hence they have the same performance. For the more general case of FIR channels, the situation is different. An MMSE equalizer can yield a higher unbiased SNR than a zero-forcing equalizer (if it exists) as demonstrated in Example 3.4.

Example 3.4 *Frequency-selective scalar channel* In this example, the channel is $C(z) = 1 + 0.95z^{-1}$. Assume that both the signal $x(n)$ and the noise $q(n)$ are zero-mean, white, and independent of each other. First we fix the signal power at $\mathcal{E}_x = 1$ and the noise power at $\mathcal{N}_0 = 0.3$. We plot the variances of $e(n)$, $e_{sig}(n)$, and $e_q(n)$ (see Eq. (3.11)) for the delay-optimized MMSE equalizer versus the equalizer order L_a. The results are shown in Fig. 3.5. Comparing these results with those of the least-squares equalizer $a_{ls}(n)$ in Example 3.2, we see that there is no noise amplification for the MMSE equalizer. The output error variance σ_e^2 is no longer dominated by the noise power $\sigma_{e_q}^2$ when L_a is large.

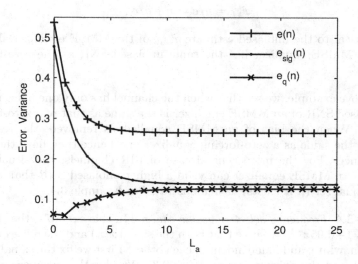

Figure 3.5. **Example 3.4.** The variances of $e(n)$, $e_{sig}(n)$, and $e_q(n)$ (defined in (3.11)) of an MMSE equalizer plotted against the equalizer orders L_a.

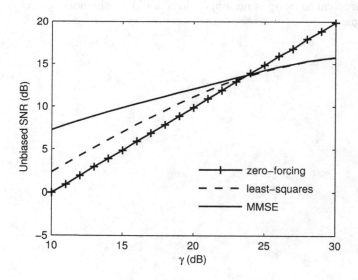

Figure 3.6. **Example 3.4.** The unbiased SNRs of the IIR zero-forcing equalizer $1/C(z)$, the least-squares equalizer $A_{ls}(z)$, and the MMSE equalizer $A_\perp(z)$. The orders of both $A_{ls}(z)$ and $A_\perp(z)$ are $L_a = 16$. Here $\gamma = \mathcal{E}_x/\mathcal{N}_0$.

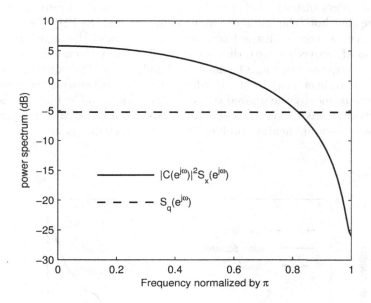

Figure 3.7. Example 3.4. The power spectra $|C(e^{j\omega})|^2 S_x(e^{j\omega})$ and $S_q(e^{j\omega})$.

Next we compare the unbiased SNRs of the MMSE equalizer, the least-squares equalizer, and the IIR zero-forcing equalizer $1/C(z)$. The results are shown in Fig. 3.6, where $\gamma = \mathcal{E}_x/\mathcal{N}_0$ and the orders of $A_\perp(z)$ and $A_{ls}(z)$ are $L_a = 16$. We see that the unbiased SNR of the MMSE equalizer is always higher than that of the least-squares equalizer, confirming the fact stated in Lemma 3.2 that the MMSE equalizer has the largest unbiased SNR among all unbiased estimates based on the same set of observations. The difference is more prominent when γ is small. This is because for a large γ, the MMSE equalizer reduces to the least-squares equalizer. Note that in the SNR region of $\gamma < 24$ dB, with a relative small order of $L_a = 16$, the unbiased SNRs of both the MMSE and least-squares equalizers are higher than that of the IIR zero-forcing equalizer even though an IIR equalizer employs all the past samples for symbol recovery. When γ is large, the ISI term becomes more significant. The MMSE and least-squares equalizers suffer from the ISI effect and thus are inferior to the zero-forcing IIR equalizer. ∎

To understand how an MMSE equalizer avoids the noise amplification problem, we examine the equalizers in Example 3.4 from the frequency domain viewpoint. In Fig. 3.7, we plot the power spectra $|C(e^{j\omega})|^2 S_x(e^{j\omega})$ (the desired signal component in the received noisy signal $y(n)$) and $S_q(e^{j\omega}) = \mathcal{N}_0$ (the noise component) for $\mathcal{E}_x = 1$ and $\mathcal{N}_0 = 0.3$. It is seen that the low-frequency content of the received samples contains mostly the signal component, whereas its high-frequency content is dominated by the noise component. Figure 3.8 shows the magnitude responses of the zero-forcing equalizer $1/C(z)$, the least-squares equalizer $A_{ls}(z)$ and the MMSE equalizer $A_\perp(z)$, where the orders of both $A_{ls}(z)$ and $A_\perp(z)$ are $L_a = 16$. From the figure, we can see that both the IIR zero-forcing equalizer and the FIR least-squares equalizer try to recover all the frequency components of the signal $x(n)$. In the process of recovering $x(n)$, they also greatly amplify the noise component in the high-frequency region. On the other hand, the MMSE equalizer $A_\perp(z)$ also tries to recover $x(n)$ in the low-frequency region because the received signal contains mostly the desired signal in this region. In the high-frequency region, where the received samples contains mostly noise, the MMSE equalizer avoids the noise amplification problem by having a smaller gain.

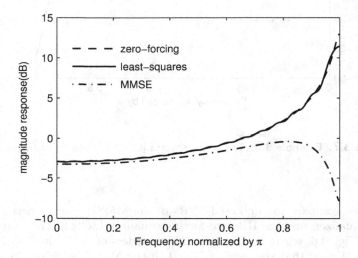

Figure 3.8. Example 3.4. The magnitude responses of the IIR zero-forcing equalizer $1/C(z)$, the least-squares equalizer $A_{ls}(z)$, and the MMSE equalizer $A_\perp(z)$. The orders of both $A_{ls}(z)$ and $A_\perp(z)$ are $L_a = 16$.

Example 3.5 *MIMO channel* Let the transmitted signal $\mathbf{x}$ and the received signal $\mathbf{y}$ be both 2×1 vectors and let the channel matrix be

$$\mathbf{C} = \begin{bmatrix} 1 & c \\ c & 1 \end{bmatrix},$$

where c is real. Suppose that $c \neq \pm 1$ so that $\mathbf{C}$ is invertible. In this case the zero-forcing receiver exists and its output is

$$\hat{\mathbf{x}} = \mathbf{x} + \mathbf{C}^{-1}\mathbf{q}.$$

Assume that $\mathbf{R}_x = \mathcal{E}_x \mathbf{I}$ and $\mathbf{R}_q = \mathcal{N}_0 \mathbf{I}$. The autocorrelation matrix of the output error is then given by $\mathbf{R}_e = \mathcal{N}_0 \mathbf{C}^{-\dagger} \mathbf{C}^{-1}$. One can calculate the unbiased

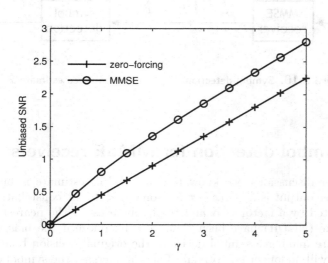

Figure 3.9. Example 3.5. Comparison of the unbiased SNRs of the zero-forcing and MMSE receivers for $c = 0.5$.

SNR of the zero-forcing receiver, which is given by

$$\beta_{k,zf} = \frac{(1-c^2)^2}{1+c^2}\gamma, \quad \text{for } k = 0, 1,$$

where the quantity $\gamma = \mathcal{E}_x/\mathcal{N}_0$. Next consider the MMSE receiver. Substituting $\mathbf{C}$ into the expression of the autocorrelation matrix $\mathbf{R}_{e,\perp}$ in (3.36), we find the two output error variances are the same, equal to

$$\sigma^2_{k,mmse} = [\mathbf{R}_{e,\perp}]_{kk} = \frac{\mathcal{N}_0(1 + \gamma^{-1} + c^2)}{(1 + \gamma^{-1} + c^2)^2 - 4c^2}, \quad \text{for } k = 0, 1.$$

The biased SNR of the MMSE receiver is

$$\beta_{k,biased} = (\gamma + 1 + \gamma c^2) - \frac{4\gamma c^2}{1 + \gamma^{-1} + c^2}.$$

The unbiased SNR of the MMSE receiver can be computed using the formula $\beta_k = \beta_{k,biased} - 1$, and it is given by

$$\beta_k = \gamma + \gamma c^2 - \frac{4\gamma c^2}{1 + \gamma^{-1} + c^2},$$

for $k = 0, 1$. Computing the difference between the unbiased SNRs of the MMSE and zero-forcing receivers, we get

$$\beta_k - \beta_{k,zf} = \frac{4\gamma c^2}{1+c^2} - \frac{4\gamma c^2}{1 + \gamma^{-1} + c^2},$$

which is always greater than zero. Unlike the scalar case in Example 3.3, the MMSE receiver has a higher unbiased SNR than the zero-forcing receiver. The two unbiased SNRs for $c = 0.5$ are plotted in Fig. 3.9. It is seen that the difference can be significant for a moderate SNR. ∎

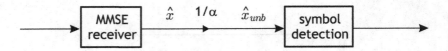

Figure 3.10. Symbol detection based on the unbiased estimate $\widehat{x}_{unb}$.

3.4   Symbol detection for MMSE receivers

From earlier discussions, we know that an MMSE estimate is biased. As the equalizer output is $\widehat{x} = \alpha x + \tau$ for some $\alpha < 1$, the constellation points are contracted by a factor of α and the boundaries for the nearest-neighbor decision rule (NNDR) have been changed. Therefore if the bias is not removed before making a symbol decision, the original decision boundaries in the NNDR will no longer be optimal. This will increase the symbol error rate (SER) as demonstrated later by Example 3.6. Thus for MMSE equalizers, it is important to remove the bias and make the symbol detection based on the unbiased estimate

$$\widehat{x}_{unb} = \widehat{x}/\alpha = x + \tau/\alpha.$$

This operation is demonstrated in Fig. 3.10. The output error of the unbiased estimate is therefore given by τ/α; it is amplified by a factor of $1/\alpha$. From (3.18), we know that α is related to the unbiased SNR β as

$$\alpha = \frac{\beta}{\beta + 1}.$$

One can see from the above formula that α will approach one as β increases. For example, $\alpha = 0.9$ if $\beta = 9$, and $\alpha = 0.99$ if $\beta = 99$. When the SNR is very high, the bias becomes negligible.

Computing SER using formulas   In Chapter 2, the SER formulas for PAM and QAM symbols are derived (see (2.12) and (2.17)). These formulas are derived under the assumption that the noise is Gaussian. From earlier discussion, we know that the quantity τ consists of both noise and ISI. Though the channel noise is, in general, Gaussian, the ISI term is a linear combination of the transmitted symbols, which are not Gaussian. However, it is known [119] that τ has an approximately Gaussian distribution when the number of transmitted symbols contributing to the ISI term is sufficiently large, e.g. 32. The Gaussian tail renders a very nice approximation of bit error rate [119, 189, 190]. As we will see in Example 3.7, this approximation is extremely good, and the formulas derived in Section 2.3 give a very accurate estimate of SER.

Example 3.6 Suppose the modulation symbol s is an equiprobable 2-bit PAM symbol with $s \in \{\pm 1, \pm 3\}$. Suppose that there is no signal processing at the transmitter so that the transmitted signal is $x = s$. One can calculate the signal power and it is given by $\sigma_x^2 = \mathcal{E}_x = 5$. Assume that the channel is AWGN. So the received signal is $y = x + q$, where the noise power is $\sigma_q^2 = \mathcal{N}_0 = 1$. The quantity γ is $\mathcal{E}_x/\mathcal{N}_0 = 5$. We consider two different receivers.

(a) *Zero-forcing equalizer* The output of a zero-forcing equalizer is $\widehat{x}_{zf} = y = x + q$. The MSE is given by

$$E[|e_{zf}|^2] = E[|q|^2] = 1.$$

From (2.12), the symbol error probability is given by

$$SER_{zf} = 3/2Q(1) = 0.238.$$

(b) *FIR MMSE equalizer* From (3.37), we have $a_\perp = \gamma/(\gamma+1) = 5/6$ and the equalizer output is

$$\widehat{x}_\perp = a_\perp y = 5/6x + 5/6q.$$

The MMSE estimate $\widehat{x}_\perp$ is biased. The MSE is given by

$$E[|e_\perp|^2] = \frac{\mathcal{N}_0}{1 + \gamma^{-1}} = 5/6,$$

which is smaller than $E[|e_{zf}|^2]$. Though it has a smaller error variance, the MMSE equalizer does not necessarily have a better SER performance than the zero-forcing equalizer. In fact if the bias is not removed before making a symbol decision, the receiver that employs an MMSE equalizer can have a worse SER performance. Suppose the receiver uses NNDR for making decisions without taking the bias into consideration. Let us compute the SER. Given that s is transmitted, $\widehat{x} = 5/6s + 5/6q$ is a Gaussian random variable with mean $m = 5/6s$ and variance $\sigma^2 = (5/6)^2$, that is,

$$P(\widehat{x}|s) = \frac{1}{\sqrt{2\pi\sigma^2}}e^{-(\widehat{x}-m)^2/(2\sigma^2)}.$$

Below we calculate the probability of symbol error when $s = 1$ is transmitted. The derivations for other cases are similar. According to NNDR, the decision $\widehat{s}$ is correct if $0 \leq \widehat{x} \leq 2$. A symbol error occurs if either $\widehat{x} < 0$ or $\widehat{x} > 2$, that is,

$$P(\widehat{s} \neq s|s = 1) = P(\widehat{x} < 0|s = 1) + P(\widehat{x} > 2|s = 1).$$

Using the Q function, we can write $P(\widehat{x} < 0|s = 1) = Q(1)$ and $P(\widehat{x} > 2|s = 1) = Q(7/5)$. Therefore

$$P(\widehat{s} \neq s|s = 1) = Q(1) + Q(7/5) = 0.2394.$$

Using a similar procedure, one can verify that $P(\widehat{s} \neq s|s = -1) = P(\widehat{s} \neq s|s = 1)$ and

$$P(\widehat{s} \neq s|s = \pm 3) = P(\widehat{x} < 2|s = 3) = Q(3/5) = 0.2743.$$

As a result, we have a symbol error rate of $SER_{mmse} = 0.2569$, which is larger than SER_{zf} in Part (a). The MMSE receiver has a worse SER performance if the bias is not corrected!

Suppose we remove the bias by dividing $\widehat{x}_\perp$ by $5/6$, the result will be $\widehat{x}/\alpha = x + q = \widehat{x}_{zf}$; it is identical to that of a zero-forcing equalizer. Thus we conclude that in this special case the MMSE and zero-forcing equalizers have the same BER performance. This example demonstrates the importance of bias removal before the symbol detection at the receiver. ∎

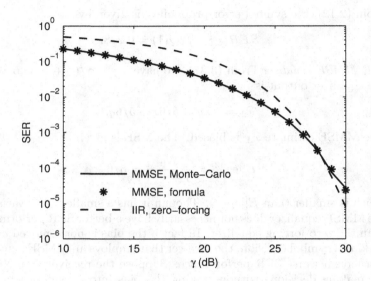

Figure 3.11. Example 3.7. Symbol error rate versus SNR $\gamma = \mathcal{E}_x/\mathcal{N}_0$.

Example 3.7 Consider the channel in Example 3.4. Suppose that $x(n)$ is an equiprobable 2-bit PAM symbol. Two different equalizers, the IIR zero-forcing equalizer and a 16th order MMSE equalizer, are considered. To obtain the SER of the system, we use Monte Carlo simulation. Figure 3.11 shows the plots of SER versus the SNR $\gamma = \mathcal{E}_x/\mathcal{N}_0$. Note that when the channel is frequency-selective, an FIR MMSE equalizer can have a smaller SER than the IIR zero-forcing equalizer for a moderate SNR γ. In the same figure, we also plot the SER values obtained by substituting the unbiased SNR of the MMSE equalizer into the SER formula in (2.12). It can be seen that the SER values obtained from the Monte Carlo experiment are indistinguishable from those obtained from the formula; the ISI term is well modeled by a Gaussian distribution. ∎

Example 3.8 In this example, we study the effect of channel zeros on the performance of the MMSE equalizers. The order of the equalizer is $L_a = 32$. The transmitted sequence $x(n)$ is a 3-bit PAM symbol and the noise $q(n)$ is

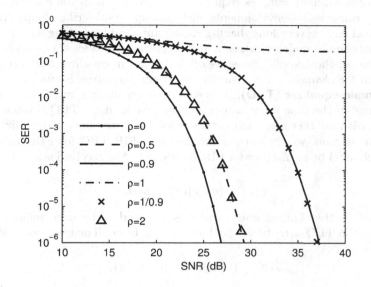

Figure 3.12. Example 3.8. Effect of channel zero on the SER performance of the MMSE equalizers.

white and Gaussian. The channel is given by

$$C(z) = \frac{1}{\sqrt{1+|\rho|^2}}(1 - \rho z^{-1}).$$

Note that the channel is normalized so that it has unit energy for all ρ. The channel has a zero at $z = \rho$. Figure 3.12 shows the SER of the delay-optimized MMSE equalizer for various positions of the channel zeros. When $\rho = 0$, the channel becomes an AWGN channel and the SER is the smallest. It is seen from the figure that when the zero moves closer to the unit circle, SER becomes worse. The system has the worst SER performance when the channel zero is on the unit circle. Moreover, note that when the channel zero is replaced with its reciprocal, the SER performance is unchanged. This phenomenon can be explained using the result in Problem 3.12.   ∎

3.5   Channel-shortening equalizers

Example 3.8 demonstrates that the performance of equalizers depends strongly on the location of channel zeros. When the channel has a zero on the unit circle, the zero-forcing equalizer is unstable and the performance of the MMSE equalizer becomes unsatisfactory. As we will see in later chapters, we can greatly alleviate this problem by using transmission schemes that insert some redundant samples known as the guard interval to the transmitted signal. The number of redundant samples required is usually equal to the channel order L_c. For transmission environments such as digital subscriber loops (DSL), the channel can be very long; having redundant samples as long as the channel order becomes impractical, as it will significantly reduce the transmission rate. Thus for these applications, we often employ an equalizer at the receiver to *shorten the channel.* Such a channel-shortening equalizer is often called the **time domain equalizer (TEQ)**. It is so named because the aim is to equalize the channel in the time domain to a shorter length, and a TEQ is often used together with another set of equalizers called frequency domain equalizers.

In this section, we shall study the design of FIR TEQ for FIR channels. Let the channel be an L_cth-order FIR filter $C(z)$. The received signal is given by

$$y(n) = (c * x)(n) + q(n),$$

where $q(n)$ is the channel noise, which is assumed to be uncorrelated with $x(n)$. Let the TEQ $a(n)$ be an L_ath-order FIR filter. Then we can write its output as

$$\widehat{x}(n) = (t * x)(n) + (a * q)(n),$$

where the "effective impulse response" $t(n)$ is given by

$$t(n) = (a * c)(n) = \sum_{k=0}^{L_a} a(k)c(n - k).$$

Unlike the previously studied equalizers, the goal of having a TEQ $a(n)$ is not to produce an output $\widehat{x}(n)$ that is close to $x(n - n_0)$ for some n_0. Its goal is to *shorten* the effective channel $t(n)$ to some predetermined target length, say $(\nu + 1)$, which is usually an integer much smaller than L_c. For an L_cth-order FIR channel and an L_ath-order FIR TEQ, the length of $(a * c)(n)$ is $(L_a + L_c + 1)$; the channel is in fact *lengthened* rather than shortened. The cascade $(a * c)(n)$ can never be a filter of length $(\nu + 1)$, but it can be a good approximation in the sense that the coefficients of $(a*c)(n)$ that fall outside a window of length $(\nu+1)$ are insignificant. By doing this we effectively shorten the long impulse response $c(n)$ to a "short" impulse response $(a * c)(n)$. We now describe the mathematical formulation of the problem. Suppose we are to design $a(n)$ so that most of the energy of $t(n)$ lies in a prescribed window of length $(\nu + 1)$, say $n_0 \leq n \leq n_0 + \nu$. Let us rewrite the equalizer output as

$$\widehat{x}(n) = \underbrace{\sum_{k=n_0}^{n_0+\nu} t(k)x(n - k)}_{\widehat{x}_d(n)} + \underbrace{\sum_{k\notin[n_0,n_0+\nu]} t(k)x(n - k)}_{\widehat{x}_u(n)} + \underbrace{(a * q)(n)}_{q_o(n)}.$$

The three quantities $\widehat{x}_d(n)$, $\widehat{x}_u(n)$, and $q_o(n)$ are, respectively, the desired signal due to the impulse response within the window, the undesired signal (i.e. the ISI term) due to the impulse response outside the window, and the noise term. Assume that the transmitted signal $x(n)$ is white with autocorrelation coefficients $r_x(k) = \mathcal{E}_x\delta(k)$ and that the noise $q(n)$ is uncorrelated with $x(n)$. Then we can write the desired signal power, the ISI power, and the noise power, respectively, as

$$P_d = \mathcal{E}_x \sum_{k=n_0}^{n_0+\nu} |t(k)|^2, \quad P_{isi} = \mathcal{E}_x \sum_{k \notin [n_0, n_0+\nu]} |t(k)|^2, \quad P_q = E[|q_o(n)|^2]. \quad (3.39)$$

The coefficients of $t(n)$ that fall inside the window contribute to the desired signal power whereas those that fall outside the window contribute to the ISI power. In what follows, we will express these three quantities in terms of the TEQ coefficients $a(n)$. Define $L = L_a + L_c$ and let the vectors $\mathbf{t}$ and $\mathbf{a}$, respectively, be defined as in (3.2). Then these vectors are related by (3.3), which we reproduce below:

$$\mathbf{t} = \mathbf{C}_{low}\mathbf{a},$$

where $\mathbf{C}_{low}$ is the $(L+1) \times (L_a+1)$ Toeplitz matrix in (3.4). To capture the desired impulse response in the prescribed window, we define an $(L+1) \times (L+1)$ diagonal matrix $\mathbf{D}_{n_0}$ whose diagonal entries are

$$[\mathbf{D}_{n_0}]_{ii} = \begin{cases} 1, & n_0 \leq n \leq n_0 + \nu; \\ 0, & \text{otherwise.} \end{cases}$$

Then the coefficients that lie inside and outside the window are, respectively, given by

$$\mathbf{D}_{n_0}\mathbf{t} = \mathbf{D}_{n_0}\mathbf{C}_{low}\mathbf{a},$$
$$(\mathbf{I} - \mathbf{D}_{n_0})\mathbf{t} = (\mathbf{I} - \mathbf{D}_{n_0})\mathbf{C}_{low}\mathbf{a}.$$

From the above expressions, we can write the three quantities in (3.39) as

$$P_d = \mathcal{E}_x \mathbf{a}^\dagger \mathbf{C}_{low}^\dagger \mathbf{D}_{n_0} \mathbf{C}_{low}\mathbf{a},$$
$$P_{isi} = \mathcal{E}_x \mathbf{a}^\dagger \mathbf{C}_{low}^\dagger (\mathbf{I} - \mathbf{D}_{n_0}) \mathbf{C}_{low}\mathbf{a}, \quad (3.40)$$
$$P_q = \mathbf{a}^\dagger \mathbf{R}_q \mathbf{a},$$

where $\mathbf{R}_q$ is the $(L_a+1) \times (L_a+1)$ autocorrelation matrix of the channel noise $q(n)$. Note that we have used $\mathbf{D}_{n_0}^\dagger \mathbf{D}_{n_0} = \mathbf{D}_{n_0}$ and $(\mathbf{I} - \mathbf{D}_{n_0})^\dagger (\mathbf{I} - \mathbf{D}_{n_0}) = (\mathbf{I} - \mathbf{D}_{n_0})$ in the above equations. With the above expressions, we now proceed to the design of the TEQ $a(n)$. There are many design criteria for TEQ. Below we consider the TEQ that maximizes the SIR or SINR depending on the availability of the noise statistics.

(1) Maximization of SIR   The signal to interference ratio (SIR) is defined as

$$SIR = \frac{P_d}{P_{isi}}.$$

Substituting the expressions for P_d and P_{isi} into the above equation, the TEQ that maximizes the SIR is given by

$$\mathbf{a}_{opt} = \arg\max_{\mathbf{a}} \frac{\mathbf{a}^\dagger \mathbf{C}_{low}^\dagger \mathbf{D}_{n_0} \mathbf{C}_{low}\mathbf{a}}{\mathbf{a}^\dagger \mathbf{C}_{low}^\dagger (\mathbf{I} - \mathbf{D}_{n_0}) \mathbf{C}_{low}\mathbf{a}}.$$

Note that the problem of finding $\mathbf{a}$ to maximize the ratio P_d/P_{isi} is equivalent to that of finding $\mathbf{a}$ to minimize P_{isi} under the constraint $P_d = 1$. The latter criteria is used to design TEQ in [90] and the optimal TEQ is called the MSSNR (maximum-shortening SNR) TEQ. Thus the SIR-maximized TEQ is the same as the MSSNR TEQ.

(2) Maximization of SINR When the noise statistics are available, we can exploit them in the design of TEQ. Define the signal to interference and noise ratio (SINR) as

$$SINR = \frac{P_d}{P_{isi} + P_q}.$$

Then the optimal TEQ $a(n)$ becomes

$$\mathbf{a}_{opt} = \arg\max_{\mathbf{a}} \frac{\mathbf{a}^\dagger \mathbf{C}_{low}^\dagger \mathbf{D}_{n_0} \mathbf{C}_{low} \mathbf{a}}{\mathbf{a}^\dagger \mathbf{C}_{low}^\dagger (\mathbf{I} - \mathbf{D}_{n_0}) \mathbf{C}_{low} \mathbf{a} + \frac{1}{\mathcal{E}_x} \mathbf{a}^\dagger \mathbf{R}_q \mathbf{a}}.$$

Note that scaling the vector $\mathbf{a}$ has no effect on the SIR and the SINR values. We can consider a unit-norm vector $\mathbf{a}$. Then the problem of finding the optimal $a(n)$ can be expressed as

$$\mathbf{a}_{opt} = \arg\max_{\mathbf{a}} \frac{\mathbf{a}^\dagger \mathbf{Q}_0 \mathbf{a}}{\mathbf{a}^\dagger \mathbf{Q}_1 \mathbf{a}}, \tag{3.41}$$

for some positive semidefinite matrix $\mathbf{Q}_0$ and positive definite matrix[4] $\mathbf{Q}_1$. The solution to the above problem can be obtained by applying the Rayleigh–Ritz principle (Appendix A). Let $\mathbf{Q}_2$ be a positive definite matrix such that $\mathbf{Q}_1^{-1} = \mathbf{Q}_2 \mathbf{Q}_2^\dagger$ and define the vector $\mathbf{b} = \mathbf{Q}_2^{-1} \mathbf{a}$. Then (3.41) can be rewritten as

$$\mathbf{a}_{opt} = \mathbf{Q}_2 \arg\max_{\mathbf{b}} \frac{\mathbf{b}^\dagger \mathbf{Q}_2^\dagger \mathbf{Q}_0 \mathbf{Q}_2 \mathbf{b}}{\mathbf{b}^\dagger \mathbf{b}}.$$

From the Rayleigh–Ritz principle, the optimal TEQ is given by $\mathbf{a}_{opt} = \mathbf{Q}_2 \mathbf{v}$, where $\mathbf{v}$ is the eigenvector of the matrix $\mathbf{Q}_2^\dagger \mathbf{Q}_0 \mathbf{Q}_2$ associated with its largest eigenvalue. As we shall see in the numerical example below, the optimal TEQ based on either objective function can effectively shorten the channel.

The Choice of n_0 The starting point n_0 of the window affects the optimal TEQ solution. To get the best n_0, one has to solve the optimal TEQ for $0 \le n_0 \le L - \nu$ and choose the value of n_0 that maximizes the objective function. In practice, it is found that the optimal n_0 is usually an integer much smaller than L_c.

Example 3.9 Let the channel $c(n)$ be an FIR filter of length 100:

$$c(n) = 0.9^n \cos\left(\frac{\pi n}{40}\right) - 0.8^n \cos\left(\frac{\pi n}{60}\right) - 0.7^n \sin\left(\frac{\pi n}{80}\right) + 0.6^n \sin\left(\frac{\pi n}{100}\right),$$

for $0 \le n < 100$. This impulse response is plotted in Fig. 3.13. Suppose that we are to design a sixth-order TEQ $a(n)$ with $\nu = 10$. TEQs optimized for the two criteria described earlier are considered. For the case of SINR, we assume that the noise is white and the SNR $\gamma = \mathcal{E}_x / \mathcal{N}_0 = 30$ dB. The resulting impulse responses $(a * c)(n)$ are shown in Fig. 3.14. It is seen that both TEQs can effectively shorten the channel with length 100 to the prescribed window of length 11. The SIR is 94.3 dB for the SIR-maximized TEQ and the SINR is 28.9 dB for the SINR-maximized TEQ. ∎

---

[4]In fact, it can be showed (see Problem 3.20) that the matrix $\mathbf{C}_{low}^\dagger (\mathbf{I} - \mathbf{D}_{n_0}) \mathbf{C}_{low}$ (hence $\mathbf{Q}_1$) is positive definite except for the degenerate case of $L_c \le \nu$, which implies that no channel shortening is needed.

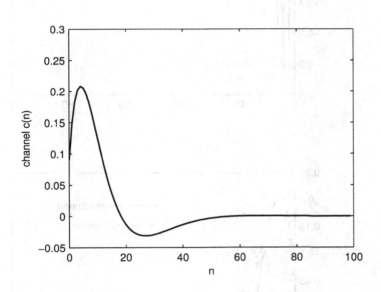

Figure 3.13. Example 3.9. Channel impulse response.

(a)

(b)

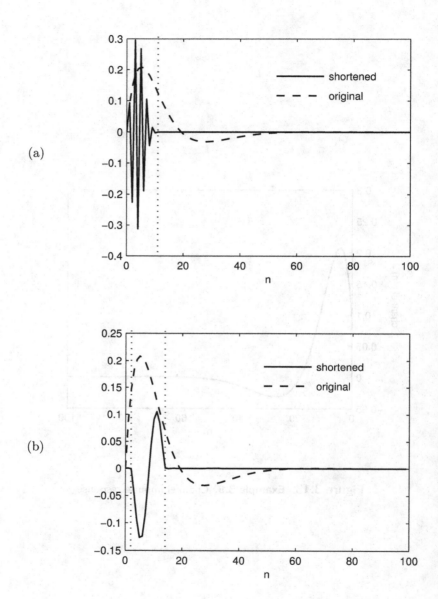

Figure 3.14. **Example 3.9.** Shortened impulse response where the TEQ $a(n)$ is designed to maximize (a) SIR, (b) $SINR$.

3.6   Concluding Remarks

In this chapter, we studied the problem of channel equalization for FIR channels. The FIR zero-forcing equalizer and the MMSE equalizer were derived. Both equalizers were obtained by simple matrix computations. A powerful tool for solving many communication problems, the orthogonality principle, was also studied. For more detailed treatments of estimation and detection problems, see [64, 65].

The idea of channel-shortening equalizers was introduced by Melsa in [90]. Since then, many design approaches have been proposed. The SIR and SINR criteria given in this chapter for designing TEQ were proposed in [37, 152]. One can also design the TEQ by incorporating the specific structure of the transmitter and receiver into the TEQ optimization. These approaches are often more complex but can result in better system performance. For further details, the reader is referred to [6, 10, 42, 82, 87, 165].

3.7   Problems

3.1 Prove that the matrix $\mathbf{C}_{low}$ in (3.4) has full rank whenever the L_cth-order channel $C(z) \neq 0$.

3.2 Let $C(z) = 1 + z^{-1}$. Find the second-order least-squares equalizer $a_{ls}(n)$. What is the optimal system delay n_0?

3.3 Let $A(z) = a(0) + a(1)z^{-1} + \cdots + a(L_a)z^{-L_a}$ be the delay-optimized least-squares equalizer of the channel $C(z) = c(0) + c(1)z^{-1} + \cdots + c(L_c)z^{-L_c}$, and let the corresponding optimal system delay be n_0.

   (a) Prove that the delay-optimized L_ath-order least-squares equalizer for the channel $C'(z) = c(L_c) + c(L_c - 1)z^{-1} + \cdots + c(0)z^{-L_c}$ is given by $A'(z) = a(L_a) + a(L_a - 1)z^{-1} + \cdots + a(0)z^{-L_a}$. What is the optimal system delay n_0'? How are the minimized MSEs for these two cases related?

   (b) What is the L_ath-order delay-optimized least-squares equalizer when the channel is $C''(z) = c(0) - c(1)z^{-1} + c(2)z^{-2} - \cdots + (-1)^{L_c}c(L_c)z^{-L_c}$? What is the optimal system delay? Is the minimized MSE for the case of $C''(z)$ the same as that of $C(z)$? Justify your answer.

   (c) Prove that $A(z^2)$ is the $2L_a$th-order delay-optimized least-squares equalizer for $C(z^2)$.

3.4 (a) Let $\mathbf{U\Lambda V}$ be the singular value decomposition of the matrix $\mathbf{C}_{low}$ in (3.4). Show that the matrix $\mathbf{B}_{ls}$ in (3.9) can be expressed as

$$\mathbf{B}_{ls} = \mathbf{U} \begin{bmatrix} \mathbf{I}_{L_a+1} & \mathbf{0} \\ \mathbf{0} & \mathbf{0} \end{bmatrix} \mathbf{U}^\dagger.$$

   (b) Express the kth diagonal entry $[\mathbf{B}_{ls}]_{kk}$ of $\mathbf{B}_{ls}$ in terms of the entries in the kth row vector of $\mathbf{U}$.

   (c) Prove that $0 \leq [\mathbf{B}_{ls}]_{kk} \leq 1$ for $0 \leq k \leq L$.

3.5 From the previous problem, we know that $0 \leq [\mathbf{B}_{ls}]_{kk} \leq 1$ for all k.

    (a) Find the conditions on the channel $c(n)$ such that $\max_k [\mathbf{B}_{ls}]_{kk} = 0$.

    (b) Find the conditions on the channel $c(n)$ such that $[\mathbf{B}_{ls}]_{kk} = 1$ for some k.

3.6 Suppose that the transmitted signal and the channel noise are both white and that they are uncorrelated. Let $e_{ls}(n) = \widehat{x}(n) - x(n - n_0)$ be the output error of an L_ath-order least-squares equalizer. Compute the output error variance $\mathcal{E}_{ls}(n_0) = E[|e_{ls}(n)|^2]$. Express your answer in terms of $\mathbf{C}_{low}$, $\mathcal{E}_x$ and $\mathcal{N}_0$.

3.7 *Orthogonality principle for the vector case* One can generalize the orthogonality principle to the vector case. Suppose we are to estimate an $M \times 1$ random vector $\mathbf{x}$ from a set of $M \times 1$ observed vectors $\mathbf{y}_0, \mathbf{y}_1, \ldots, \mathbf{y}_{K-1}$. The linear estimate is

$$\widehat{\mathbf{x}} = \mathbf{A}_0 \mathbf{y}_0 + \mathbf{A}_1 \mathbf{y}_1 + \cdots + \mathbf{A}_{K-1} \mathbf{y}_{K-1}.$$

Show that the estimate $\widehat{\mathbf{x}}_\perp$ is optimal if and only if its error vector $\mathbf{e}_\perp = \widehat{\mathbf{x}}_\perp - \mathbf{x}$ satisfies $E[\mathbf{e}_\perp \mathbf{y}_i^\dagger] = \mathbf{0}$, for $i = 0, 1, \ldots, K - 1$.

3.8 Suppose we are to estimate x_i using the observation variables $y_i = a_i x_i + q_i$. Assume that x_i are uncorrelated with q_i for $0 \leq i, j \leq K - 1$, and the autocorrelation matrices of $\mathbf{x}$ and $\mathbf{q}$ are both diagonal (not necessarily equal to a scaled identity matrix). Show that the vector MMSE estimation problem reduces to K scalar MMSE estimation problems.

3.9 Let the $M \times N$ matrix $\mathbf{A}$ be the optimal linear estimator of an $M \times 1$ vector $\mathbf{s}$ given an $N \times 1$ observation vector $\mathbf{r}$. Suppose now we are to estimate $\mathbf{s}$ using $\mathbf{Br}$, where $\mathbf{B}$ is an invertible matrix. Prove that the optimal estimator in this case is $\mathbf{AB}^{-1}$ and the mean square error remains the same. (This shows that, without loss of generality, applying an invertible matrix to the observation vector does not change the minimized MSE of an MMSE equalizer.)

3.10 Verify that the minimized MSE for an MMSE equalizer can be expressed as in (3.28).

3.11 Let $C(z) = 1 + z^{-1}$. Suppose both the transmitted signal and channel noise are white with powers $\mathcal{E}_x$ and $\mathcal{N}_0$, and they are uncorrelated. Find the second-order MMSE equalizer $a_\perp(n)$. If the channel is changed to $C'(z) = 1 - z^{-1}$, how is the new MMSE equalizer $a'_\perp(n)$ related to $a_\perp(n)$?

3.12 Suppose both the transmitted signal and channel noise are white with powers $\mathcal{E}_x$ and $\mathcal{N}_0$, and they are uncorrelated. Let the channel be $C(z) = \sum_{n=0}^{L_c} c(n) z^{-n}$ and the L_ath-order delay-optimized MMSE equalizer be $a_\perp(n)$ with the optimal decision delay given by n_0.

    (a) Express σ_e^2 (the output error variance due to the channel noise) in terms of $\mathcal{N}_0$ and $a_\perp(n)$.

(b) Prove that the delay-optimized L_ath-order MMSE equalizer for the reversed channel $C'(z) = c(L_c) + c(L_c - 1)z^{-1} + \cdots + c(0)z^{-L_c}$ is given by $a'_\perp(n) = a_\perp(L_a - n)$. What is the optimal decision delay n'_0? How are the minimized MSEs for these two cases related?

(c) What is the delay-optimized L_ath-order MMSE equalizer $a''_\perp(n)$ when the channel is $C''(z) = c(0) - c(1)z^{-1} + c(2)z^{-2} - \cdots + (-1)^{L_c}c(L_c)z^{-L_c}$? What is the optimal system delay? Is the minimized MSE for the case of $C''(z)$ the same as that of $C(z)$? Justify your answer.

(d) What is the $2L_a$th-order delay-optimized equalizer when the channel is $C(z^2)$? Justify your answer.

3.13 Explain why the quantities

$$\min_{n_0} \mathcal{E}_\perp(n_0) \quad \text{and} \quad \min_{n_0} \mathcal{E}_{d,ls}(n_0)$$

given in (3.10) and (3.29) cannot increase with L_a, the order of the equalizer.

3.14 Suppose that we have a two-input one-output channel whose transfer matrix is $\mathbf{c} = [1 \ 1]$. The input vector $\mathbf{x}$ is 2×1 with $\mathbf{R}_x = \mathcal{E}_x \mathbf{I}_2$ and the noise q has variance $\mathcal{N}_0$. Find the MMSE equalizer and the minimized MSE trace($\mathbf{R}_e$).

3.15 Consider the matrix $\mathbf{B}_\perp$ given in (3.31). Show that $[\mathbf{B}_\perp]_{kk} = 1$ for some k only if the channel $c(n)$ has only one nonzero tap and the noise $q(n)$ is zero. (Hint: $[\mathbf{B}_\perp]_{kk} = 1$ for some k if and only if $\mathcal{E}_\perp(k) = 0$ for some k.)

3.16 Let $\mathbf{C} = [c(0) \ c(1) \ \ldots \ c(L_c)]$, $\mathbf{x} = [x(n) \ x(n-1) \ \ldots \ x(N-L_c)]^T$, $\mathbf{y} = y(n)$, and $\mathbf{q} = q(n)$. Show that the MMSE equalizer in (3.35) reduces to that in (3.26).

3.17 *MMSE equalizers for linear phase channels* A matrix $\mathbf{J}$ is said to be a reversal matrix if all its entries are zero except the anti-diagonal entries. For example, a 3×3 reversal matrix is given by

$$\mathbf{J} = \begin{bmatrix} 0 & 0 & 1 \\ 0 & 1 & 0 \\ 1 & 0 & 0 \end{bmatrix}.$$

Let the input signal $x(n)$, the channel $c(n)$, and the noise $q(n)$ be real in Fig. 3.1.

(a) Let $\mathbf{x}$ and $\mathbf{q}$ be the vectors defined in (3.25). Show that their autocorrelation matrices satisfy $\mathbf{J}\mathbf{R}_x\mathbf{J} = \mathbf{R}_x$ and $\mathbf{J}\mathbf{R}_q\mathbf{J} = \mathbf{R}_q$.

(b) Show that if $c(n)$ is symmetric, i.e. $c(L_c - n) = c(n)$, then $\mathbf{J}\mathbf{C}_{low}\mathbf{J} = \mathbf{C}_{low}$, where $\mathbf{C}_{low}$ is the $(L+1) \times (L_a+1)$ Toeplitz matrix defined in (3.4).

(c) Suppose that the order L_a of the equalizer is such that $L = L_a + L_c$ is even. Let the system delay be $n_0 = L/2$. Show that the MMSE equalizer $a_\perp(n)$ is symmetric when $c(n)$ is symmetric.

(d) What would $a_\perp(n)$ be if $c(n)$ is anti-symmetric (i.e. $c(L_c - n) = -c(n)$)?

3.18 Let $\mathbf{Q}$ be an $m \times m$ positive definite matrix so that $\mathbf{Q} = \mathbf{SS}^\dagger$ for some invertible $\mathbf{S}$. Consider the following matrix:

$$\mathbf{P} = \mathbf{A}^\dagger \left[\mathbf{AA}^\dagger + \mathbf{Q} \right]^{-1} \mathbf{A},$$

where $\mathbf{A}$ is an $m \times n$ matrix. By following the procedure described below, we can prove that all the diagonal entries of $\mathbf{P}$ are strictly smaller than one.

(a) Show that $\mathbf{P}$ can be expressed as $\mathbf{B}^\dagger \left[\mathbf{BB}^\dagger + \mathbf{I}_m \right]^{-1} \mathbf{B}$, where $\mathbf{B} = \mathbf{S}^{-1}\mathbf{A}$.

(b) For any $m \times n$ matrix $\mathbf{B}$, verify the identity

$$\mathbf{B}[\mathbf{B}^\dagger\mathbf{B} + \mathbf{I}_n] = [\mathbf{BB}^\dagger + \mathbf{I}_m]\mathbf{B},$$

which proves that

$$\left[\mathbf{BB}^\dagger + \mathbf{I}_m \right]^{-1} \mathbf{B} = \mathbf{B} \left[\mathbf{B}^\dagger\mathbf{B} + \mathbf{I}_n \right]^{-1}$$

for any $\mathbf{B}$.

(c) Verify the following identity:

$$\mathbf{B}^\dagger\mathbf{B} \left[\mathbf{B}^\dagger\mathbf{B} + \mathbf{I}_n \right]^{-1} + \left[\mathbf{B}^\dagger\mathbf{B} + \mathbf{I}_n \right]^{-1} = \mathbf{I}.$$

(d) Using the results in (a), (b), and (c), and the fact that the matrix $\left[\mathbf{B}^\dagger\mathbf{B} + \mathbf{I}_n \right]^{-1}$ is positive definite, show that the diagonal entries of $\mathbf{P}$ are strictly smaller than one.

3.19 *Single-input multi-output channel (SIMO)* Consider the following SIMO channel:

$$\mathbf{y} = \mathbf{c}x + \mathbf{q},$$

where x is the scalar transmitted signal, $\mathbf{y}$ is the $N \times 1$ received vector, and $\mathbf{q}$ is the noise vector which is assumed to be uncorrelated with x. Let the equalizer output be

$$\widehat{x} = \mathbf{a}^T\mathbf{c}x + \mathbf{a}^T\mathbf{q}.$$

Define the unbiased SNR

$$\beta = \frac{|\mathbf{a}^T\mathbf{c}|^2\sigma_x^2}{\mathbf{a}^T\mathbf{R}_q\mathbf{a}^*}.$$

Assume that the noise autocorrelation matrix $\mathbf{R}_q$ is positive definite and let $\mathbf{R}_q = \mathbf{QQ}^\dagger$.

(a) Find the equalizer $\mathbf{a}_{opt}^T$ that maximizes the unbiased SNR β. Express your answer in terms of $\mathbf{c}$ and $\mathbf{Q}$.

(b) Let $\mathbf{a}_\perp^T$ be the MMSE equalizer. How is $\mathbf{a}_{opt}^T$ related to $\mathbf{a}_\perp^T$?

3.20 Show that the matrix $\mathbf{C}_{low}^{\dagger}(\mathbf{I}-\mathbf{D}_{n_0})\mathbf{C}_{low}$ in (3.40) can be singular only if the channel order $L_c \le \nu$. This implies that the ISI can never be completely removed by the TEQ unless the original channel $c(n)$ is already shorter than or equal to the target channel length. (Hint: When $L_c > \nu$, can we find a nonzero TEQ $a(n)$ such that $(\mathbf{I}-\mathbf{D}_{n_0})\mathbf{t}=\mathbf{0}$?)

3.21 Let the channel be $C(z) = 1+z^{-1}+z^{-2}+z^{-3}+z^{-4}$. Let $\nu = 2$ and $n_0 = 2$. Assume that both signal $x(n)$ and noise $q(n)$ are white with power $\mathcal{E}_x$ and $\mathcal{N}_0$, respectively. Find a first-order TEQ $A(z) = a(0)+a(1)z^{-1}$ that maximizes (a) SIR, (b) $SINR$.

3.22 *An MMSE criterion for designing TEQs* [6] Consider the block diagram shown in Fig. P3.22. A different way of designing the TEQ $a(n)$ is to find $a(n)$ and $t_d(n)$ so that the mean squared error

$$E[|w(n) - y(n)|^2] \tag{3.42}$$

is minimized. The sequence $t_d(n)$ is called the target impulse response (TIR). This design problem is split into three steps as described below. (Assume that the signal $x(n)$ and noise $q(n)$ are uncorrelated, and the signal $x(n)$ is white with signal power $\mathcal{E}_x$.)

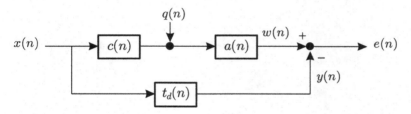

Figure P3.22. The TEQ that minimizes the output mean squared error.

(a) Let $t_d(n)$ be FIR with nonzero coefficients for $n_0 \le n \le n_0 + \nu$. Given $t_d(n)$, find $a(n)$ so that the mean squared error

$$\mathcal{E}(t_d) = E[|w(n) - (x * t_d)(n)|^2]$$

is minimized. Express your answer in terms of the TIR vector $\mathbf{t}_d = [t_d(n_0)\ t_d(n_0+1)\ \ldots\ t_d(n_0+\nu)]^T$, the matrix $\mathbf{C}_{low}$, and the noise autocorrelation matrix $\mathbf{R}_q$.

(b) Let $\mathcal{E}_0(t_d)$ be the minimized mean squared error. Express $\mathcal{E}_0(t_d)$ in terms of $\mathbf{t}_d$ and $\mathbf{C}_{low}$.

(c) Find a unit-norm $\mathbf{t}_d$ so that $\mathcal{E}_0(t_d)$ is minimized. What is the corresponding optimal TEQ $a(n)$?

3.23 Let the system parameters be the same as those in Problem 3.21. Use the method described in Problem 3.22 to design an optimal first-order TEQ $A(z) = a(0)+a(1)z^{-1}$. Is the resultant TEQ identical to the TEQ designed in Problem 3.21(b)?

3.24 In this problem, we would like to prove that the optimal TEQ obtained
by using the MMSE criteria described in Problem 3.22 is a scaled version
of the optimal TEQ that maximizes the SINR. This can be done by
following the steps given below.

(a) Let $a_\perp(n)$ and $t_{d,\perp}(n)$ be, respectively, the optimal TEQ and de-
sired response obtained in Problem 3.22(c). Prove that $t_{d,\perp}(n) = (a_\perp * c)(n)$ for $n_0 \leq n \leq n_0 + \nu$. (Hint: Show that if $t_{d,\perp}(n) \neq (a_\perp * c)(n)$ for some $n_0 \leq n \leq n_0 + \nu$, then we can find another $t_d(n)$ with a smaller mean squared error.)

(b) Using the result from (a), show that finding the $a(n)$ and $t_d(n)$ that
minimize the mean squared error defined in (3.42) is equivalent to
finding the $a(n)$ that minimizes $P_{isi} + P_q$.

(c) Prove that the optimal $a(n)$ that minimizes $P_{isi} + P_q$, under the
constraint that $P_d = 1$, differs from the TEQ that maximizes the
SINR only by a scale factor.

As the TEQ performance is not changed by scaling, this proves that the
MMSE TEQ and SINR-maximized TEQ have the same performance.

4

Fundamentals of multirate signal processing

Multirate signal processing has been found to be very useful in modern communication systems. In this chapter, we introduce basic multirate concepts which are fundamental to many of our discussions in future chapters.

4.1 Multirate building blocks

The decimator and expander are two fundamental building blocks in multirate signal processing systems. The M-fold **decimator**, shown in Fig. 4.1(a), is defined by the input-output relation

$$y_d(n) = x(Mn), \tag{4.1}$$

which means that every Mth sample of the input is retained. This is demonstrated in the figure for $M = 3$. Note that the samples are renumbered by the decimator such that

$$y_d(0) = x(0), \quad y_d(1) = x(3), \quad y_d(2) = x(6), \quad \ldots,$$

and so forth. The decimator is also known as the downsampler. In general, decimation results in a loss of information. But if a signal is bandlimited in a certain way then it can be decimated without loss of information, as we shall see in Section 4.2.

The M-fold **expander**, or upsampler, shown in Fig. 4.2(a), is defined in the time domain by the input-output relation

$$y_e(n) = \begin{cases} x(n/M), & n = \text{multiple of } M; \\ 0, & \text{otherwise.} \end{cases} \tag{4.2}$$

Thus the expander merely inserts zero-valued samples in a systematic manner, namely $M - 1$ zeros are inserted between adjacent samples of $x(n)$. This is demonstrated in Figs. 4.2(b) and (c) for $M = 3$. Evidently there is no loss of information due to the expanding operation. From the definition of the expander we see that

$$y_e(Mn) = x(n),$$

71

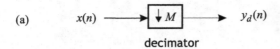

(a)

decimator

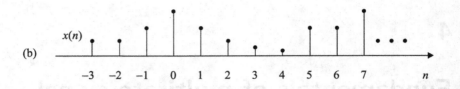

(b)

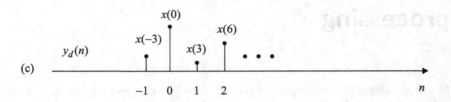

(c)

Figure 4.1. (a) The M-fold decimator or downsampler. (b),(c) Examples of input and output signals with $M = 3$.

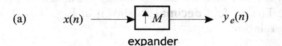

(a)

expander

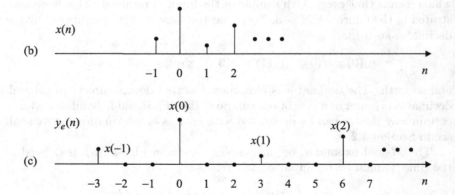

(b)

(c)

Figure 4.2. (a) The M-fold expander or upsampler. (b),(c) Examples of input and output signals with $M = 3$.

which shows that we can recover the input $x(n)$ from $y_e(n)$ simply by M-fold decimation. The expander is usually followed by a digital filter called an interpolation filter which makes it more interesting (Section 4.3).

4.1.1 Transform domain formulas

In the z-domain the input–output relation for the expander can be written as

$$Y_e(z) = X(z^M) \qquad \text{(expander)}. \tag{4.3}$$

To see this note that $y_e(n) = 0$ when n is not a multiple of M, and we get

$$Y_e(z) = \sum_{n=-\infty}^{\infty} y_e(Mn) z^{-Mn} = \sum_{n=-\infty}^{\infty} x(n) z^{-Mn} = X(z^M),$$

where we have used the fact that $y_e(Mn) = x(n)$. Next, for the decimator we claim that the input–output relation in the z-domain can be written as

$$Y_d(z) = \frac{1}{M} \sum_{\ell=0}^{M-1} X(z^{1/M} W^\ell) \qquad \text{(decimator)}, \tag{4.4}$$

where the quantity W is defined as

$$W = e^{-j2\pi/M}. \tag{4.5}$$

Note that W is an Mth root of unity, that is $W^M = 1$. We shall defer the proof of the above expression to Section 4.4.

It is important to understand (4.3) and (4.4) in terms of the frequency variable ω. Substituting $z = e^{j\omega}$ we have

$$Y_e(e^{j\omega}) = X(e^{j\omega M}) \qquad \text{(expander)} \tag{4.6}$$

and

$$Y_d(e^{j\omega}) = \frac{1}{M} \sum_{\ell=0}^{M-1} X(e^{j(\omega - 2\pi\ell)/M}) \qquad \text{(decimator)}. \tag{4.7}$$

Thus, for the expander, $Y_e(e^{j\omega})$ is obtained by *squeezing* or shrinking the plot of $X(e^{j\omega})$ by a factor of M (Fig. 4.3). There are M squeezed copies of $X(e^{j\omega})$ in a period of 2π. The extra $M - 1$ copies are called images. For the decimator, the output $Y_d(e^{j\omega})$ is a sum of M terms. The zeroth term contains $X(e^{j\omega/M})$, which is simply an M-fold *stretched version* of $X(e^{j\omega})$. The first term is

$$X(e^{j(\omega - 2\pi)/M})$$

which is the stretched version shifted by 2π. More generally, the ℓth term $X(e^{j(\omega - 2\pi\ell)/M})$ is the stretched version shifted by $2\pi\ell$. Figure 4.4 demonstrates this for $M = 3$.

The shifted versions in general overlap with the original stretched version. This is called the **aliasing effect**, and it is similar to aliasing created by undersampling a bandlimited signal [105]. Since the transform domain expressions for decimation are rather complicated, we often abbreviate them as follows:

$$X(e^{j\omega})\Big|_{\downarrow M} \quad \text{or} \quad \left[X(e^{j\omega})\right]_{\downarrow M} \triangleq \frac{1}{M} \sum_{\ell=0}^{M-1} X(e^{j(\omega - 2\pi\ell)/M}); \tag{4.8}$$

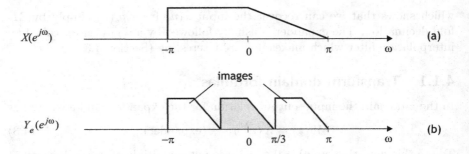

Figure 4.3. Fourier transforms of (a) input to the expander and (b) output of the expander for $M = 3$. In general there are $M - 1$ images.

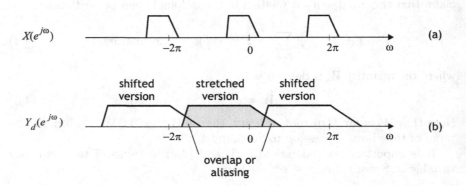

Figure 4.4. Fourier transforms of (a) input to the decimator and (b) output of the decimator for $M = 3$. In general there are $M - 1$ shifted versions of the stretched version.

$$X(z)\Big|_{\downarrow M} \quad \text{or} \quad \big[X(z)\big]_{\downarrow M} \triangleq \frac{1}{M} \sum_{\ell=0}^{M-1} X(z^{1/M} W^{\ell}). \qquad (4.9)$$

Similarly, the notation $[x(n)]_{\downarrow M}$ is often used in the time domain instead of $x(Mn)$. For the expander we use similar notations, namely $[x(n)]_{\uparrow M}$ and $[X(z)]_{\uparrow M}$. These notations are especially useful in the simplification of complicated mathematical expressions containing multirate operators.

Remark In many signal processing applications, such as subband coding and subband adaptive filtering, it is often desirable to have a multirate system that is free from the aliasing effect. Therefore in these applications, the multirate system is usually designed in such a way that the aliasing effect is eliminated or minimized. However, in most modern communication systems, an input signal often passes through an M-fold expander before it goes through an M-fold decimator. As we will see next, such a cascade system of an expander followed by a decimator does not create aliasing.

4.1.2   Multirate identities

In many applications, decimators and expanders are used in cascade with linear time invariant systems. Since the decimator and expander are linear, but time-varying building blocks, they cannot in general be interchanged with other systems in the cascade. However, under some conditions a restricted amount of movement is possible. For example, the two systems shown in Fig. 4.5(a) are equivalent and so are the two systems shown in Fig. 4.5(b). These are known as **noble identities**. For convenience we shall refer to them as the first and the second noble identity, respectively. A proof of these identities is as follows.

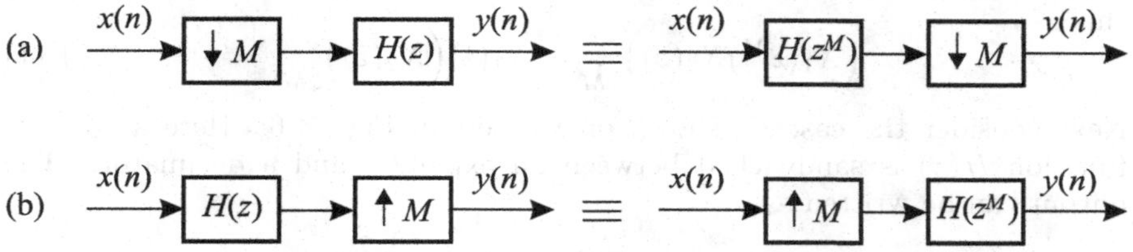

Figure 4.5. Noble identities. (a) The first noble identity, and (b) the second noble identity.

Proof. The filter $H(z^M)$ on the right in Fig. 4.5(a) has the output $X(z)H(z^M)$. The output is therefore the decimated version

$$Y(z) = [X(z)H(z^M)]_{\downarrow M} = \frac{1}{M} \sum_{k=0}^{M-1} X(z^{1/M}W^k)H\left((z^{1/M}W^k)^M\right),$$

where $W = e^{-j2\pi/M}$. By using the fact that $W^M = 1$, we find

$$Y(z) = \frac{H(z)}{M} \sum_{k=0}^{M-1} X(z^{1/M}W^k) = H(z)\left(X(z)\right)_{\downarrow M}.$$

The right-hand side is precisely the output of the system on the left side of Fig. 4.5(a). This proves the first identity. For the second identity note that the output of the system on the left side of Fig. 4.5(b) is

$$Y(z) = \left(X(z)H(z)\right)_{\uparrow M} = X(z^M)H(z^M),$$

which is precisely the output of the system on the right side. This establishes the second identity. ∎

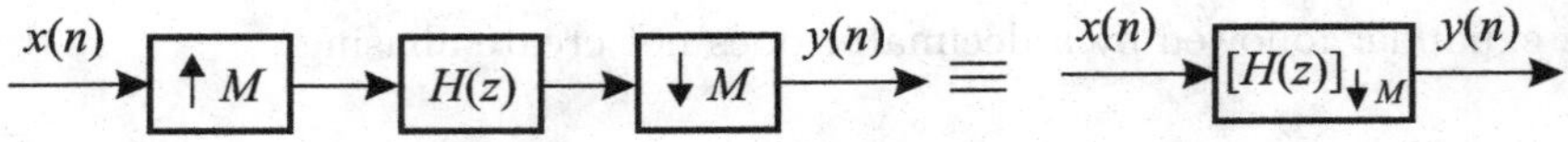

Figure 4.6. The polyphase identity.

Note that the first noble identity can also be written in the form

$$\Big(H(z^M)X(z) \Big)_{\downarrow M} = H(z) \Big(X(z) \Big)_{\downarrow M}. \tag{4.10}$$

As this is valid for any two z-transforms $X(z)$ and $H(z)$, it is useful to write this as

$$\Big(X_1(z^M)X_2(z) \Big)_{\downarrow M} = X_1(z) \Big(X_2(z) \Big)_{\downarrow M}. \tag{4.11}$$

Next consider the cascade shown on the left in Fig. 4.6. Here a transfer function $H(z)$ is sandwiched between an expander and a decimator. The output can be written as

$$Y(z) = \Big(H(z)X(z^M) \Big)_{\downarrow M} = X(z) \Big(H(z) \Big)_{\downarrow M}, \tag{4.12}$$

where we have used (4.11). Thus the system is equivalent to a simple transfer function $[H(z)]_{\downarrow M}$, which represents an LTI system with the decimated impulse response $h(Mn)$. This proves the identity in Fig. 4.6, which is called the **polyphase identity**. Note that it is simply the first noble identity with the roles of $X(z)$ and $H(z)$ interchanged. As we will see in future chapters, the polyphase identity is very useful for the applications of multirate signal processing in communication systems. We summarize the results in the following.

Theorem 4.1 Polyphase Identity. Consider the cascade system on the left in Fig. 4.6. The system is LTI with transfer function $[H(z)]_{\downarrow M}$ as shown on the right of the figure. Or equivalently in the time domain its impulse response is $h(Mn)$.   ∎

4.1.3   Blocking and unblocking

The multirate building blocks can be used to describe schematically two basic operations called blocking and unblocking that appear frequently in many modern communication systems. Consider Fig. 4.7(a). Here the vector $\mathbf{x}(n)$ is made from successive samples of $x(n)$. If the original samples $x(n)$ are spaced apart by T seconds, then the samples of $\mathbf{x}(n)$ are spaced apart by MT seconds, as shown in the figure. Note that the vectors $\mathbf{x}(n)$ and $\mathbf{x}(n+1)$ have no common samples. Thus the samples of the vector $\mathbf{x}(n)$ represent successive nonoverlapping blocks of the input stream $x(n)$ (Fig. 4.8). We therefore say that $\mathbf{x}(n)$ is an M-blocked version or just a *blocked version* of $x(n)$. The integer M is the block size. In other words, a serial input data stream is converted to a set of parallel data streams. For this reason, the blocking operation is also known as **serial to parallel conversion**, denoted by $S/P(M)$. Whenever the details of blocking are not relevant, we shall use the schematic diagram of Fig. 4.7(b).

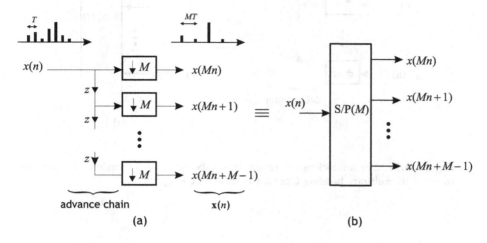

(a)   (b)

Figure 4.7. The blocking operation or serial to parallel conversion: (a) implementation with multirate building blocks; (b) schematic diagram.

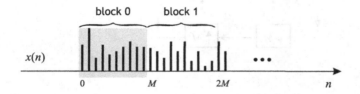

Figure 4.8. The blocking operation demonstrated in the time domain.

The inverse of the blocking operation is unblocking, which is shown in Fig. 4.9(a). Here a set of M signals is *interleaved* by the combination of expanders and the delay chain. We say that $x(n)$ is the *unblocked* version of the set $\{x_k(n)\}$. We also say that the signals $x_k(n)$ are *time-multiplexed* to get $x(n)$. The input sequence $x_k(n)$ is related to $x(n)$ by

$$x_k(n) = x(Mn + k).$$

For obvious reasons, the unblocking operation is also regarded as a **parallel to serial conversion**, and its schematic diagram is shown in Fig. 4.9(b).

If the blocking and unblocking systems are connected in cascade (in either order) it is clear that we get identity. This is shown in Fig. 4.10. This *trivial multirate identity* is surprisingly useful, as we shall see later.

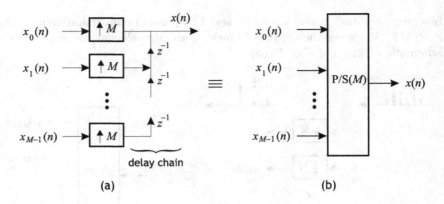

Figure 4.9. The unblocking operation, or parallel to serial conversion: (a) implementation with multirate building blocks; (b) schematic diagram.

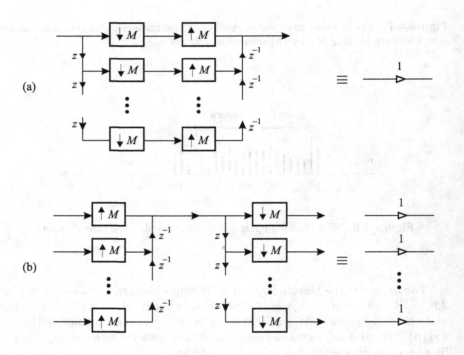

Figure 4.10. Two systems which are equivalent to identity systems. (a) Blocking followed by unblocking; (b) unblocking followed by blocking.

4.2 Decimation filters

Consider the frequency domain relation (4.7) for a decimator. This was demonstrated in Fig. 4.4. Note that the stretched version $X(e^{j\omega/M})$ in general overlaps with some of the shifted versions. The original input $x(n)$ cannot in general be recovered from the decimated output $y_d(n) = x(Mn)$. This overlap is commonly referred to as **aliasing**. However, when the original signal $x(n)$ is **bandlimited** to the region

$$-\pi/M \le \omega < \pi/M,$$

then there is no aliasing because the shifted versions do not overlap. In this case we can recover the higher-rate signal $x(n)$ from its decimated version $y_d(n)$.

A signal $x(n)$ need not be a bandlimited lowpass signal in order to be recoverable from its M-fold decimated version. For example, consider the bandpass signals with Fourier transform shown in Fig. 4.11. The total bandwidth is $2\pi/M$. It can be verified (Problem 4.2) that if the signal in Fig. 4.11(a) is decimated, then the stretched version $X(e^{j\omega/M})$ does not overlap with the shifted versions $X(e^{j(\omega-2\pi\ell)/M})$. For the signal in Fig. 4.11(b), decimation by M will not generate any aliasing if and only if $\omega_0 = k\pi/M$. In these cases, we can therefore recover $x(n)$ from $x(Mn)$ by using an ideal bandpass filter with passband region coinciding with the nonzero part of the signal. When a signal can be decimated by M without creating any aliasing, it is said to satisfy the aliasfree(M) condition. The most general aliasfree(M) condition can be found in [159].

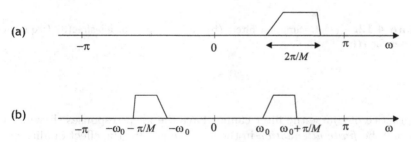

Figure 4.11. Bandpass signals with total bandwidth $2\pi/M$. (a) Complex signal; (b) real signal.

Given an arbitrary signal $x(n)$ we have to pass it through an appropriate bandlimiting filter $H(z)$ so that there is no aliasing upon decimation by M. This is shown in Fig. 4.12(a). The preconditioning filter $H(z)$ is called the **decimation filter**. It can be a lowpass, bandpass, or highpass, as demonstrated in Figs. 4.12(b)–(d) for $M = 3$. The output $y(n)$ is related to the input $x(n)$ by

$$y(n) = \sum_k h(k)x(Mn - k) = \sum_k x(k)h(Mn - k).$$

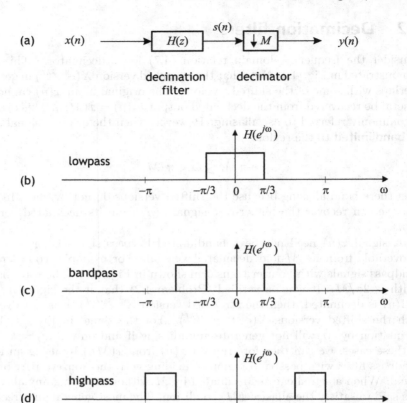

Figure 4.12. (a) Decimation filter; (b)–(d) examples of aliasfree(3) frequency responses for $H(z)$.

In practice of course the filter cannot have the ideal responses shown. There can only be *finite attenuation* in the stopbands. But the effect of aliasing can be made arbitrarily small by making this attenuation large.

4.3 Interpolation filters

An interpolation filter is a discrete-time filter used at the output of an expander, as shown in Fig. 4.13(a). This combination of an expander followed by an interpolation filter is said to be a discrete-time **interpolator**. From its definition we know that the expander inserts zero-valued samples between adjacent samples of the input. The purpose of the interpolation filter is to replace these zeros with a weighted average of the input samples, as we shall explain later.

First consider what happens in the frequency domain. Recall that the expander has M squeezed copies (images) of the input Fourier transform

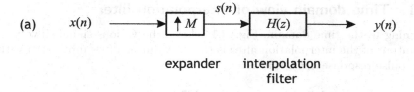

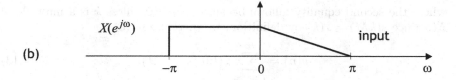

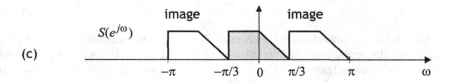

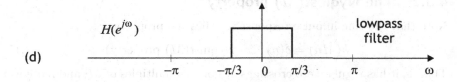

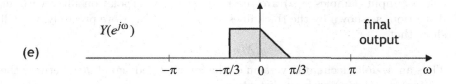

Figure 4.13. (a) Discrete-time interpolator; (b)–(e) explanation of its operation in the frequency domain for $M = 3$.

$X(e^{j\omega})$, as shown in Fig. 4.13(b). The interpolation filter retains one out of these M copies of the Fourier transform. This is demonstrated in Fig. 4.13 for the case when the filter is lowpass and $M = 3$. Thus the interpolated output $Y(e^{j\omega})$ is simply a squeezed version of $X(e^{j\omega})$ with the images removed.

4.3.1   Time domain view of interpolation filter

Returning to the time domain, Fig. 4.14 shows the various signals involved. The output of the interpolation filter is the convolution of its input $s(n)$ with the impulse response $h(n)$, that is

$$y(n) = \sum_{k=-\infty}^{\infty} s(k)h(n-k) = \sum_{k=-\infty}^{\infty} s(kM)h(n-kM),$$

where the second equality follows because $s(k) = 0$ unless k is a multiple of M. Since $s(kM) = x(k)$ by definition, we therefore have

$$y(n) = \sum_{k=-\infty}^{\infty} x(k)h(n-kM). \tag{4.13}$$

Suppose $H(e^{j\omega})$ is ideal lowpass. The impulse response $h(n)$ will be the sinc function

$$h(n) = \frac{\sin(\pi n/M)}{\pi n/M}. \tag{4.14}$$

Figure 4.14(c) shows $h(n)$ for the case $M = 3$. The convolution $(s * h)(n)$ results in the replacement of the zero-valued samples of $s(n)$ with *interpolated values*, as demonstrated in Fig. 4.14(d). The output $y(n)$ has a smooth lowpass appearance because of the filtering.

4.3.2   The Nyquist(M) property

Note that the sinc impulse response satisfies the property

$$h(Mn) = \delta(n) \qquad \text{(Nyquist(M) property)}. \tag{4.15}$$

That is, it has regular zero crossings at nonzero multiples of M, and moreover $h(0) = 1$. A filter $H(z)$ satisfying this property is called a Nyquist(M) filter, or just a **Nyquist filter**. We will show that this property ensures that the original input samples $x(n)$ are retained by the interpolation filter without distortion, as shown by the thick lines in Fig. 4.14(d). More precisely, we will show that

$$y(Mn) = x(n).$$

Thus an M-fold decimated version of the interpolated signal $y(n)$ returns the original signal $x(n)$ without distortion.

Proof. The fact that the Nyquist(M) condition ensures that $y(Mn) = x(n)$ can be verified directly from Eq. (4.13). Setting $n = \ell M$, this equation can be rewritten as follows:

$$y(\ell M) = \sum_{k=-\infty}^{\infty} x(k)h(\ell M - kM) = \sum_{k=-\infty}^{\infty} x(k)g(\ell - k),$$

where $g(n) \stackrel{\triangle}{=} h(Mn)$. We have proved that the sequence $y(nM)$ is equal to the convolution $(x * g)(n)$. So it follows that $y(nM) = x(n)$ for all inputs $x(n)$ if and only if the Nyquist condition $h(Mn) = \delta(n)$ is satisfied.   ∎

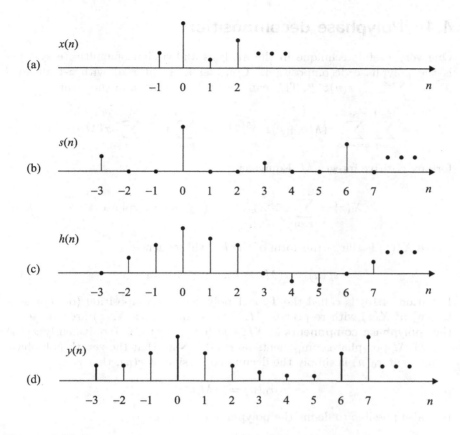

Figure 4.14. Operation of the interpolation filter in the time domain. (a) Input signal; (b) output of the expander; (c) impulse response of the interpolation filter; (d) the final interpolated output.

In practice, the interpolation filter is a realizable system with *finite attenuation* in stopbands. The images are therefore not completely suppressed. It is, however, still possible to satisfy the Nyquist property exactly, thus ensuring that original samples $x(n)$ are preserved. There are many methods for the design of Nyquist filters. One simple approach is the window method. Starting with the ideal sinc function in (4.14), one can choose any window function $w(n)$, such as a Blackman, Hanning, or Kaiser window. Then the following is an FIR Nyquist filter:

$$h(n) = \frac{\sin(\pi n/M)}{\pi n}\, w(n).$$

We can also obtain FIR Nyquist filters by incorporating the Nyquist constraint into standard filter design methods, e.g. the eigenfilter method [114, 151, 156].

4.4   Polyphase decomposition

One very useful technique in the analysis and design of multirate systems is the polyphase decomposition. Consider a signal $x(n)$ with z-transform $X(z) = \sum_{n=-\infty}^{\infty} x(n)z^{-n}$. This can always be rewritten in the form

$$X(z) = \sum_{k=0}^{M-1} \sum_{n=-\infty}^{\infty} x(Mn+k)z^{-(Mn+k)} = \sum_{k=0}^{M-1} z^{-k} \sum_{n=-\infty}^{\infty} x(Mn+k)z^{-Mn}$$

for any positive integer M. Equivalently,

$$X(z) = \sum_{k=0}^{M-1} z^{-k}X_k(z^M) \quad \text{(Type 1 polyphase)} \tag{4.16}$$

where $X_k(z)$ is the z-transform of the kth subsequence

$$x_k(n) = x(Mn+k), \quad 0 \le k \le M-1. \tag{4.17}$$

Equation (4.16) is called the Type 1 polyphase decomposition (or representation) of $X(z)$ with respect to M. The M quantities $X_k(z)$ are said to be the **polyphase components** of $X(z)$ with respect to M (equivalently $x_k(n)$ are the M polyphase components of $x(n)$).[1] Note that the zeroth polyphase component $x_0(n)$ is simply the decimated version of $x(n)$, that is

$$x_0(n) = x(Mn).$$

It is also possible to define the polyphase decomposition

$$X(z) = \sum_{k=0}^{M-1} z^k \widehat{X}_k(z^M) \quad \text{(Type 2 polyphase)}, \tag{4.18}$$

where the Type 2 polyphase components $\widehat{X}_k(z)$ are z-transforms of $\widehat{x}_k(n) = x(Mn-k)$ instead of $x(Mn+k)$. Figure 4.15 shows how the polyphase components can be represented in terms of delays and decimators. A time domain example for Type 1 polyphase decomposition is shown in Fig. 4.16 for $M = 3$; note how the samples of the polyphase components are numbered with consecutive integers. We can also apply the polyphase decomposition to a transfer function $H(z)$. The Type 1 and Type 2 polyphase decompositions of $H(z)$ are, respectively,

$$H(z) = \begin{cases} \sum_{k=0}^{M-1} z^{-k} G_k(z^M), & \text{(Type 1)}; \\ \\ \sum_{k=0}^{M-1} z^k S_k(z^M), & \text{(Type 2)}. \end{cases} \tag{4.19}$$

The usefulness of polyphase decompositions will become clear later. With the polyphase representation, we are now ready to prove the transform domain formula of the decimator output.

---

[1] Clearly $X_k(z)$ depends on M, so a notation such as $X_k^{(M)}(z)$ would have been more appropriate; but the quantity M is almost always clear from the context. So we shall use the simpler notation $X_k(z)$.

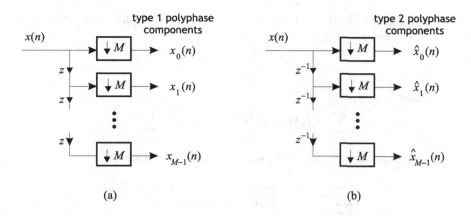

Figure 4.15. Simple block diagram interpretation of polyphase components. (a) Type 1 polyphase components $x_k(n)$; (b) Type 2 polyphase components $\widehat{x}_k(n)$.

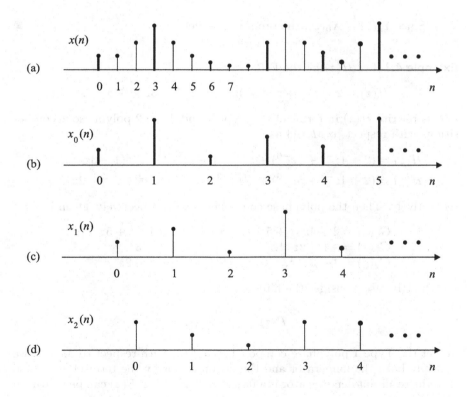

Figure 4.16. A signal $x(n)$ and its three polyphase components $x_k(n)$, for $M = 3$. Note carefully how the samples are renumbered.

Proof of (4.4). Replacing z with zW^ℓ in the polyphase representation (4.16) we get

$$X(zW^\ell) = \sum_{k=0}^{M-1} z^{-k} W^{-k\ell} X_k(z^M).$$

Thus

$$\sum_{\ell=0}^{M-1} X(zW^\ell) = \sum_{k=0}^{M-1} z^{-k} X_k(z^M) \sum_{\ell=0}^{M-1} W^{-k\ell}. \qquad (4.20)$$

Using the fact $W = e^{j2\pi/M}$, one can verify the following:

$$\sum_{\ell=0}^{M-1} W^{-k\ell} = \begin{cases} M, & \text{if } k \text{ is a multiple of } M; \\ 0, & \text{otherwise.} \end{cases}$$

Substituting the above expression into (4.20), we get

$$\sum_{\ell=0}^{M-1} X(z^{1/M} W^\ell) = M X_0(z).$$

Since $Y_d(z) = X_0(z)$, the proof is complete. ∎

Example 4.1 Let $H(z)$ be the FIR filter

$$H(z) = 2 - z^{-1} + z^{-2} + 4z^{-3} - 2z^{-4} - 3z^{-5} + 5z^{-6}.$$

Let us rewrite $H(z)$ in terms of the Type 1 and Type 2 polyphase decompositions with respect to $M = 3$ as

$$H(z) = (2 + 4z^{-3} + 5z^{-6}) + z^{-1}(-1 - 2z^{-3}) + z^{-2}(1 - 3z^{-3}),$$
$$H(z) = (2 + 4z^{-3} + 5z^{-6}) + z^1(z^{-3} - 3z^{-6}) + z^2(-z^{-3} - 2z^{-6}),$$

respectively. Then the polyphase components are, respectively, given by

$$\begin{aligned} G_0(z) &= 2 + 4z^{-1} + 5z^{-2}, & S_0(z) &= 2 + 4z^{-1} + 5z^{-2}, \\ G_1(z) &= -1 - 2z^{-1}, & S_1(z) &= z^{-1} - 3z^{-2}, \\ G_2(z) &= 1 - 3z^{-1}, & S_2(z) &= -z^{-1} - 2z^{-2}. \end{aligned}$$

For the IIR case, consider the filter

$$F(z) = \frac{1 - 2z^{-1}}{1 + z^{-1}}.$$

To get the Type 1 polyphase components of $F(z)$ with respect to $M = 2$, we multiply both the numerator and the denominator by the term $(1 - z^{-1})$ so that the resultant denominator is a function of z^2. Thus $F(z)$ can be rewritten as

$$F(z) = \frac{1 - 3z^{-1} + 2z^{-2}}{1 - z^{-2}} = \frac{1 + 2z^{-2}}{1 - z^{-2}} + z^{-1} \frac{-3}{1 - z^{-2}}.$$

From the above equation, we immediately get the Type 1 polyphase components as

$$G_0(z) = \frac{1 + 2z^{-1}}{1 - z^{-1}}, \quad G_1(z) = \frac{-3}{1 - z^{-1}}.$$

Similarly we can get the Type 2 polyphase components of $F(z)$ as

$$S_0(z) = \frac{1 + 2z^{-1}}{1 - z^{-1}}, \quad S_1(z) = \frac{-3z^{-1}}{1 - z^{-1}}.$$

∎

Observe that when $H(z)$ is an FIR filter having L coefficients, the total number of coefficients in the M polyphase components is also L. The same is not true for the IIR case. If we implement the IIR polyphase components $S_i(z)$ (or $G_i(z)$), the total number of multipliers is larger than that in $F(z)$.

4.4.1 Decimation and interpolation filters

First consider the decimation filter reproduced in Fig. 4.17(a). Note that one out of every M output samples are discarded. It is therefore wasteful to compute all the outputs of $H(z)$. The structure for $H(z)$ can be rearranged in such a way that none of the discarded samples are computed in the first place. For this we can use the polyphase decomposition. Consider the Type 2 polyphase decomposition

$$H(z) = \sum_{k=0}^{M-1} z^k S_k(z^M).$$

We can redraw Fig. 4.17(a) as in Fig. 4.17(b). The first noble identity (Fig. 4.5) can now be applied to redraw this in the form of Fig. 4.17(c), which is called the **polyphase implementation** of the decimation filter. Note that none of the computed results are discarded, and furthermore that all the computations (filtering with $S_k(z)$) are implemented at the lower rate, i.e. after the decimator. The polyphase implementation is therefore computationally very efficient. Note from Fig. 4.17 that the Type 2 polyphase components $S_k(z)$ of the filter are operating individually on the Type 1 polyphase components $x_k(n)$ of the input $x(n)$ (compare Fig. 4.15(a)) before the results are added to produce $y(n)$. Equivalently the polyphase components $S_k(z)$ operate on the blocked version $\mathbf{x}(n)$ of the input $x(n)$.

Next consider the interpolation filter shown in Fig. 4.18(a). The zero-valued samples inserted by the expander enter the multipliers $f(k)$ and result in wasted computation. Furthermore these multipliers operate at the higher rate (they only have T/M seconds per computation), where T is the separation between input samples. To avoid such inefficient computations we represent the filter in its Type 1 polyphase form:

$$F(z) = \sum_{k=0}^{M-1} z^{-k} G_k(z^M).$$

Then the system can be redrawn as in Fig. 4.18(b). By using the second noble identity (Fig. 4.5(b)) this can be simplified to the form shown in Fig. 4.18(c). In this implementation there are no zero-valued samples entering the filters $G_k(z)$. Furthermore all the computations are taking place at the lower rate (i.e. before the expanders). Observe that the output $y(n)$ is the *interleaved*

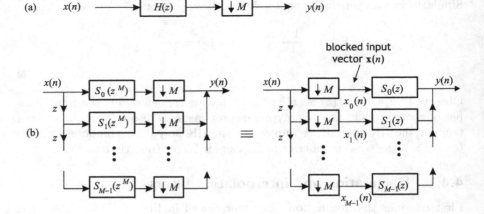

Figure 4.17. (a) Decimation filter; (b) Type 2 polyphase representation; (c) simplification with the help of the first noble identity.

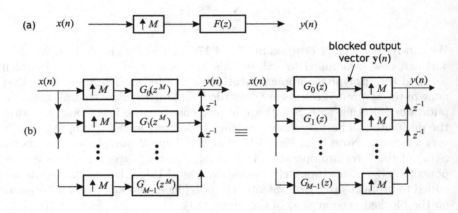

Figure 4.18. (a) Interpolation filter, (b) Type 1 polyphase representation, and (c) simplification with the help of the second noble identity.

version of the outputs of $G_k(z)$. Thus each Type 1 polyphase component $G_k(z)$ of the filter computes the corresponding Type 1 polyphase component of the output $y(n)$, and these results are interleaved to get $y(n)$.

In order to maximize the efficiency of computation in multirate systems, we *move all decimators as far to the left as possible* and *move all expanders as far to the right as possible*. In this way, the computational building blocks operate at the lowest possible rate.

4.4.2   Synthesis filter banks

We next consider the system shown in Fig. 4.19. In this system, there are M interpolation filters with inputs $s_k(n)$, and their outputs are added. In general, the interpolation ratio N can be different from the number of branches M (usually $N > M$ in communications and $N \leq M$ in other applications such as signal compression and adaptive filtering). The system is called a **synthesis filter bank** because it combines a set of signals $s_k(n)$ into a single signal $x(n)$. We now show how this filter bank can be expressed in polyphase form, which leads to a more efficient implementation.

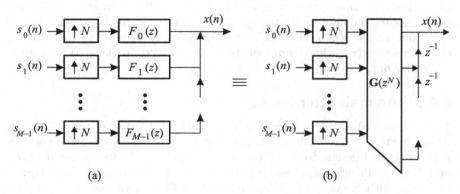

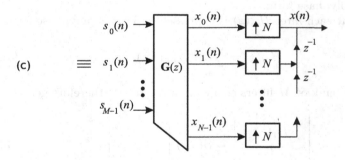

Figure 4.19. (a) Synthesis filter bank; (b) polyphase version; (c) simplification with the use of the second noble identity.

Assume that each filter $F_m(z)$ is expressed in Type 1 polyphase form with respect to N:

$$F_m(z) = \sum_{k=0}^{N-1} z^{-k} G_{km}(z^N). \qquad (4.21)$$

Expressing the bank of M filters as a row vector we therefore get

$$\begin{bmatrix} F_0(z) & F_1(z) & \cdots & F_{M-1}(z) \end{bmatrix} = \begin{bmatrix} 1 & z^{-1} & \cdots & z^{-(N-1)} \end{bmatrix} \mathbf{G}(z^N), \qquad (4.22)$$

where $\mathbf{G}(z)$ is the $N \times M$ matrix of polyphase components:

$$\mathbf{G}(z) = \begin{bmatrix} G_{00}(z) & G_{01}(z) & \cdots & G_{0,M-1}(z) \\ G_{10}(z) & G_{11}(z) & \cdots & G_{1,M-1}(z) \\ \vdots & \vdots & \ddots & \vdots \\ G_{N-1,0}(z) & G_{N-1,1}(z) & \cdots & G_{N-1,M-1}(z) \end{bmatrix}. \tag{4.23}$$

We say that $\mathbf{G}(z)$ is the **polyphase matrix** of the synthesis filter bank. Equation (4.22) is said to be the polyphase representation of the filter bank, and can be represented as in Fig. 4.19(b).   With the help of the second noble identity we can further redraw it as in Fig. 4.19(c), which is called the polyphase implementation of the synthesis filter bank. In this representation the input vector $\mathbf{s}(n)$ is filtered by the MIMO system $\mathbf{G}(z)$ to produce the output vector $\mathbf{x}(n)$, whose components are interleaved (unblocked) to produce the scalar output $x(n)$.

4.4.3   Analysis filter banks

Now consider the system shown in Fig. 4.20(a). In this system, there are M decimation filters with a common input $x(n)$. The decimation ratio N is in general different from M (again usually $N > M$ in communications and $N \leq M$ in other applications such as signal compression and adaptive filtering). Such a system is called an **analysis filter bank** because it splits a signal $x(n)$ into M components. We now show how this filter bank can be expressed in polyphase form.

Assume that each filter $H_m(z)$ is expressed in Type 2 polyphase form:

$$H_m(z) = \sum_{k=0}^{N-1} z^k S_{mk}(z^N). \tag{4.24}$$

Expressing the bank of M filters as a column vector we therefore get

$$\begin{bmatrix} H_0(z) \\ H_1(z) \\ \vdots \\ H_{M-1}(z) \end{bmatrix} = \mathbf{S}(z^N) \begin{bmatrix} 1 \\ z \\ \vdots \\ z^{(N-1)} \end{bmatrix}, \tag{4.25}$$

where $\mathbf{S}(z)$ is the $M \times N$ matrix of polyphase components:

$$\mathbf{S}(z) = \begin{bmatrix} S_{00}(z) & S_{01}(z) & \cdots & S_{0,N-1}(z) \\ S_{10}(z) & S_{11}(z) & \cdots & S_{1,N-1}(z) \\ \vdots & \vdots & \ddots & \vdots \\ S_{M-1,0}(z) & S_{M-1,1}(z) & \cdots & S_{M-1,N-1}(z) \end{bmatrix}. \tag{4.26}$$

We say that $\mathbf{S}(z)$ is the   **polyphase matrix** of the analysis filter bank. Equation (4.25) is said to be the polyphase representation of the filter bank, and can be represented as in Fig. 4.20(b). With the help of the first noble identity this polyphase representation can be redrawn as shown in Fig. 4.20(c). In this representation the input $x(n)$ is first blocked into its Type 1 polyphase components $x_k(n)$ and these are filtered by the MIMO system $\mathbf{S}(z)$ to produce the output vector.

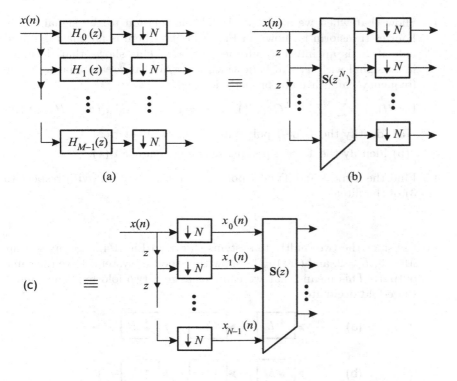

Figure 4.20. (a) Analysis filter bank; (b) polyphase version; (c) simplification with the use of the first noble identity.

4.5 Concluding remarks

Filter bank systems have been successfully applied to many areas such as signal compression, subband adaptive filtering, and so forth. Readers interested in a more detailed treatment of these systems are referred to [4, 30, 43, 49, 146, 159, 169].   In this book, we focus on the application of multirate signal processing in modern communication systems. In the next few chapters we shall find that many recent designs of communication systems can be efficiently represented in terms of the multirate filter bank language introduced in this chapter.

4.6 Problems

4.1 Determine whether the decimator and expander have one or more of the following properties: (a) linearity; (b) causality; (c) time invariance.

4.2 Show that the signal in Fig. 4.11(a) satisfies the aliasfree(M) condition. Suggest a circuit to recover the original signals from their decimator outputs.

4.3 Show that when we apply an M-fold decimation to the signal whose Fourier transform is shown in Fig. 4.11 (b), there is no aliasing if and only if ω_0 is an integer multiple of π/M. This shows that decimating a signal with a total bandwidth of $2\pi/M$ can create aliasing if the frequency bands are not properly located.

4.4 Let $H(z) = \sum_{k=0}^{M-1} z^{-k} G_k(z^M)$, $X(z) = 1+z^{-1}$, and $Y(z) = H(z)X(z)$.

(a) Identify the Type 1 polyphase components of $Y(z)$.

(b) Identify the Type 2 polyphase components of $Y(z)$.

4.5 Find the Type 1 and Type 2 polyphase representation (with respect to 3) of the filter

$$F(z) = \frac{1 - z^{-1}}{1 + z^{-1}}.$$

4.6 Consider the two multirate systems shown in Fig. P4.6. Suppose that $M = 3$, $L = 2$, and $C(z) = 1$. Show that the two systems have the same output. This means that we can exchange a two-fold expander and a three-fold decimator.

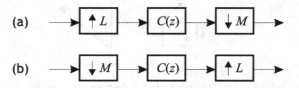

Figure P4.6. Two cascades of multirate building blocks.

4.7 Suppose that $L = 2$ and $M = 3$, and the transfer function $C(z)$ in Fig. P4.6 is a delay z^{-1}. Do the two systems have the same output? Justify your answer.

4.8 Let $M = N = 2$ and let the synthesis filters $F_0(z)$ and $F_1(z)$ of Fig. 4.19(a) be related by $F_1(z) = F_0(-z)$. How are the Type 2 polyphase components of these filters related? Exploiting this relation, find an implementation (in the polyphase form) of the synthesis filter bank with the smallest number of multipliers.

4.9 Let $M = N = 2$ and let the analysis filters $H_0(z)$ and $H_1(z)$ of Fig. 4.20(a) be related by $H_1(z) = H_0(-z)$. How are the Type 1 polyphase components of these filters related? Exploiting this relation, find an implementation (in the polyphase form) of the analysis filter bank with the smallest number of multipliers.

4.10 Consider the systems in Fig. P4.10. Let $H_k(z)$ and $F_k(z)$ be the following ideal filters:

$$H_k(e^{j\omega}) = F_k(e^{j\omega}) = \begin{cases} 1, & \frac{k\pi}{M} \le |\omega| < \frac{(k+1)\pi}{M}, \\ 0, & \text{otherwise,} \end{cases}$$

for $k = 0, 1, \ldots, M - 1$.

   (a) Determine $U_k(e^{j\omega})$ in Fig. P4.10(a) in terms of $X(e^{j\omega})$.

   (b) Determine $Y(e^{j\omega})$ in Fig. P4.10(b) in terms of $X(e^{j\omega})$.

(a)

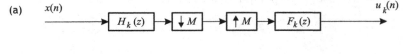

(b)

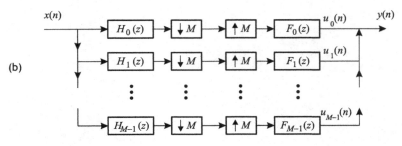

Figure P4.10. Two multirate systems.

4.11 *DFT banks.* Consider the analysis filter bank in Fig. 4.20. Suppose that $N = M$ and the analysis filters $H_m(z)$ are related by $H_m(z) = H_0(zW^m)$, for $m = 1, 2, \ldots, M - 1$. Such a filter bank is called a DFT bank. DFT banks can be implemented efficiently using the polyphase structure and they are widely used in many applications.

   (a) Suppose that $H_0(z)$ is an ideal lowpass filter with the passband $[-\pi/M, \pi/M]$. Plot the magnitude responses of $H_k(z)$ for $0 \leq k \leq M - 1$.

   (b) Let the Type 2 polyphase decomposition of $H_0(z)$ with respect to M be

$$H_0(z) = \sum_{k=0}^{M-1} z^k S_k(z^M).$$

Write the Type 2 polyphase decomposition of $H_m(z)$ in terms of $S_k(z)$.

   (c) Let $\mathbf{h}(z) = [H_0(z)\ H_1(z)\ \ldots\ H_{M-1}(z)]^T$. Show that

$$\mathbf{h}(z) = \mathbf{W}\mathbf{D}(z^M)\mathbf{e}(z),$$

where $\mathbf{W}$ is the $M \times M$ normalized DFT matrix with (k, l)th entry $(1/\sqrt{M})e^{-j2\pi kl/M}$, $\mathbf{e}(z) = [1\ z\ \ldots\ z^{M-1}]^T$, and $\mathbf{D}(z)$ is a diagonal matrix. What are the diagonal entries of $\mathbf{D}(z)$?

   (d) Suppose that $H_0(z)$ is an Lth-order FIR filter. Using the result in (c), suggest a low-cost implementation of the analysis filter bank shown in Fig. 4.20(c). Show that the implementation cost of the analysis filter bank is equal to that of $H_0(z)$ plus an $M \times M$

DFT matrix, which can be implemented efficiently using FFT algorithms.

Similarly, when the synthesis filters have the frequency shift property, the filter bank is called a DFT bank and it can also be implemented efficiently using the polyphase structure.

5

Multirate formulation of communication systems

In this chapter, we give the formulation of some modern communication systems using the multirate building blocks. Under the multirate framework, we show how to insert redundancy at the transmitter side to help the task of equalization. These formulations are fundamental to many of our discussions in future chapters. We also show that the multirate framework can be used to simplify the study of the so-called fractionally spaced equalizer system in which the receiver take samples at a higher rate.

5.1   Filter bank transceivers

Figure 5.1 shows a system called the **filter bank transceiver** (*trans*mitter and re*ceiver*). Such a system is also known as a transmultiplexer. This system and its variations will be central to many of our discussions in this book. From the figure we realize that the transmitter is a synthesis filter bank whereas the receiver is an analysis filter bank (see Section 4.4). The filters $F_k(z)$ are known as the **transmitting filters** and $H_k(z)$ are called the **receiving filters**. When all these filters are FIR, we say that the transceiver is an FIR filter bank transceiver; otherwise it is called an IIR filter bank transceiver. The M input signals $s_k(n)$ are usually symbol streams (such as PAM or QAM signals, see Section 2.2). These could be symbols generated by different users that share the same communication channel. Or they could be different parts of the signals generated by one user. The number M is also called the number of subchannels or bands. The symbol streams $s_k(n)$ are passed through the transmitting filters $F_k(z)$ and the results are added to form the transmitted signal

$$x(n) = \sum_{k=0}^{M-1} \sum_{i} s_k(i) f_k(n - iM).$$

Figure 5.2 shows the action of the transmitting filter $F_k(z)$ on the kth input signal $s_k(n)$. It takes each sample of $s_k(n)$ and "puts a pulse $f_k(n)$ around it." Thus the filters $F_k(z)$ are also called *pulse-shaping filters*. Figure 5.2(b) shows an example of the shaped output when $F_k(z)$ is a lowpass filter. The

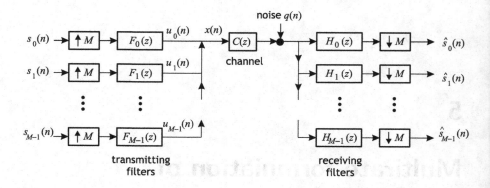

Figure 5.1. The M-band filter bank transceiver system.

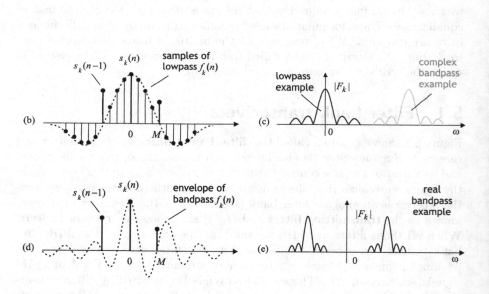

Figure 5.2. (a) Pulse shaping filter; (b) example where $F_k(z)$ is lowpass; (c) examples of lowpass and bandpass $F_k(z)$; (d) example where $F_k(z)$ is real-coefficient bandpass; (e) response of a real bandpass filter.

filter $F_k(z)$ can also be a complex bandpass filter, as shown in part (c), or a real bandpass filter, as shown in part (e). Part (d) shows an example of the shaped signal when $f_k(n)$ is a real-coefficient bandpass filter.

In Fig. 5.1, the output $x(n)$ of the transmitting filter bank is sent over an LTI channel with transfer function $C(z)$ and additive noise $q(n)$. At the receiver end, the filters $H_k(z)$ have the task of separating the signals and reducing them to the original rates by decimation. The transmitted symbols

$s_k(n)$ are then identified from the signals $\widehat{s}_k(n)$, which in general are distorted versions of the symbols $s_k(n)$ because of the combined effects of channel, noise, and the filters. The general goal is to identify the symbols $s_k(n)$ accurately in the presence of these distortions. The choice of filters $\{F_k(z), H_m(z)\}$ depends on the details of the specific instances of application. We will discuss details of this throughout the book.

5.1.1 The multiplexing operation

We can always regard the set of signals $s_k(n)$ as the polyphase components of a hypothetical signal $s(n)$, as demonstrated in Fig. 5.3(a) for $M = 3$. That is, we can regard $s_k(n)$ as components of a **time-multiplexed** signal $s(n)$. On the other hand, with the filters $F_k(z)$ chosen as a contiguous set of bandpass filters (Fig. 5.3(b)), we can regard the transmitted signal $x(n)$ as a **frequency-multiplexed** version of the set of signals $s_k(n)$. To see this consider an arbitrary $s_1(n)$ with a Fourier transform as demonstrated in Fig. 5.3(c). The expander squeezes this Fourier transform by a factor of M, and the interpolation filter $F_1(z)$ retains only one copy (Fig. 5.3(d)). Thus the output $x(n)$ of the synthesis bank has a Fourier transform as shown in Fig. 5.3(e); it is just a concatenation of squeezed versions of $S_k(e^{j\omega})$ for $0 \leq k \leq M - 1$. Thus $x(n)$ is the frequency-multiplexed version of $\{s_k(n)\}$.

The earliest applications of the transmultiplexers were indeed conversions from **time-division multiplexing (TDM)** to **frequency-division multiplexing (FDM)** of telephone signals prior to their transmission [11, 99]. The received signal is then converted from an FDM back to a TDM format. So transmultiplexers are often referred to as TDM to FDM (and vice versa) converters. The transmultiplexer in Fig. 5.1 is said to be perfect if, in the absence of noise, $\widehat{s}_k(n) = s_k(n)$ for $0 \leq k < M$. The theory of FIR perfect transmultiplexers was first studied in [168], and the condition of perfect transmultiplexer was derived for the case of $C(z) = 1$. When the channel $C(z)$ is frequency-selective, the design of perfect transmultiplexers becomes much more complicated. As we shall see in Chapter 10, no FIR perfect transmultiplexers exist when the channel is frequency-selective. To aid the design of transmultiplexer systems, redundant samples are often introduced to the transmitting side. This topic will be investigated next.

5.1.2 Redundancy in filter bank transceivers

In wideband communications, the channel $C(z)$ in general introduces some degree of distortion to the transmitted signal. To aid the task of equalization, redundancy is usually embedded in the transmitted signals. To do this we can use a filter bank transceiver, of which the number of subchannels or bands is smaller than the interpolation ratio. Figure 5.4 shows such a system, where $N > M$. To understand how redundancy is embedded in the transmitted sequence $x(n)$, suppose that the sampling period of the D/C converter is T.

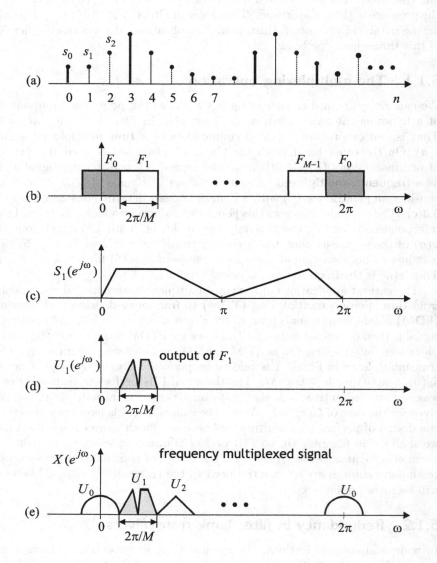

Figure 5.3. The frequency-multiplexing operation performed by the transmultiplexer. (a) Original TDM signal; (b) filter bank response; (c) Fourier transform of a signal $s_1(n)$; (d) output of the corresponding filter $F_1(z)$; (e) FDM signal $X(e^{j\omega})$.

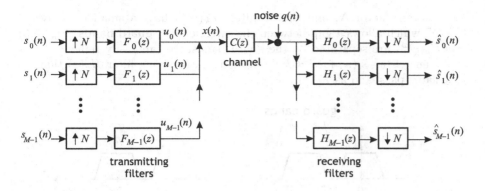

Figure 5.4. The M-band or M-subchannel filter bank transceiver system with redundancy ($N > M$).

Then, due to the N-fold expander, the modulation symbols $s_k(n)$ are spaced apart by NT seconds. There are M information carrying symbols in one input block, but the transmitter sends out N samples of $x(n)$ over the channel in NT seconds. In other words, for every block of M input symbols, the transmitting filter bank sends out N samples of $x(n)$. The number of redundant samples inserted is $(N-M)$ per input block. If each modulation symbol $s_k(n)$ carries b bits of information, then Mb bits are sent for every N samples of $x(n)$ transmitted. The transmission rate of the system is given by

$$\mathcal{R} = \frac{M}{N} \frac{b}{T} \qquad \text{bits per second (bps).}$$

The transmission rate has been reduced by a factor of N/M due to the introduction of redundant samples. To compensate for this reduction, we can increase the sampling frequency of the D/C and C/D converters by a factor of N/M. This means that the channel bandwidth has to be increased by a factor of N/M. For this reason the ratio

$$\zeta = N/M$$

is also known as the **bandwidth expansion ratio.** In practice, the number of redundant samples per block $(N-M)$ is usually much smaller than M so that ζ is close to one.

We can also understand from a frequency domain viewpoint how the inserted redundant samples could help in the equalization process. Note that if the filters $F_k(z)$ in Fig. 5.3(b) are not perfectly bandlimited to a width of $2\pi/M$ then the Fourier transforms $U_k(e^{j\omega})$ will contain leakage from the multiple images created by the expanders. Thus the spectra $U_k(e^{j\omega})$ will overlap, and cannot in general be separated perfectly by the receiving filters $H_k(z)$. Thus the kth output $\widehat{s}_k(n)$ contains contributions from $s_i(n)$ for $i \neq k$, which results in interference. A simple way to avoid this is to have nonoverlapping $U_k(e^{j\omega})$. This can be achieved by making the interpolation ratio N larger than the number of subchannels M. The spectra of $s_k(n)$ are now squeezed by a larger factor N, and the M filters $F_k(e^{j\omega})$ have a smaller bandwidth $2\pi/N$, which allows a gap between the filters called **guard bands** (Fig. 5.5). As $U_k(e^{j\omega})$ are separated by the guard bands, we can easily remove the interference from $s_i(n)$ for $i \neq k$ by designing a receiving filter $H_k(z)$ that has the same passband as $F_k(z)$.

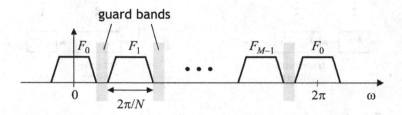

Figure 5.5. Illustration of guard bands in a redundant transceiver.

Filter bank transceivers with $N = M$ are called **minimal transceivers**, and those with $N > M$ are called **redundant transceivers**. In practice the filters are not perfectly bandlimiting filters; in fact, filters of fairly small orders are used. In such systems the introduction of redundancy is still useful for the purpose of channel equalization. This will be explained in more detail in future chapters.

5.1.3   Types of distortion in transceivers

The received signals $\widehat{s}_k(n)$ in general are different from $s_k(n)$ for several reasons.

(1) First, there is **inter subchannel interference**. This means that $\widehat{s}_k(n)$ is affected not only by $s_k(n)$, but also by $s_m(n), m \neq k$.

(2) Second, for any fixed k, the signal $\widehat{s}_k(n)$ depends also on $s_k(n-m)$ for $m \neq 0$. This is due to the effect of filtering created by the channel and the various filters. This is called **intra subchannel interference**.

(3) Finally there is additive channel noise.[1]

The task at the receiver is to minimize the effects of these distortions so that the transmitted symbols $s_k(n)$ can be detected from $\widehat{s}_k(n)$ with an acceptable low probability of error. We will see later that this task is easier when there is redundancy (i.e. $N > M$). We shall present a quantitative study of inter subchannel and intra subchannel interferences in Section 5.2.

---

[1]There are other sources of errors, such as the nonlinear distortion in analog circuits, synchronization error, carrier frequency offset, and so on, which are not considered here.

5.2　Analysis of filter bank transceivers

In this section we show that the filter bank transceiver can be represented in a form that does not require the use of multirate building blocks. Such a representation gives the channel $C(z)$ in the form of a special transfer matrix called the pseudocirculant. This alternative mathematical formulation is very useful in the theoretical study of communication systems. Below we will carry out the analysis for the redundant filter bank transceivers ($N \geq M$). The minimal transceiver case can be obtained by setting $N = M$.

5.2.1　ISI-free filter bank transceivers

Consider the path from $s_m(n)$ to $\widehat{s}_k(n)$ in Fig. 5.4. In the absence of channel noise this path is simply a transfer function sandwiched between an expander and a decimator, as shown in Fig. 5.6(a). By using the polyphase identity (Fig. 4.6 and Theorem 4.1), we see that this path has the transfer function

$$T_{km}(z) = [H_k(z)C(z)F_m(z)]_{\downarrow N}. \tag{5.1}$$

Ignoring the channel noise for a moment, the transceiver can be described

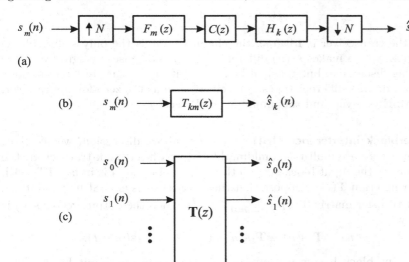

(a)

(b)

(c)

Figure 5.6.　(a) The path from the mth input to the kth output of the transceiver; (b) the equivalent system with noise ignored; (c) matrix representation of the filter bank transceiver in Fig. 5.4, with noise ignored.

as a simple transfer matrix $\mathbf{T}(z)$. The off-diagonal elements of the matrix, namely $T_{km}(z), k \neq m$, represent the path from the mth input to the kth output. If the matrix is diagonal, that is,

$$T_{km}(z) = 0, \quad k \neq m,$$

then there is *no inter subchannel interference*. With inter subchannel interference eliminated, each diagonal element $T_{kk}(z)$ represents the transfer function from $s_k(n)$ to $\widehat{s}_k(n)$. If, in addition,

$$T_{kk}(z) = 1,$$

then $\widehat{s}_k(n)$ is affected by only a single sample of $s_k(n)$ (e.g. $s_k(n-1)$ does not affect it). There is *no intra subchannel interference* in the kth path. Thus the system is free from both interferences if

$$\mathbf{T}(z) = \mathbf{I},$$

and we see that there is **perfect symbol recovery** or the **ISI-free property**

$$\widehat{s}_k(n) = s_k(n) \tag{5.2}$$

in the absence of noise. More generally in an ISI-free transceiver, $\mathbf{T}(z)$ is allowed to be a diagonal matrix with diagonal elements $\alpha_k z^{-n_k}$ for some integer n_k and some $\alpha_k \neq 0$, so that

$$\widehat{s}_k(n) = \alpha_k s_k(n - n_k).$$

When the transceiver is ISI-free, the channel noise is the only source of imperfection, which makes $\widehat{s}_k(n)$ different from $s_k(n)$. Noise was ignored in the preceding discussion, but we shall handle it with great care in later chapters where we discuss different types of receivers such as the zero-forcing receiver, the MMSE receiver, and the minimum BER receiver.

Interblock interference (IBI)   From the above discussion, we see that, although there are multirate building blocks in a filter bank transceiver, the system from the input block $\mathbf{s}(n)$ to the output block $\widehat{\mathbf{s}}(n)$ is in fact LTI with transfer function $\mathbf{T}(z)$. Suppose that the transceiver is not ISI-free and it has an FIR transfer matrix $\mathbf{T}(z) = \sum_{k=0}^{J} \mathbf{T}_k z^{-k}$. Then the output vector $\widehat{\mathbf{s}}(n)$ is given by

$$\widehat{\mathbf{s}}(n) = \mathbf{T}_0 \mathbf{s}(n) + \mathbf{T}_1 \mathbf{s}(n-1) + \cdots + \mathbf{T}_J \mathbf{s}(n-J).$$

The output block becomes a mixture of more than one input block. This phenomenon is called **interblock interference (IBI)**. When $\mathbf{T}(z) = \mathbf{T}_0$, the output block $\widehat{\mathbf{s}}(n)$ depends only on the input block $\mathbf{s}(n)$. In this case, we say that the transceiver is IBI-free.

Interchanging the transmitting and receiving filters   Suppose we interchange the transmitting filters $F_k(z)$ and the receiving filters $H_k(z)$ and let $\mathbf{T}'(z)$ be the transfer matrix of the new transceiver system. Then, from (5.1), we have

$$T'_{km}(z) = [F_k(z)C(z)H_m(z)]_{\downarrow N} = T_{mk}(z).$$

In other words, $\mathbf{T}'(z) = \mathbf{T}^T(z)$. In particular, this implies that *interchanging the transmitting and receiving filters does not affect the IBI-free and ISI-free properties*.

5.2.2 Polyphase approach

From the preceding discussions, we know that, in the absence of channel noise, the transceiver is an MIMO (multi-input multi-output) LTI system with transfer matrix $\mathbf{T}(z)$, even though it has multirate building blocks. Below we will redraw the filter bank transceiver in a form that does not require the use of any multirate building blocks. To derive such a representation, we implement the transmitting and receiving filter banks using the polyphase structures (Section 4.4). Ignoring the channel noise, the transceiver in Fig. 5.4 can be redrawn as in Fig. 5.7. The $N \times M$ matrix $\mathbf{G}(z)$ is known as the **transmitting** (polyphase) **matrix**, whereas the $M \times N$ matrix $\mathbf{S}(z)$ is called the **receiving** (polyphase) **matrix**. These matrices are related to the transmitting and receiving filters by

$$\mathbf{G}(z) = \begin{bmatrix} G_{00}(z) & G_{01}(z) & \cdots & G_{0,M-1}(z) \\ G_{10}(z) & G_{11}(z) & \cdots & G_{1,M-1}(z) \\ \vdots & \vdots & \ddots & \vdots \\ G_{N-1,0}(z) & G_{N-1,1}(z) & \cdots & G_{N-1,M-1}(z) \end{bmatrix}$$

and

$$\mathbf{S}(z) = \begin{bmatrix} S_{00}(z) & S_{01}(z) & \cdots & S_{0,N-1}(z) \\ S_{10}(z) & S_{11}(z) & \cdots & S_{1,N-1}(z) \\ \vdots & \vdots & \ddots & \vdots \\ S_{M-1,0}(z) & S_{M-1,1}(z) & \cdots & S_{M-1,N-1}(z) \end{bmatrix},$$

where $G_{km}(z)$ is the kth Type 1 polyphase component of $F_m(z)$ and $S_{km}(z)$ is the mth Type 2 polyphase component of $H_k(z)$. The transmitting and receiving filters are, respectively, related to the transmitting and receiving polyphase matrices as follows:

$$\begin{bmatrix} F_0(z) & F_1(z) & \cdots & F_{M-1}(z) \end{bmatrix} = \begin{bmatrix} 1 & z^{-1} & \cdots & z^{-N+1} \end{bmatrix} \mathbf{G}(z^N)$$

and

$$\begin{bmatrix} H_0(z) \\ H_1(z) \\ \vdots \\ H_{M-1}(z) \end{bmatrix} = \mathbf{S}(z^N) \begin{bmatrix} 1 \\ z \\ \vdots \\ z^{N-1} \end{bmatrix}.$$

Consider the system shown in the gray box in Fig. 5.7. We will see later that this is an LTI system with an $N \times N$ transfer matrix $\mathbf{C}_{ps}(z)$ which depends only on the scalar transfer function $C(z)$ and the integer N, but not the transceiver used. Because the matrix $\mathbf{C}_{ps}(z)$ operates on the blocked version of the channel input $x(n)$ to produce the blocked version of the channel output $r(n)$, we say that $\mathbf{C}_{ps}(z)$ is the **blocked version of the channel** $C(z)$. We will see in Section 5.3 that the transfer matrix $\mathbf{C}_{ps}(z)$ has a form called the pseudocirculant form. The subscript "ps" is a reminder of "pseudocirculant." Thus the entire transceiver system can be drawn as a cascade of MIMO systems, as shown in Fig. 5.8. From the figure we see that the overall system from the input vector $\mathbf{s}(n)$ to the output vector $\hat{\mathbf{s}}(n)$ has the transfer matrix

$$\mathbf{T}(z) = \mathbf{S}(z)\mathbf{C}_{ps}(z)\mathbf{G}(z).$$

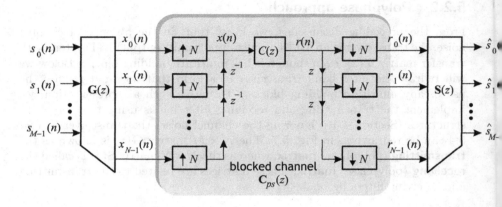

Figure 5.7. The M-band transceiver system in polyphase form.

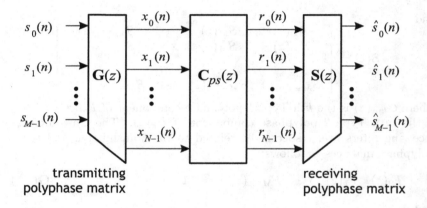

Figure 5.8. The M-band transceiver system in matrix form with transmitting and receiving polyphase matrices indicated.

In general, $\mathbf{T}(z)$ is a matrix dependent on z. From this we can come to a number of conclusions:

(1) inter subchannel interference is eliminated if and only if $\mathbf{S}(z)\mathbf{C}_{ps}(z)\mathbf{G}(z)$ is diagonal;

(2) interblock interference is eliminated if and only if $\mathbf{S}(z)\mathbf{C}_{ps}(z)\mathbf{G}(z)$ is a constant matrix;

(3) the transceiver has the ISI-free property if and only if $\mathbf{S}(z)\mathbf{C}_{ps}(z)\mathbf{G}(z) = \mathbf{I}$. This condition is equivalent to $\widehat{\mathbf{s}}(n) = \mathbf{s}(n)$ in the absence of noise. More generally, since $\widehat{s}_k(n) = \alpha_k s_k(n - n_k)$ is acceptable for nonzero α_k, the product $\mathbf{S}(z)\mathbf{C}_{ps}(z)\mathbf{G}(z)$ can be a diagonal matrix with nonzero diagonal delay elements.

It turns out that the ISI-free condition is easier to satisfy with a redundant filter bank transceiver $(N > M)$, as we shall elaborate later.

Block transceivers When the polyphase matrices $\mathbf{G}(z)$ and $\mathbf{S}(z)$ are both constant matrices independent of z, the filter bank transceiver is called a block transceiver. In this case, the processing at both the transmitter and receiver is done in a block by block manner:

$$\mathbf{x}(n) = \mathbf{G}\mathbf{s}(n) \quad \text{and} \quad \widehat{\mathbf{s}}(n) = \mathbf{S}\mathbf{r}(n).$$

As the polyphase components are constant, the lengths of the transmitting and receiving filters do not exceed the interpolation ratio N. The systems to be discussed in Chapters 6, 7, and 8 are such block transceivers. In practice, the polyphase matrices are often chosen to be matrices that allow a low-cost implementation, e.g. DFT matrices. Block transceivers with DFT polyphase matrices are referred to as DFT-based transceivers. Many modern communication systems fall into this category (see Chapter 6).

5.2.3 Channel-independent ISI-free filter bank transceivers

Generally speaking, an ISI-free filter bank transceiver is channel-dependent. That is, the transmitting filters $F_k(z)$ and the receiving filters $H_k(z)$ depend on the channel coefficients $c(n)$. In many applications, especially those involving wireless communications, the channels are often time-varying. When the channel $c(n)$ changes, we have to redesign $F_k(z)$ and $H_k(z)$ so that the transceiver remains ISI-free. For these applications, it is desirable to have ISI-free transceiver systems that are channel-independent. In particular, we would like to have a transceiver that can be ISI-free for any νth-order channel. From earlier discussions, we know that the transfer function from the mth input $s_m(n)$ to the kth output is given by $T_{km}(z)$ in (5.1). For an FIR channel of order ν, we can write $T_{km}(z)$ as

$$T_{km}(z) = \sum_{n=0}^{\nu} c(n) \left[z^{-n} H_k(z) F_m(z) \right]_{\downarrow N}.$$

The channel-independent ISI-free property implies that for any $c(n)$, the transfer function $T_{km}(z) = \alpha_k z^{-D_k} \delta(m - k)$. This is true if and only if

$$\left[z^{-n} H_k(z) F_m(z) \right]_{\downarrow N} = \rho_{kn} z^{-D_k} \delta(m - k),$$

for $n = 0, 1, \ldots, \nu$. Substituting this result into the expression for $T_{km}(z)$, we find that the kth subchannel gain is given by

$$\alpha_k = \sum_{n=0}^{\nu} c(n) \rho_{kn}.$$

In Chapter 6, we will study a number of transceiver systems with such a channel-independent ISI-free property. We will see that these systems belong to the class of DFT-based block transceivers. The first non-DFT-based block transceivers with such a property were derived by Scaglione and Giannakis, and the system is called a Lagrange–Vandermonde (LV) transceiver [132]. In a LV transceiver, the transmit filters are Lagrange interpolation polynomials, whereas the receive filters are Vandermonde filters. A dual system called a Vandermonde–Lagrange (VL) transceiver, where the transmit filters are Vandermonde filters and the receive filters are Lagrange filters, was derived in [133]. Both the LV and VL systems include the widely known DFT-based transceivers as a special case. In [46], the LV and VL transceivers were generalized to case of multiuser transmission. It was later shown [84] that the LV and VL systems are the only solutions of block transceivers with the channel-independent ISI-free property. For filter bank transceivers with transmitting and receiving filters longer than N, the solution of channel-independent ISI-free transceivers is still unknown. In [117], a method has been proposed for the design of near-ISI-free channel-independent filter bank transceivers. Filter bank transceivers with high signal to interference ratio and good frequency response have been successfully designed therein.

5.3   Pseudocirculant and circulant matrices

In this section, we shall study two special classes of matrices, known as pseudocirculant and circulant matrices, which appear frequently in the study of transceiver systems. These matrices are very useful in the discussion of transceiver systems using a matrix representation.

5.3.1   Pseudocirculants and blocked versions of scalar systems

Definition 5.1   Pseudocirculant matrices An $N \times N$ matrix $\mathbf{C}_{ps}(z)$ is said to be a pseudocirculant if any column (except the leftmost column which is arbitrary) is obtained from the preceding column by performing the following operations: (a) shift down by one element, (b) recirculate the spilled element to the top, and (c) multiply the recirculated element by the delay operator z^{-1}. ∎

Let $\begin{bmatrix} C_0(z) & C_1(z) & \cdots & C_{N-1}(z) \end{bmatrix}^T$ be the first column of $\mathbf{C}_{ps}(z)$. Then one can obtain $\mathbf{C}_{ps}(z)$ by following the mechanism described in the above definition. Following the procedure, we can find the following general form of the pseudocirculant matrix $\mathbf{C}_{ps}(z)$:

$$
\begin{bmatrix}
C_0(z) & z^{-1}C_{N-1}(z) & z^{-1}C_{N-2}(z) & \cdots & z^{-1}C_2(z) & z^{-1}C_1(z) \\
C_1(z) & C_0(z) & z^{-1}C_{N-1}(z) & \cdots & z^{-1}C_3(z) & z^{-1}C_2(z) \\
C_2(z) & C_1(z) & C_0(z) & \cdots & z^{-1}C_4(z) & z^{-1}C_3(z) \\
\vdots & \vdots & \vdots & \vdots & \ddots & \vdots \\
C_{N-2}(z) & C_{N-3}(z) & C_{N-4}(z) & \cdots & C_0(z) & z^{-1}C_{N-1}(z) \\
C_{N-1}(z) & C_{N-2}(z) & C_{N-3}(z) & \cdots & C_1(z) & C_0(z)
\end{bmatrix}.
$$

$$(5.3)$$

Note that along any line parallel to the main diagonal, all elements are identical. Thus a pseudocirculant is also a Toeplitz matrix. Other minor variations are possible in the definition of pseudocirculants. For example, we can shift a column *up* and then circulate the spilled element with z^{-1}. Definition 5.1 will be used throughout this book.

Pseudocirculant matrices were originally introduced in the filter bank literature in the context of aliasfree filter banks and blocked LTI systems (see [157] and [159]). In our context, their importance arises from the fact that the blocked channel matrix $\mathbf{C}_{ps}(z)$ is a pseudocirculant matrix of exactly the form defined above. This is stated in the following theorem.

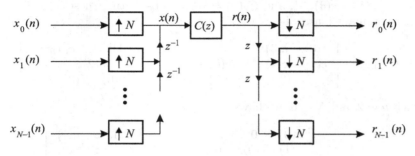

Figure 5.9. $N \times N$ blocked version of a channel $C(z)$.

Theorem 5.1 Blocked versions and pseudocirculants Consider the N-input N-output system shown in Fig. 5.9. Let the Type 1 polyphase representation of the channel be

$$C(z) = C_0(z^N) + C_1(z^N)z^{-1} + \cdots + C_{N-1}(z^N)z^{-N+1}.$$

Then the $N \times N$ system is LTI and its transfer matrix is the pseudocirculant matrix given by $\mathbf{C}_{ps}(z)$ in (5.3). ∎

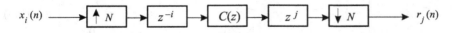

Figure 5.10. Path from $x_i(n)$ to $r_j(n)$ of the blocked system in Fig. 5.9.

A sketch of the proof is as follows. Let us look at the system from $x_i(n)$ to $r_j(n)$ in Fig. 5.9. It is simply $z^{j-i}C(z)$ sandwiched between an N-fold expander and an N-fold decimator (Fig. 5.10). According to the polyphase identity (Fig. 4.6), this system is LTI with transfer function $\left[z^{j-i}C(z)\right]_{\downarrow N}$. Using the polyphase decomposition of $C(z)$, it is not difficult to show (Problem 5.2) that

$$\left[z^{j-i}C(z)\right]_{\downarrow N} = \begin{cases} C_{j-i}(z), & \text{if } j - i \geq 0; \\ z^{-1}C_{N+j-i}(z), & \text{otherwise,} \end{cases} \tag{5.4}$$

which proves that the transfer matrix of the N-input N-output system in Fig. 5.9 has the pseudocirculant form $\mathbf{C}_{ps}(z)$ in (5.3).

Assume that the channel is FIR with order ν:

$$C(z) = \sum_{i=0}^{\nu} c(i)z^{-i}.$$

For example, when $\nu = 4$ and $N = 3$ we have $C_0(z) = c(0) + c(3)z^{-1}$, $C_1(z) = c(1) + c(4)z^{-1}$, and $C_2(z) = c(2)$. Substituting these into (5.3), the pseudocirculant matrix is

$$\mathbf{C}_{ps}(z) = \begin{bmatrix} c(0) + c(3)z^{-1} & c(2)z^{-1} & c(1)z^{-1} + c(4)z^{-2} \\ c(1) + c(4)z^{-1} & c(0) + c(3)z^{-1} & c(2)z^{-1} \\ c(2) & c(1) + c(4)z^{-1} & c(0) + c(3)z^{-1} \end{bmatrix}.$$

When $\nu = 2$ and $N = 6$, we have

$$\mathbf{C}_{ps}(z) = \begin{bmatrix} c(0) & 0 & 0 & 0 & z^{-1}c(2) & z^{-1}c(1) \\ c(1) & c(0) & 0 & 0 & 0 & z^{-1}c(2) \\ c(2) & c(1) & c(0) & 0 & 0 & 0 \\ 0 & c(2) & c(1) & c(0) & 0 & 0 \\ 0 & 0 & c(2) & c(1) & c(0) & 0 \\ 0 & 0 & & c(2) & c(1) & c(0) \end{bmatrix}.$$

In many practical applications, the block size N is usually much larger than the channel order. The polyphase components of $C(z)$ with respect to N are all constant (independent of z), and only the first $(\nu + 1)$ polyphase components are nonzero. In this case, $\mathbf{C}_{ps}(z)$ has the special form

$$\mathbf{C}_{ps}(z) = \begin{bmatrix} c(0) & 0 & \cdots & 0 & z^{-1}c(\nu) & \cdots & z^{-1}c(1) \\ c(1) & c(0) & & & & \ddots & \vdots \\ \vdots & & \ddots & & & & z^{-1}c(\nu) \\ c(\nu) & & & & & & 0 \\ \vdots & & & & & & \vdots \\ & & & c(0) & 0 & \cdots & 0 \\ & & & c(1) & c(0) & & \vdots \\ & & & \vdots & & \ddots & \\ 0 & & & c(\nu) & c(\nu - 1) & \cdots & c(0) \end{bmatrix}. \quad (5.5)$$

Observe that only the $\nu \times \nu$ submatrix at the top right corner depends on z. We will exploit this property of $\mathbf{C}_{ps}(z)$ for IBI elimination. Some deeper properties of pseudocirculant matrices will be explored in Chapter 10.

5.3.2   Circulants and circular convolutions

If we substitute $z = 1$ into a pseudocirculant matrix $\mathbf{C}_{ps}(z)$, we get a constant matrix $\mathbf{C}_{ps}(1)$. It has a special form known as circulant, which is defined below.

Definition 5.2   Circulant matrices An $M \times M$ matrix $\mathbf{C}_{circ}$ is said to be a circulant if any column (except the leftmost column which is arbitrary) is obtained from the preceding column by performing the following operations: (a) shift down by one element and (b) recirculate the spilled element to the top.   ∎

Similar to the case of pseudocirculant matrices, a circulant matrix is completely determined by any of its columns or any of its rows. Letting its first column be $\left[\; c(0) \quad c(1) \quad \cdots \quad c(M-1) \;\right]^{T}$, the $M \times M$ circulant matrix is

$$
\mathbf{C}_{circ} = \begin{bmatrix}
c(0) & c(M-1) & \cdots & c(2) & c(1) \\
c(1) & c(0) & \cdots & c(3) & c(2) \\
\vdots & & \ddots & & \vdots \\
c(M-2) & c(M-3) & & c(0) & c(M-1) \\
c(M-1) & c(M-2) & \cdots & c(1) & c(0)
\end{bmatrix}.
$$

Observe that, apart from the leftmost column, all the other columns are obtained by shifting *down* the previous column. This matrix is therefore called a circulant.[2] It is not difficult to verify (Problem 5.5) that a matrix is circulant if and only if its (k, m)th entry is given by

$$
[\mathbf{C}_{circ}]_{km} = c((k-m))_M, \tag{5.6}
$$

where the argument $((k-m))_M$ is interpreted modulo M, which is an integer between 0 and $(M-1)$. Comparing the pseudocirculant and circulant matrices, we find that pseudocirculant matrices have a form similar to circulant matrices, except that the recirculated element is multiplied by the delay operator z^{-1}. This explains the name "pseudocirculant."

An important property of circulants is that they can be **diagonalized** by the DFT matrix. The $M \times M$ DFT matrix $\mathbf{W}$ is a matrix whose (k, m)th entry is given by[3]

$$
[\mathbf{W}]_{km} = \frac{W^{km}}{\sqrt{M}}, \tag{5.7}
$$

where $W = e^{-j2\pi/M}$ is the Mth root of unity. First consider the product $\mathbf{C}_{circ}\mathbf{W}^{\dagger}$, where the superscript $\dagger$ denotes transpose conjugation. The (k, ℓ)th element of this product is

$$
[\mathbf{C}_{circ}\mathbf{W}^{\dagger}]_{k\ell} = \frac{1}{\sqrt{M}} \sum_{m=0}^{M-1} c((k-m))_M W^{-m\ell}.
$$

Making a change of variable and using the fact that $W^M = 1$, we can rewrite the above expression as

$$
[\mathbf{C}_{circ}\mathbf{W}^{\dagger}]_{k\ell} = \frac{1}{\sqrt{M}} \sum_{m=0}^{M-1} c(m)W^{-(k-m)\ell} = \frac{W^{-k\ell}}{\sqrt{M}}\left(\sum_{m=0}^{M-1} c(m)W^{m\ell}\right).
$$

---

[2] Or, more specifically, a *down* circulant (to distinguish from an *up* circulant which is defined in an obvious way) In this book the term *circulant* always refers to a down circulant.

[3] For convenience, we have normalized the DFT matrix so that it is unitary, $\mathbf{W}^{\dagger}\mathbf{W} = \mathbf{I}_M$; the definition is slightly different from the conventional one, for which the (k, m)th element is W^{km}.

On the right-hand side of the above expression, the quantity inside the bracket is, in fact, the ℓth DFT coefficient of $c(n)$:

$$C_\ell = \sum_{n=0}^{M-1} c(n) W^{n\ell}. \tag{5.8}$$

Note that the DFT coefficients are defined in the usual manner, although the DFT matrix $\mathbf{W}$ is normalized. Using the above expression, we can write

$$\left[\mathbf{C}_{circ}\mathbf{W}^\dagger\right]_{k\ell} = \frac{W^{-k\ell}}{\sqrt{M}} C_\ell.$$

Writing the above expression for $0 \le k \le M-1$, we get the ℓth column of the product $\mathbf{C}_{circ}\mathbf{W}^\dagger$:

$$\left[\mathbf{C}_{circ}\mathbf{W}^\dagger\right]_\ell = \frac{1}{\sqrt{M}} \begin{bmatrix} 1 \\ W^{-\ell} \\ \vdots \\ W^{-(M-1)\ell} \end{bmatrix} C_\ell = C_\ell \left[\mathbf{W}^\dagger\right]_\ell,$$

where $\left[\mathbf{W}^\dagger\right]_\ell = 1/\sqrt{M} \begin{bmatrix} 1 & W^{-\ell} & \cdots & W^{-(M-1)\ell} \end{bmatrix}^T$ is the ℓth column of $\mathbf{W}^\dagger$. In other words, the columns of the IDFT matrix $\mathbf{W}^\dagger$ are the eigenvectors of $\mathbf{C}_{circ}$ and the corresponding eigenvalues are the DFT coefficients of $c(n)$. This is true *regardless of the values of* $c(n)$. Collecting all the M columns for $0 \le \ell \le M-1$, we have proved

$$\mathbf{C}_{circ}\mathbf{W}^\dagger = \mathbf{W}^\dagger\mathbf{\Gamma},$$

where $\mathbf{\Gamma}$ is the diagonal matrix

$$\mathbf{\Gamma} = \begin{bmatrix} C_0 & 0 & \cdots & 0 \\ 0 & C_1 & \cdots & 0 \\ \vdots & \vdots & \ddots & \vdots \\ 0 & 0 & \cdots & C_{M-1} \end{bmatrix}. \tag{5.9}$$

Using the fact that $\mathbf{W}^\dagger\mathbf{W} = \mathbf{I}$, we get

$$\mathbf{W}\mathbf{C}_{circ}\mathbf{W}^\dagger = \mathbf{\Gamma},$$

or equivalently

$$\mathbf{C}_{circ} = \mathbf{W}^\dagger\mathbf{\Gamma}\mathbf{W}; \tag{5.10}$$

that is, circulant matrices are diagonalized by DFT matrices. Conversely, given any matrix of the form $\mathbf{C} = \mathbf{W}^\dagger\mathbf{\Gamma}\mathbf{W}$, we can write its (k, m)th element as

$$[\mathbf{C}]_{km} = \sum_{\ell=0}^{M-1} C_\ell \frac{W^{-k\ell}}{\sqrt{M}} \frac{W^{\ell m}}{\sqrt{M}} = \frac{1}{M} \sum_{\ell=0}^{M-1} C_\ell W^{-(k-m)\ell}.$$

Define the sequence $c(n)$ as the IDFT coefficients of C_k. That is,

$$c(n) = \frac{1}{M} \sum_{\ell=0}^{M-1} C_\ell W^{-n\ell}.$$

Then we can write $[\mathbf{C}]_{km} = c(k - m)$. As the sequence $c(n)$ is periodic with period M, we have

$$[\mathbf{C}]_{km} = c((k - m))_M,$$

which proves that the matrix $\mathbf{C}$ is circulant. As a matrix of the form $\mathbf{W}^\dagger \mathbf{\Gamma} \mathbf{W}$ is circulant, its diagonal entries are the same and they are equal to $c(0)$, which is the average of C_ℓ, the diagonal entries of $\mathbf{\Gamma}$. This fact will be used repeatedly later. Summarizing, we have proved the following theorem.

Theorem 5.2 Diagonalization of circulants If $\mathbf{C}_{circ}$ is a circulant matrix, it can be diagonalized as in (5.10), where $\mathbf{\Gamma}$ is the diagonal matrix (5.9) of the DFT coefficients C_k defined in (5.8), and $\mathbf{W}$ is the $M \times M$ normalized DFT matrix defined in (5.7). *Conversely*, a product of the form $\mathbf{W}^\dagger \mathbf{\Gamma} \mathbf{W}$ is necessarily a circulant, and its leftmost column is $[c(0) \ c(1) \ \cdots \ c(M-1)]^T$, where $c(n) = 1/M \sum_{k=0}^{M-1} C_k W^{-kn}$ are the inverse DFT coefficients of the diagonal elements C_k of $\mathbf{\Gamma}$. ∎

Circulant matrices and circular convolutions Let the two $M \times 1$ vectors

$$\mathbf{r} = \begin{bmatrix} r(0) & r(1) & \cdots & r(M-1) \end{bmatrix}^T,$$
$$\mathbf{s} = \begin{bmatrix} s(0) & s(1) & \cdots & s(M-1) \end{bmatrix}^T$$

be such that $\mathbf{r} = \mathbf{C}_{circ}\mathbf{s}$. Then, using (5.6), we can write the nth entry of $\mathbf{r}$ as

$$r(n) = \sum_{\ell=0}^{M-1} [\mathbf{C}_{circ}]_{n\ell} \, s(\ell) = \sum_{\ell=0}^{M-1} c((n - \ell))_M s(\ell).$$

That is, the sequence $r(n)$ is the *circular convolution* of the sequences $s(n)$ and $c(n)$, which corresponds to the entries in the first column of $\mathbf{C}_{circ}$. Substituting (5.10) into the expression $\mathbf{r} = \mathbf{C}_{circ}\mathbf{s}$ and rearranging the terms, we get

$$\sqrt{M}\mathbf{W}\mathbf{r} = \mathbf{\Gamma}\left(\sqrt{M}\mathbf{W}\mathbf{s}\right). \tag{5.11}$$

Note that the entries of the vectors $\sqrt{M}\mathbf{W}\mathbf{r}$ and $\sqrt{M}\mathbf{W}\mathbf{s}$ are, respectively, the M-point DFT coefficients of $r(n)$ and $s(n)$. Equation (5.11) means that a circular convolution in the time domain becomes a pointwise multiplication operation in the DFT domain, which is simply the circular convolution theorem [105].

5.4 Redundancy for IBI elimination

In a filter bank transceiver the channel $c(n)$ in general introduces interblock interference (IBI). In this section, we will show that by inserting enough redundant samples in the transmitted sequence, IBI can be avoided or eliminated completely. In the following discussion, we assume that the number of redundant samples inserted per input block is ν and the channel is causal FIR of order $\leq \nu$, i.e.

$$C(z) = \sum_{n=0}^{\nu} c(n)z^{-n}. \tag{5.12}$$

The last few coefficients are zero if the order of $c(n)$ is less than ν. We will also assume that there is no channel noise, i.e. $q(n) = 0$. The effect of noise will be studied in more detail in future chapters.

5.4.1 Zero-padded systems

One form of redundancy is the insertion of zeros in the transmitted sequence. This is known as **zero padding**. Figure 5.11(a) shows a system that employs the zero-padding scheme to transmit a sequence $s(n)$. Figures 5.11(b)–(d) illustrate how the zero-padded system works. At the transmitter, the input $s(n)$ is first partitioned into blocks of length M (Fig. 5.11(b)). Then ν zeros are inserted at the end of each block to obtain the zero-padded result $x(n)$ (Fig. 5.11(c)). Thus the new block size N is given by

$$N = M + \nu.$$

The zero-padded sequence $x(n)$ is sent over the channel. The ν zeros will become nonzero due to filtering by $C(z)$. However, because the channel has at most order ν, the N received samples in the current block of $r(n)$ depend only on the M transmitted symbols in the current block of $s(n)$ (Fig. 5.11(d)); there is no IBI. It should be emphasized that IBI is completely avoided regardless of what the impulse response $c(n)$ is so long as its order does not exceed ν.

Matrix formulation of the zero-padded systems

The zero-padded system in Fig. 5.11(a) can be represented using multirate building blocks. This gives rise to a useful representation that can be easily incorporated later in a matrix formulation. The insertion of zeros can be described using decimators and expanders as in Fig. 5.12. First the input sequence $s(n)$ is partitioned into blocks of size M using the advance chain and decimators. The nth block $\mathbf{s}(n)$ is related to the sequence $s(n)$ by[4]

$$\mathbf{s}(n) = \begin{bmatrix} s(nM) & s(nM+1) & \cdots & s(nM+M-1) \end{bmatrix}^T.$$

Each block $\mathbf{s}(n)$ is padded by ν zeros to form $\mathbf{x}(n)$ of size N. Then $\mathbf{x}(n)$ is interleaved using the expander and the delay chain to form the sequence $x(n)$, which is transmitted over the channel $c(n)$. To derive the matrix representation, we block the received sequence $r(n)$ into blocks of size N using the advance chain and decimators as in Fig. 5.12. The nth block $\mathbf{r}(n)$ is given by

$$\mathbf{r}(n) = \begin{bmatrix} r(nN) & r(nN+1) & \cdots & r(nN+N-1) \end{bmatrix}^T.$$

Let us look at the N-input N-output system from $\mathbf{x}(n)$ to $\mathbf{r}(n)$. From Theorem 5.1, we know that it is simply the $N \times N$ blocked version $\mathbf{C}_{ps}(z)$ of the channel $c(n)$. Thus we have

$$\mathbf{r}_z(z) = \mathbf{C}_{ps}(z)\mathbf{x}_z(z) = \mathbf{C}_{ps}(z) \begin{bmatrix} \mathbf{s}_z(z) \\ \mathbf{0} \end{bmatrix},$$

---

[4]By convention, when a sequence $a(n)$ is partitioned into blocks of size P, the kth block begins at $a(kP)$ and ends at $a(kP + P - 1)$. This convention will be used throughout the book.

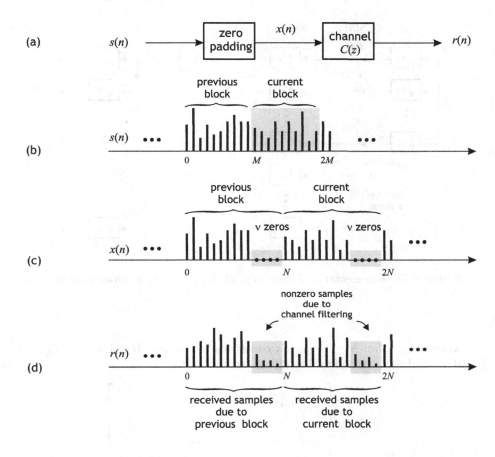

Figure 5.11. Zero-padded system. (a) Transmission scheme employing zero padding; (b)–(d) illustration of how IBI is avoided in the zero-padded system.

where the vectors $\mathbf{r}_z(z)$, $\mathbf{x}_z(z)$, and $\mathbf{s}_z(z)$ are the z-transforms of the vectors $\mathbf{r}(n)$, $\mathbf{x}(n)$, and $\mathbf{s}(n)$ respectively. Recall that the matrix $\mathbf{C}_{ps}(z)$ is an $N \times N$ pseudocirculant matrix. Because the block size satisfies $N = M + \nu > \nu$, the matrix $\mathbf{C}_{ps}(z)$ has the simple form in (5.5). Only the last ν columns of $\mathbf{C}_{ps}(z)$ are dependent on z. Thus we can write

$$\mathbf{r}_z(z) = \mathbf{C}_{low}\mathbf{s}_z(z),$$

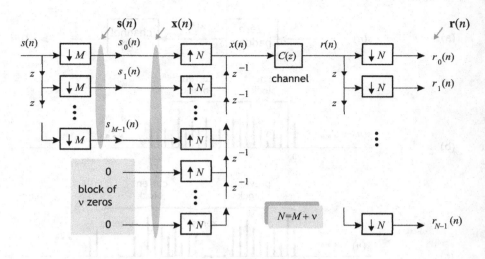

Figure 5.12. Representation of the zero-padded system using multirate building blocks.

where $\mathbf{C}_{low}$ is an $N \times M$ matrix consisting of the first M columns of $\mathbf{C}_{ps}(z)$. It is the lower triangular Toeplitz matrix given by

$$
\mathbf{C}_{low} = \begin{bmatrix}
c(0) & 0 & \cdots & 0 & \cdots & 0 \\
c(1) & c(0) & & & & 0 \\
\vdots & & \ddots & & & \vdots \\
c(\nu) & & & c(0) & & \\
0 & \ddots & & & \ddots & \\
\vdots & & c(\nu) & & \cdots & c(0) \\
0 & \cdots & 0 & c(\nu) & & c(1) \\
\vdots & & \vdots & & \ddots & \vdots \\
0 & \cdots & 0 & & & c(\nu)
\end{bmatrix}. \qquad (5.13)
$$

As $\mathbf{C}_{low}$ is independent of z, in the time domain we can write

$$\mathbf{r}(n) = \mathbf{C}_{low}\mathbf{s}(n);$$

the nth received block $\mathbf{r}(n)$ depends only on the nth input block $\mathbf{s}(n)$, but not other input blocks. *There is no IBI.* This holds for *any* causal FIR channel $C(z)$ with order less than or equal to ν. As a result, symbol recovery can be done in a block by block manner. Whenever the details of blocking and unblocking are not relevant to our discussion, we will adopt the schematic plot shown in Fig. 5.13, which is a more compact description of Fig. 5.12. The legend under the zero-padding box is meant to indicate that the bandwidth expansion factor is N/M (Section 5.1.2); that is, $N - M$ zeros are inserted per block of size M.

As the zero-padded system is free from IBI, symbol recovery can be achieved by choosing any left inverse of the $N \times M$ matrix $\mathbf{C}_{low}$. The rank of $\mathbf{C}_{low}$

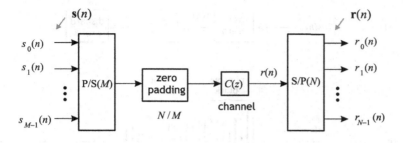

Figure 5.13. Schematic plot of the zero-padded system.

is M as long as the coefficients $c(n)$ are not all zero (Problem 3.1). We can always recover the input vector $\mathbf{s}(n)$ from the received vector $\mathbf{r}(n)$ unless the channel $C(z)$ is identically zero. One possible zero-forcing solution is

$$\left(\mathbf{C}_{low}^{\dagger} \mathbf{C}_{low}\right)^{-1} \mathbf{C}_{low}^{\dagger},$$

which in general has an implementation complexity of $\mathcal{O}(MN)$. Other solutions that have a much lower implementation cost will be explored in Chapter 6.

5.4.2 Cyclic-prefixed systems

Another way of dealing with IBI is to allow a certain amount of IBI during transmission and ensure the received samples corrupted by IBI are removed at the receiver. For an FIR channel of order ν, only the first ν samples of each received block are corrupted by the previous transmitted block. Therefore IBI can be removed by discarding the first ν samples of each received block. This strategy gives rise to another form of redundancy for IBI elimination, namely **cyclic prefixing**. It is the most popular form of redundancy employed in modern communication systems. We shall explain this below.

Figure 5.14(a) shows a system that employs the cyclic-prefixing scheme to transmit a sequence $s(n)$. Figures 5.14(b)–(e) illustrate how the cyclic-prefixed system works for the case of $\nu = 4$. At the transmitter, the input $s(n)$ is partitioned into blocks of size M, as in Fig. 5.14(b). Then we copy the last ν samples at the end of each block and place them at the beginning (Fig. 5.14(c)). This construction implicitly means $M \geq \nu$. The new sequence $x(n)$ can be considered to have blocks of length

$$N = M + \nu,$$

with the first ν samples identical to the last ν samples. The set of ν samples at the beginning is referred to as a **cyclic prefix (CP)**. The integer ν is called the CP length. The cyclic-prefixed sequence $x(n)$ is then sent over the channel $c(n)$. As the channel has order ν, the first ν samples of each received block are corrupted by the previous transmitted block (Fig. 5.14(d)). These

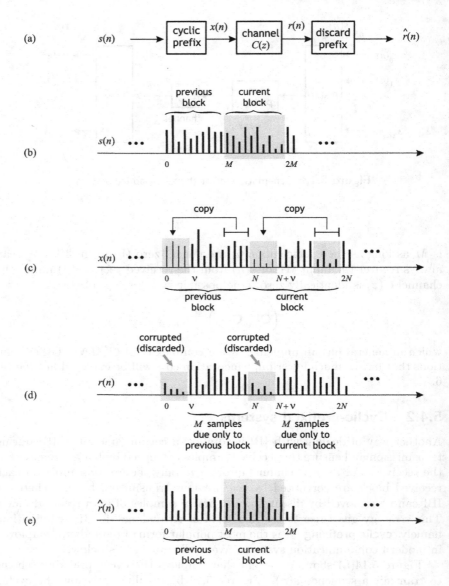

Figure 5.14. The Cyclic-prefixed system. (a) Transmission scheme employing cyclic prefixing; (b)–(e) illustration of how the cyclic prefix is added and how IBI is eliminated in the cyclic-prefixed system.

contaminated samples are discarded, and this operation is denoted by a box labeled "discard prefix" (Fig. 5.14(a)). After discarding the prefix, we obtain the new sequence $\hat{r}(n)$, which can be viewed as a sequence consisting of blocks of length M (Fig. 5.14(e)). Note that each received block $\hat{\mathbf{r}}(n)$ depends only on the corresponding transmitted block of the input signal $\mathbf{s}(n)$. IBI is completely eliminated regardless of what the impulse response $c(n)$ is so long as its order does not exceed ν.

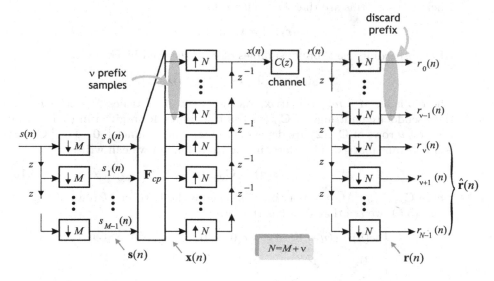

Figure 5.15. The cyclic-prefixed system shown in terms of multirate building block notation.

Matrix formulation of the cyclic-prefixed systems

The cyclic-prefixed system in Fig. 5.14(a) can also be represented using multirate building blocks. Again this representation gives rise to a useful formulation that can be conveniently included in a matrix framework. Using multirate building blocks, the cyclic-prefixed system in Fig. 5.14(a) can be redrawn as Fig. 5.15. The input sequence $s(n)$ is first blocked into blocks of size M using the advance chain and decimators. The input block $\mathbf{s}(n)$ is then multiplied by an $N \times M$ matrix $\mathbf{F}_{cp}$ defined as

$$\mathbf{F}_{cp} = \begin{bmatrix} \mathbf{0} & \mathbf{I}_\nu \\ \mathbf{I}_{M-\nu} & \mathbf{0} \\ \mathbf{0} & \mathbf{I}_\nu \end{bmatrix}. \tag{5.14}$$

The result is an $N \times 1$ vector

$$\mathbf{x}(n) = \mathbf{F}_{cp}\mathbf{s}(n).$$

It is not difficult to verify that multiplying $\mathbf{s}(n)$ by $\mathbf{F}_{cp}$ is equivalent to adding a cyclic prefix of length ν. The vector $\mathbf{x}(n)$ is the cyclic-prefixed version of $\mathbf{s}(n)$ and it is interleaved to obtain the cyclic-prefixed sequence $x(n)$ as in Fig. 5.15. At the receiver, the received sequence $r(n)$ is blocked into $N \times 1$ vectors $\mathbf{r}(n)$. After removing the first ν samples in each vector $\mathbf{r}(n)$ (discarding the prefix), we obtain $M \times 1$ output vectors $\widehat{\mathbf{r}}(n)$ as indicated in Fig. 5.15. Note that the system from $\mathbf{x}(n)$ to $\mathbf{r}(n)$ is the blocked version $\mathbf{C}_{ps}(z)$ of the channel. Their z-transforms are therefore related by

$$\mathbf{r}_z(z) = \mathbf{C}_{ps}(z)\mathbf{x}_z(z).$$

After discarding the ν prefix samples of each received block, we have

$$\widehat{\mathbf{r}}_z(z) = \begin{bmatrix} \mathbf{0} & \mathbf{I}_M \end{bmatrix} \mathbf{C}_{ps}(z)\mathbf{x}_z(z),$$

where $\mathbf{0}$ is an $M \times \nu$ zero matrix. As the block size satisfies $N = M + \nu > \nu$, the pseudocirculant matrix $\mathbf{C}_{ps}(z)$ has the simple form given in (5.5). Only the first ν rows of $\mathbf{C}_{ps}(z)$ are dependent on z. The product $\begin{bmatrix} \mathbf{0} & \mathbf{I}_M \end{bmatrix} \mathbf{C}_{ps}(z)$ is a constant matrix. Therefore, in the time domain we can write

$$\widehat{\mathbf{r}}(n) = \mathbf{C}_{up}\mathbf{x}(n), \tag{5.15}$$

where $\mathbf{C}_{up}$ is an $M \times N$ matrix consisting of the bottom M rows of $\mathbf{C}_{ps}(z)$. From (5.5), we see that $\mathbf{C}_{up}$ has the form:

$$\mathbf{C}_{up} = \begin{bmatrix} c(\nu) & c(\nu-1) & \cdots & c(0) & 0 & \cdots & 0 & \cdots & 0 \\ 0 & c(\nu) & \ddots & & \ddots & & \vdots & & \vdots \\ \vdots & & \ddots & & & c(0) & 0 & \cdots & 0 \\ 0 & & & c(\nu) & & & c(0) & & \\ \vdots & & & & \ddots & & & \ddots & \\ 0 & 0 & \cdots & & c(\nu) & \cdots & c(1) & c(0) \end{bmatrix}. \tag{5.16}$$

The subscript "up" serves as a reminder that the matrix is an upper triangular Toeplitz matrix. Substituting the cyclic-prefix relation $\mathbf{x}(n) = \mathbf{F}_{cp}\mathbf{s}(n)$ into (5.15), we get

$$\widehat{\mathbf{r}}(n) = \mathbf{C}_{up}\mathbf{F}_{cp}\mathbf{s}(n).$$

The nth output block $\widehat{\mathbf{r}}(n)$ depends only on the nth input block $\mathbf{s}(n)$. *IBI is eliminated* by the operation of discarding the prefix. The output and input vectors are related by the $M \times M$ product matrix $(\mathbf{C}_{up}\mathbf{F}_{cp})$. It turns out that this product is a circulant matrix. To see this, observe that multiplying $\mathbf{C}_{up}$ with $\mathbf{F}_{cp}$ from the right is equivalent to adding the first ν columns of $\mathbf{C}_{up}$ to the last ν columns. This operation will transform the $M \times N$ upper triangular Toeplitz matrix $\mathbf{C}_{up}$ into an $M \times M$ circulant matrix. To understand this, let us consider the example of $\nu = 3$ and $M = 4$. In this case, the $M \times N$ matrix $\mathbf{C}_{up}$ is

$$\mathbf{C}_{up} = \begin{bmatrix} c(3) & c(2) & c(1) & | & c(0) & 0 & 0 & 0 \\ 0 & c(3) & c(2) & | & c(1) & c(0) & 0 & 0 \\ 0 & 0 & c(3) & | & c(2) & c(1) & c(0) & 0 \\ 0 & 0 & 0 & | & c(3) & c(2) & c(1) & c(0) \end{bmatrix}.$$

Adding the first three columns to the last three columns, we get

$$\mathbf{C}_{up}\mathbf{F}_{cp} = \begin{bmatrix} c(0) & c(3) & c(2) & c(1) \\ c(1) & c(0) & c(3) & c(2) \\ c(2) & c(1) & c(0) & c(3) \\ c(3) & c(2) & c(1) & c(0) \end{bmatrix}, \tag{5.17}$$

which is an $M \times M$ circulant matrix. For general M and ν that satisfy $M \geq \nu$, the relation

$$\widehat{\mathbf{r}}(n) = \mathbf{C}_{circ}\mathbf{s}(n) \tag{5.18}$$

continues to hold. In this case, the first column of the circulant matrix $\mathbf{C}_{circ}$ is the same as the νth column of $\mathbf{C}_{up}$. In practice, M is usually much larger than ν, and in this case many of the entries of $\mathbf{C}_{circ}$ are zero:

$$\mathbf{C}_{circ} = \begin{bmatrix} c(0) & 0 & \cdots & \cdots & c(\nu) & \cdots & c(1) \\ c(1) & c(0) & & & & \ddots & \vdots \\ \vdots & & \ddots & & & & c(\nu) \\ c(\nu) & & \ddots & \ddots & & & 0 \\ 0 & \ddots & & \ddots & c(0) & & \vdots \\ \vdots & \ddots & \ddots & & & \ddots & 0 \\ 0 & \cdots & 0 & c(\nu) & \cdots & c(1) & c(0) \end{bmatrix}. \tag{5.19}$$

When $M = \nu$, the matrix $\mathbf{C}$ is still circulant but it is slightly different from the above expression (Problem 5.16).

From Section 5.4.2, we know that the multiplication of a circulant matrix and a vector can be interpreted as a circular convolution. Equation (5.18) means that the entries of $\widehat{\mathbf{r}}(n)$ represent the circular convolution of the entries of $\mathbf{s}(n)$ and the channel impulse response. The removal of the cyclic prefix at the receiver eliminates IBI and *the insertion of the cyclic prefix at the transmitter converts a linear convolution to a circular one.* For symbol recovery, we can simply invert the matrix $\mathbf{C}_{circ}$ (provided that it is invertible) to get

$$\mathbf{s}(n) = \mathbf{C}_{circ}^{-1}\widehat{\mathbf{r}}(n).$$

From Theorem 5.3, we know that $\mathbf{C}_{circ}$ is invertible if and only if the diagonal matrix $\mathbf{\Gamma}$ is invertible. The diagonal entries of $\mathbf{\Gamma}$ are the M-point DFT coefficients of the channel impulse response $c(n)$. Thus *zero-forcing equalization exists for a CP system if and only if the channel $C(z)$ does not have zeros at the DFT frequencies $2k\pi/M$.* Whenever the details of blocking and unblocking are not relevant to our discussion, we will adopt the schematic plot of the cyclic-prefixed system shown in Fig. 5.16. Again the ratio N/M is the bandwidth expansion ratio; that is, $(N - M)$ CP samples are inserted per block of size M. In Chapter 6, we will introduce a number of transceiver systems that employ the cyclic-prefixing scheme.

5.4.3 Summary and comparison

The zero-padded (ZP) system and the cyclic-prefixed (CP) system are very similar, but there are also some differences. Here is a summary and comparison of the two systems.

(1) *Bandwidth efficiency.* Both systems have the same bandwidth expansion ratio:

$$\zeta = \frac{M + \nu}{M} = \frac{N}{M}.$$

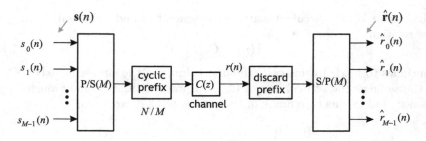

Figure 5.16. Schematic plot of the cyclic-prefixed system.

In practice, due to bandwidth efficiency the block size M is usually much larger than ν. Also note that, in order to eliminate IBI, the number ν has to be at least as large as the channel order. For some applications, such as DSLs, the channels can have very long impulse responses. In this case, a *channel-shortening filter* is usually employed at the receiver to shorten the channels so that the equivalent channel is well approximated by a much shorter filter. The design of channel-shortening filters was described in Section 3.5.

(2) *Transmission power.* In a ZP system the redundant samples are zeros, whereas in a CP system the redundant samples are the last ν samples in each block, which are in general nonzero. As the zero samples have zero power, the ZP system needs a smaller transmission power than the CP system for the same input block. The difference is small as the block size M is usually much larger than ν.

(3) *Symbol recovery.* In a ZP system, there is no IBI. All the N samples in the received block $\mathbf{r}(n)$ can be utilized for symbol recovery. On the other hand, in a CP system only the M samples in the vector $\hat{\mathbf{r}}(n)$ can be used for symbol recovery. So when there is no constraint on the complexity of the receiver, the ZP system will outperform the CP system. However, when the receivers of both systems are implemented with roughly the same complexity, their performances are almost the same (Chapter 6).

(4) *Channel equalization.* In both systems, the equalization of FIR channels can be achieved by inverting a constant matrix provided that the transmitter inserts enough redundant samples. *No IIR filtering is needed.* For ZP systems, zero-forcing equalizers always exist as long as the channel is not identically zero. On the other hand, for CP systems the existence of zero-forcing equalizers depends on the locations of the channel zeros.

For the purpose of channel equalization, the ZP system in general has a better performance than the CP system. However, the inserted CP samples can be exploited at the receiver for solving other issues, such as timing synchronization [100] and channel shortening [87]. A more detailed comparison of the ZP and CP systems can be found in [96].

5.4.4 IBI-free systems with reduced redundancy

Both the ZP and CP systems are IBI-free for any causal FIR channels provided that the number of inserted redundant samples per block, ν, is larger than or equal to the channel order. A ZP system *avoids* IBI by inserting enough zeros between successive blocks. On the other hand, in a CP system the channel does incur IBI on successive blocks, but the receiver *eliminates* IBI by discarding the CP samples. Instead of inserting ν zeros, one can also append fewer zeros at the transmitter, and the IBI due to insufficient redundancy can be eliminated by removing the contaminated samples at the receiver. It was shown in [76] that, by doing so, we can have IBI-free block transceivers with fewer redundant samples per block. Below we will demonstrate how to achieve this.

Let ν be order of the channel $C(z)$. Suppose that we append ρ (with $\rho \leq \nu$) zeros to each transmission block $\mathbf{s}(n)$ of M samples. Let $\mathbf{r}(n)$ denote the received block of $N = M + \rho$ samples. Then the z-transforms of $\mathbf{r}(n)$ and $\mathbf{s}(n)$ are related by

$$\mathbf{r}_z(z) = \mathbf{C}_{ps}(z) \begin{bmatrix} \mathbf{s}_z(z) \\ \mathbf{0} \end{bmatrix},$$

where $\mathbf{C}_{ps}(z)$ is the $N \times N$ pseudocirculant matrix in (5.5) (assuming that $N = M + \rho > \nu$). Using the facts that $\mathbf{0}$ is an $\rho \times 1$ zero vector and $\mathbf{C}_{ps}(z)$ has the special form in (5.5), only the first $(\nu - \rho)$ entries of $\mathbf{r}_z(z)$ are dependent on z. In other words, the last $(N - \nu + \rho)$ received samples in $\mathbf{r}(n)$ depend only on the current transmitted block $\mathbf{s}(n)$. Therefore by discarding the first $(\nu - \rho)$ of each received block, we will have an IBI-free system as shown in Fig. 5.17. Let $\hat{\mathbf{r}}(n)$ be the vector consisting of the last $(N - \nu + \rho)$ entries of $\mathbf{r}(n)$. Then we have

$$\hat{\mathbf{r}}(n) = \mathbf{C}_\rho \mathbf{s}(n),$$

where $\mathbf{C}_\rho$ is the $(N - \nu + \rho) \times M$ Toeplitz matrix with first row $[c(\nu - \rho)\ c(\nu - \rho - 1)\ \ldots\ c(0)\ 0\ \ldots\ 0]$ and first column $[c(\nu - \rho)\ c(\nu - \rho + 1)\ \ldots\ c(\nu)\ 0\ \ldots\ 0]^T$. To ensure the existence of a zero-forcing receiver, the rank of $\mathbf{C}_\rho$ should be at least M. Thus one necessary condition is

$$N - \nu + \rho \geq M.$$

Using $N = M + \rho$, the above condition can be rewritten as $2\rho - \nu \geq 0$. Thus we can conclude that the minimum redundancy needed for the existence of ISI-free block transceivers is

$$\rho_{min} = \begin{cases} \nu/2, & \text{for even } \nu; \\ (\nu + 1)/2, & \text{for odd } \nu. \end{cases}$$

The number of redundant samples needed is about *half of that in the ZP and CP systems*. Like the CP system, the existence of a zero-forcing equalizer depends on the channel $c(n)$. The block transceivers with minimum redundancy were introduced in [76]. One major drawback of such transceivers was their high complexity. Unlike the CP and the ZP systems, which have low complexity DFT-based implementations (as we will show in Chapter 6), the cost of implementing the left inverse of the Toeplitz matrix $\mathbf{C}_\rho$ is high. A reduced-complexity implementation of these minimum redundancy block transceivers

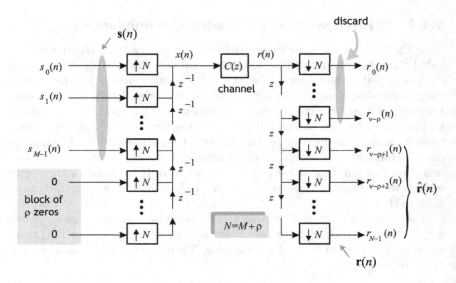

Figure 5.17. IBI-free system with reduced redundancy.

was given in [26]. The minimum redundancy given in this subsection is for block transceivers where the transmitting and receiving polyphase matrices $\mathbf{G}(z)$ and $\mathbf{S}(z)$ are constant matrices independent of z. If we allow $\mathbf{G}(z)$ and $\mathbf{S}(z)$ to have memory (z-dependent), then the redundancy can be even lower (Chapter 10).

5.5   Fractionally spaced equalizer systems

In modern communication systems, the receiver sometimes takes samples of the received continuous-time signal at a higher rate to enhance the system performance. Figure 5.18 shows such a system, where the transmitter sends the sequence $x(n)$ at a rate of $1/T$ (Hz) but the receiver takes the samples $r(n)$ at a rate of N/T (Hz). The sampling rate of the receiver is N times higher than that of the transmitter. We will assume N is an integer in our discussion. If we look at the symbol spacing, the input signal $x(n)$ is spaced apart by T seconds, whereas the received signal $r(n)$ is spaced apart by T/N seconds. Hence the system is known as the **fractionally spaced equalizer (FSE)** system. When the transmitter and receiver have the same sampling rate, that is $N = 1$, the FSE system reduces to the **symbol spaced equalizer (SSE)** system, which has been discussed in detail in Section 2.1. We know that the SSE system can be described by an equivalent discrete-time LTI system. We shall derive that the FSE system also has a discrete-time equivalent system. As the sampling rates are different at the transmitter and receiver, the equivalent system is no longer LTI, and multirate building blocks are needed.

Let us compare Fig. 5.18 with Fig. 2.3. We find that the signals $w_a(t)$ in these two figures are identical and they are given by (2.1). For convenience,

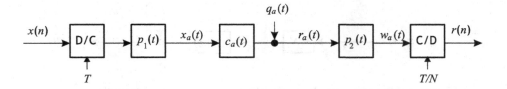

Figure 5.18. Transmission system with oversampling at the receiver.

we reproduce the expression for $w_a(t)$ below:

$$w_a(t) = \sum_{k=-\infty}^{\infty} x(k)c_e(t - kT) + q_e(t),$$

where $c_e(t) = (p_1 * c_a * p_2)(t)$ and $q_e(t) = (q_a * p_2)(t)$ are the effective channel and noise, respectively. When $w_a(t)$ is uniformly sampled every T/N seconds as in Fig. 5.18, we obtain the discrete-time signal

$$r(n) = \sum_{k=-\infty}^{\infty} x(k)c_e(nT/N - kT) + q_e(nT/N).$$

Define the discrete-time equivalent channel and noise, respectively, as follows:

$$c(n) = (p_1 * c_a * p_2)(t)\Big|_{t=nT/N}, \qquad (5.20)$$

$$q(n) = (q_a * p_2)(t)\Big|_{t=nT/N}. \qquad (5.21)$$

Then we can rewrite $r(n)$ as

$$r(n) = \sum_{k=-\infty}^{\infty} x(k)c(n - kN) + q(n).$$

The first term simply represents the interpolation filter operating on the input $x(n)$. In terms of discrete-time signal processing notations, the preceding equation has the beautiful interpretation shown in Fig. 5.19, where

$$C(z) = \sum_n c(n)z^{-n}.$$

Let us compare the discrete-time equivalent channel and noise of the FSE system with those of the SSE system derived in Section 2.1. For an SSE system where the receiver takes samples at the sampling rate $1/T$, the discrete-time equivalent channel and noise are, respectively, given by

$$c_{sse}(n) = (p_1 * c_a * p_2)(t)\Big|_{t=nT},$$

$$q_{sse}(n) = (q_a * p_2)(t)\Big|_{t=nT}.$$

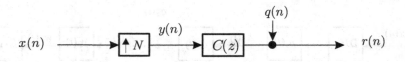

Figure 5.19.   Discrete-time equivalent model of the communication system (in Fig. 5.18) with oversampling at the receiver.

From (5.20) and (5.21), we find that the two discrete-time models are related by

$$c_{sse}(n) = c(Nn) = [c(n)]_{\downarrow N} \quad \text{and} \quad q_{sse}(n) = q(Nn) = [q(n)]_{\downarrow N} \, ;$$

the channel impulse response and noise of the SSE system are, respectively, the N-fold decimated versions of the channel and noise of the associated FSE system.

5.5.1   Zero-forcing FSE systems

In an FSE system, the received samples $r(n)$ are spaced apart by T/N seconds whereas the input samples $x(n)$ are spaced apart by T seconds. The received data rate is N times the input data rate. To recover $x(n)$ from $r(n)$, we can use an N-fold decimation filter. This is shown in Fig. 5.20. The simplest way to equalize the effect of the channel is to choose the equalizer as $H(z) = 1/C(z)$. Then the system is zero-forcing and the output error is simply $q(n)$ passed through $1/C(z)$ and decimated by N. A very important observation here is that if we want to cancel out the effect of $C(z)$ completely, it is not necessary to use $1/C(z)$ as the equalizer because the output is decimated anyway. From the polyphase identity (Fig. 4.6), we know that in the absence of noise the system from $x(n)$ to $\hat{x}(n)$ is in fact LTI with transfer function

$$T(z) = \Big[C(z)H(z) \Big]_{\downarrow N}.$$

Recall from Chapter 4 that $[C(z)H(z)]_{\downarrow N}$ is simply the zeroth polyphase component of the product $C(z)H(z)$. Thus the system in Fig. 5.20 is ISI-free when

$$\Big[C(z)H(z) \Big]_{\downarrow N} = 1.$$

Or, equivalently in the time domain, $t(n) = (c * h)(Nn) = \delta(n)$, which means that the convolution $(c * h)(n)$ satisfies the Nyquist(N) property (Section 4.3.2). In other words, the system is ISI-free if and only if $h(n)$ is such that the impulse response $(c * h)(n)$ has the zero-crossing property.[5] More

---

[5] When two transfer functions $C(z)$ and $H(z)$ are such that $[C(z)H(z)]_{\downarrow N} = 1$, they are called biorthogonal partners. Further details about the theory and applications of biorthogonal partners and their extensions can be found in [160] and [171].

generally, if we allow a delay in the system output, the zero-forcing condition becomes

$$T(z) = \Big[C(z)H(z)\Big]_{\downarrow N} = z^{-n_0},$$

for some non-negative integer n_0. In the absence of noise, the output is $\widehat{x}(n) = x(n - n_0)$.

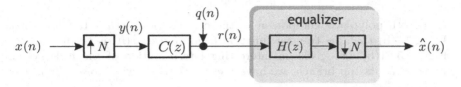

Figure 5.20. Decimation filter for FSE channel equalization.

Note that for the FSE system, we do not need $C(z)H(z) = 1$ to achieve zero-forcing equalization. Only the zeroth polyphase component of the product $C(z)H(z)$ needs to be unity; all the other polyphase components can be *arbitrary*. Thus channel inversion can be avoided by applying oversampling at the receiver. In fact, for most FIR channels, perfect channel equalization can be achieved with an FIR filter $H(z)$, as we will see in the following.

5.5.2   Polyphase approach

Let us decompose the channel $C(z)$ and the filter $H(z)$ using Type 1 and 2 polyphase representations respectively:

$$C(z) = \sum_{k=0}^{N-1} z^{-k} G_k(z^N),$$

$$H(z) = \sum_{k=0}^{N-1} z^{k} S_k(z^N).$$

Then the zeroth polyphase component of the product $C(z)H(z)$ is given by

$$T(z) = \sum_{k=0}^{N-1} G_k(z)S_k(z). \tag{5.22}$$

Given the channel $C(z)$, the equalizer is zero-forcing if and only if the polyphase components $S_k(z)$ satisfy

$$\sum_{k=0}^{N-1} G_k(z)S_k(z) = z^{-n_0}. \tag{5.23}$$

A pictorial view of (5.22) We can also derive the transfer function $T(z)$ in (5.22) pictorially. Using the polyphase implementation of the decimation and interpolation filters in Section 4.4, we can redraw the system in Fig. 5.20 as in Fig. 5.21(a). Recall from Fig. 4.10(b) that the multi-input multi-output (MIMO) system in the gray box is an identity system. Thus Fig. 5.21(a) can be redrawn as Fig. 5.21(b), where

$$q_k(n) = q(Nn + k)$$

is the kth polyphase component of the channel noise $q(n)$. It follows from Fig. 5.21(b) that the transfer function from $x(n)$ to $\hat{x}(n)$ is given by (5.22). Figure 5.21(b) also gives an interesting interpretation of the FSE system. Looking at its top branch, we have a channel $G_0(z)$ with noise $q_0(n)$. Note that

$$G_0(z) = [C(z)]_{\downarrow N} \quad \text{and} \quad q_0(n) = [q(n)]_{\downarrow N};$$

these quantities will be the channel and noise of an SSE system, respectively, if the receiver takes samples $r(n)$ at the symbol spacing. On the other hand, in an FSE system we have N branches. *The FSE receiver has N copies of the transmitted signal, each going through a different "channel" $G_k(z)$.* As a result, the receiver of an FSE system can recover $x(n)$ using an FIR equalizer $H(z)$ for most FIR channels $C(z)$, as stated by Theorem 5.3.

Theorem 5.3 Existence of an FIR zero-forcing equalizer. Given an FIR channel $C(z) = \sum_{n=0}^{\nu} c(n)z^{-n}$, there exists an FIR $H(z)$ such that

$$T(z) = \Big[C(z)H(z)\Big]_{\downarrow N} = 1$$

if and only if the polyphase components $G_i(z)$ of $C(z)$ do not have any common factor of the form $(1 - \alpha z^{-1})$, $\alpha \neq 0$. ∎

Proof. From (5.22) we know that $T(z) = \sum_{i=0}^{N-1} G_i(z)S_i(z)$. When all $G_i(z)$ have a nontrivial common factor $(1 - \alpha z^{-1})$, $\alpha \neq 0$, it is clear that $T(z)$ will also have such a common factor if $S_i(z)$ are FIR. Because the polyphase components $S_i(z)$ of FIR $H(z)$ are also FIR, we conclude that there does not exist any FIR $H(z)$ such that $T(z) = 1$ when all $G_i(z)$ have a nontrivial common factor $(1 - \alpha z^{-1})$, $\alpha \neq 0$. Conversely, suppose that $G_i(z)$ do not have such a factor. Then Euclid's theorem states that there exist causal FIR filters $S_i'(z)$ such that

$$\sum_{i=0}^{N-1} G_i(z)S_i'(z) = z^{-K},$$

where z^{-K} is the greatest common factor of all the polyphase components $G_i(z)$. Letting $S_i(z) = z^K S_i'(z)$, then $H(z) = \sum_{i=0}^{N-1} z^i S_i(z^N)$ is FIR and it satisfies $[C(z)H(z)]_{\downarrow N} = 1$. ∎

In practice, it rarely happens that all the polyphase components $G_i(z)$ of a channel $C(z)$ have a nontrivial common factor $(1 - \alpha z^{-1})$. If it does happen,

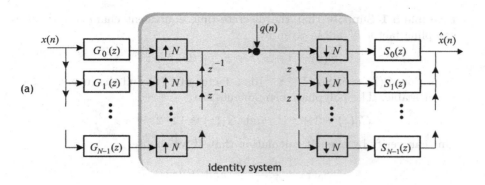

(a)

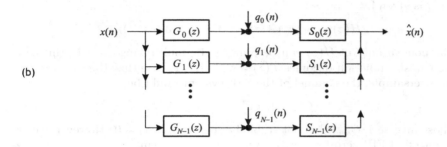

(b)

Figure 5.21. (a) Oversampled channel $C(z)$ and fractionally spaced equalizer H(z); (b) polyphase representation; (c) simplified equivalent.

then an FIR zero-forcing solution does not exist and one needs an IIR filter. Let $Q(z)$ be the greatest common factor of all $G_i(z)$. Then Euclid's theorem says that we can find FIR $S_i'(z)$ such that

$$\sum_{i=0}^{N-1} G_i(z)S_i'(z) = Q(z).$$

Then the zero-forcing condition can be satisfied if we choose

$$S_i(z) = S_i'(z)/Q(z).$$

In this case, the equalizer $H(z) = \sum_{i=0}^{N-1} z^i S_i(z^N)$ is an IIR filter. If $Q(z)$ does not have any zeros on the unit circle, then $H(z)$ is stable. When all the zeros of $Q(z)$ are inside the unit circle, $H(z)$ is both causal and stable.

Causal equalizer The transfer function $H(z) = \sum_{i=0}^{N-1} z^i S_i(z^N)$ can be noncausal even when all $S_i(z)$ are causal. To obtain a causal equalizer, one can multiply a noncausal $H(z)$ by z^{-kN}. Note that multiplying a zero-forcing equalizer $H(z)$ by z^{-m} can destroy its zero-forcing property when m is not a multiple of N. This is because $[z^{-m}C(z)H(z)]_{\downarrow N}$ is not a delayed version of $[C(z)H(z)]_{\downarrow N}$ when m is not a multiple of N.

Example 5.1 Suppose that the discrete-time equivalent channel with over-sampling factor $N = 2$ is given by

$$C(z) = 1 + z^{-1} + z^{-2} - 2z^{-3} - z^{-5}.$$

Note that $C(1) = 0$. So $(1 - z^{-1})$ is a factor of $C(z)$ and the inverse $1/C(z)$ is not stable. The polyphase components are

$$G_0(z) = 1 + z^{-1} \quad \text{and} \quad G_1(z) = 1 - 2z^{-1} - z^{-2}.$$

One can verify by explicit calculation that, if we choose

$$S_0(z) = \frac{3 - z^{-1}}{4} \quad \text{and} \quad S_1(z) = \frac{1}{4},$$

then we will have $G_0(z)S_0(z) + G_1(z)S_1(z) = 1$. The zero-forcing FIR filter $H(z)$ is given by

$$H(z) = S_0(z^2) + zS_1(z^2) = (z + 3 - z^{-2})/4.$$

The noncausality of $H(z)$ can be removed by multiplying z^{-2}. In this case, the transfer function becomes $T(z) = z^{-1}$. In the case that the receiver does not oversample, the channel of the SSE system would be

$$C_{sse}(z) = G_0(z) = 1 + z^{-1},$$

whose inverse $1/C_{sse}(z)$ is not stable because $C_{sse}(-1) = 0$. However, we are able to find FIR zero-forcing $H(z)$ for the FSE system.   ∎

Nonuniqueness of the zero-forcing FIR equalizer   Whenever it exists, the zero-forcing FIR equalizer in an FSE system is not unique. To see this, consider the case of $N = 2$. Suppose $H(z) = S_0(z^2) + zS_1(z^2)$ is zero-forcing so that

$$G_0(z)S_0(z) + G_1(z)S_1(z) = 1.$$

Note now that, given any transfer function $A(z)$, the following identity holds trivially:

$$G_0(z)\Big[G_1(z)A(z)\Big] + G_1(z)\Big[-G_0(z)A(z)\Big] = 0.$$

This means that if we define a new transfer function $\widehat{H}(z)$ with polyphase components

$$\widehat{S}_0(z) = S_0(z) + G_1(z)A(z) \quad \text{and} \quad \widehat{S}_1(z) = S_1(z) - G_0(z)A(z),$$

then $T(z) = G_0(z)\widehat{S}_0(z) + G_1(z)\widehat{S}_1(z)$ is also equal to unity. The filter $\widehat{H}(z)$ is a valid zero-forcing FIR filter as well. Since $A(z)$ is arbitrary, there exists infinitely many zero-forcing FIR filters. In the absence of channel noise, all zero-forcing equalizers are equivalent. This is not the case when there is noise. In practice, one can optimize $A(z)$ to minimize the power of the output noise. Further details of this idea can be found in [171].

In the above discussions, the oversampling factor N is assumed to be an integer. The idea of representing an FSE system using multirate building blocks can also be extended to the case of rational oversampling factors. An FSE system with a rational oversampling factor can also be described using multirate building blocks. More details on the equalizer design for such an FSE system can be found in [172].

5.6   Concluding remarks

The connection between a perfect reconstruction filter bank and a perfect transmultiplexer was first recognized by Vetterli [168]. Multirate signal processing for channel equalization was reported in [121]. The application of nonmaximally decimated (redundant) filter banks to precoding was first proposed by Xia [182, 185]. Optimization of filter bank precoders is studied in [135]. Some tutorials on this topic can be found in [3, 161, 162]. More topics on filter bank transceivers will be explored later in the book. The matrix formulations of the ZP and CP systems are useful for the designs of a number of block transceiver systems (Chapters 6, 7, and 8). The filter bank structure will greatly facilitate the design of transceiver systems with better frequency responses (Chapter 9). The multirate theory will be applied later to the study of the existence of zero-forcing FIR filter bank transceivers (Chapter 10).

5.7   Problems

5.1 Consider the filter bank transceiver system shown in Fig. 5.1. Suppose that the transmitting filter $F_1(z)$ is a real bandpass filter with frequency response given by

$$F_1(e^{j\omega}) = \begin{cases} 1, & \omega \in [\pi/M, \ 2\pi/M] \cup [-2\pi/M, \ -\pi/M]; \\ 0, & \text{otherwise}, \end{cases}$$

and the Fourier transform of the signal $s_1(n)$ is as given in Fig. 5.3(c). Draw the Fourier transform of the output $u_1(n)$ of $F_1(z)$.

5.2 Let $\mathbf{C}_{ps}(z)$ be the $N \times N$ blocked version of the channel $C(z)$. Show that its (i, j)th entry $[\mathbf{C}_{ps}(z)]_{ij}$ is given by (5.4).

5.3 Suppose that we block a scalar channel $C(z)$ as in Fig. P5.3. We know that when $p = 1$, the $N \times N$ system from $\mathbf{s}(n)$ to $\widehat{\mathbf{s}}(n)$ is LTI with transfer matrix $\mathbf{C}_{ps}(z)$. In this problem, we explore the case of $p > 1$. Let the polyphase decomposition of the channel be $C(z) = C_0(z^N) + z^{-1}C_1(z^N) + \cdots + z^{-N+1}C_{N-1}(z^N)$.

   (a) Let $N = 3$ and $p = 2$. Find the 3×3 transfer matrix $\mathbf{T}(z)$. Express your answer in terms of $C_i(z)$.

   (b) Repeat (a) for $N = p = 3$. What can we say about the rank of $\mathbf{T}(z)$?

   (c) Repeat (a) for $N = 6$ and $p = 2$. What is the rank of $\mathbf{T}(z)$?

   (d) More generally, let the greatest common divisor (gcd) of N and p be $\gcd(N, p) = K$ and let $N/K = Q$. Prove that the rank of $\mathbf{T}(z)$ is at most Q.

5.4 Suppose that we block a scalar channel $C(z)$, as we did in Fig. P 5.3. Let $\mathbf{T}(z)$ be its $N \times N$ transfer matrix. Assume that the integers p and N are coprime. Let the polyphase decomposition of the channel be $C(z) =$

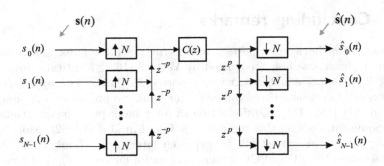

Figure P 5.3. Blocking of a scalar channel using more general delay and advance chains.

$C_0(z^N) + z^{-1}C_1(z^N) + \cdots + z^{-N+1}C_{N-1}(z^N)$. For $k = 0, 1, \ldots, N-1$, define

$$kp = q_k N + r_k,$$

where q_k and r_k are, respectively, the quotient and remainder of kp divided by N. These integers satisfy $q_k \geq 0$ and $0 \leq r_k \leq N-1$.

(a) Prove that $r_k + r_{N-k} = N$ and $q_k + q_{N-k} = p - 1$.

(b) Show that there exists an $N \times N$ permutation $\mathbf{P}$ such that

$$\begin{bmatrix} 0 \\ 1 \\ \vdots \\ N-1 \end{bmatrix} = \mathbf{P} \begin{bmatrix} r_0 \\ r_1 \\ \vdots \\ r_{N-1} \end{bmatrix}.$$

(c) Find the first column and the first row of $\mathbf{T}(z)$. Write your answer in terms of $C_i(z)$, r_i, and q_i.

(d) Show that $\mathbf{T}(z)$ is a Toeplitz matrix. That is,

$$[\mathbf{T}(z)]_{kl} = \begin{cases} [\mathbf{T}(z)]_{0,l-k}, & \text{if } l \geq k; \\ [\mathbf{T}(z)]_{k-l,0}, & \text{if } k > l. \end{cases}$$

Using the result in (c), write $\mathbf{T}(z)$ in terms of $C_i(z)$, r_i, and q_i.

(e) Show that

$$\mathbf{T}(z) = \mathbf{D}(z)\mathbf{P}\mathbf{C}_{ps}(z)\mathbf{P}^T\mathbf{D}(z^{-1}),$$

where $\mathbf{C}_{ps}(z)$ is the pseudocirculant matrix given in (5.3), $\mathbf{P}$ is a permutation matrix, and $\mathbf{D}(z)$ is a diagonal matrix with $[\mathbf{D}(z)]_{kk} = z^{n_k}$ for some integer n_k.

This problem shows that by blocking a scalar channel using a delay chain and an advance chain with z^{-p}, the resulting transfer matrix $\mathbf{T}(z)$ is closely related to $\mathbf{C}_{ps}(z)$. For example, all the columns of $\mathbf{T}(z)$ can be obtained from its first column by following a mechanism that is very similar to that in Definition 5.1, except that the recirculated element is multiplied by z^{-p} rather than z^{-1}.

5.5 Prove that a matrix $\mathbf{C}_{circ}$ satisfies Definition 5.2 if and only if its (k, m)th entry satisfies (5.6).

5.6 Show that the inverse (if it exists) of a circulant matrix is also circulant.

5.7 Is the product of two $N \times N$ circulant matrices also circulant? Justify your answer.

5.8 *Variations of circulant matrices.* An $M \times M$ matrix $\mathbf{C}_{up,circ}$ is said to be *up circulant* if any column of $\mathbf{C}_{up,circ}$ is obtained from the preceding column by performing an *up* shift followed by recirculation of the element that spills over. Note that these matrices can also be obtained by applying a left shift to the rows followed by recirculation. Hence they are also called *left circulant*. In view of this, circulant matrices defined in Definition 5.2 are also known as the *down* or *right* circulant matrices.

   (a) Give an example of a 5×5 up circulant matrix.
   (b) Let $\mathbf{C}_0 = \mathbf{C}_1 \mathbf{J}$, where $\mathbf{J}$ is the $M \times M$ reversal matrix. For example, a 3×3 reversal matrix is

$$\mathbf{J} = \begin{bmatrix} 0 & 0 & 1 \\ 0 & 1 & 0 \\ 1 & 0 & 0 \end{bmatrix}.$$

   Show that $\mathbf{C}_0$ is a down circulant matrix if and only if $\mathbf{C}_1$ is an up circulant matrix.

5.9 Using the results of Problem 5.8(b), modify Theorem 5.2 so that it is true for up circulant matrices. Prove your result.

5.10 Show that the matrix $\mathbf{A} = \mathbf{W}\mathbf{D}\mathbf{W}^\dagger$, where $\mathbf{D}$ is a diagonal matrix, is circulant.

5.11 In Subsection 5.3.2, we focus only on circulant matrices that are independent of z. One can generalize Definition 5.2 to include circulant matrices $\mathbf{C}_{circ}(z)$ in which all the entries of $\mathbf{C}_{circ}(z)$ can be a function of z. A 3×3 example is

$$\begin{bmatrix} C_0(z) & C_2(z) & C_1(z) \\ C_1(z) & C_0(z) & C_2(z) \\ C_2(z) & C_1(z) & C_0(z) \end{bmatrix}.$$

Can we express $\mathbf{C}_{circ}(z)$ as

$$\mathbf{C}_{circ}(z) = \mathbf{P}^{-1}\mathbf{\Gamma}(z)\mathbf{P},$$

for some diagonal matrix $\mathbf{\Gamma}(z)$? If we can, what is $\mathbf{P}$ and how is $\mathbf{\Gamma}(z)$ related to the entries $C_i(z)$ in the first column of $\mathbf{C}_{circ}(z)$?

5.12 In Fig. 5.9, we use a delay chain to block the input and an advance chain to block the output of the channel. Suppose instead that an advance chain and a delay chain are used to block the input and output of a channel $C(z)$, respectively. Find the transfer matrix $\mathbf{C}'_{ps}(z)$ of the blocked system.

5.13 Let two $N \times N$ matrices $\mathbf{A}(z)$ and $\mathbf{B}(z)$ be related by

$$\mathbf{B}(z) = \mathbf{\Lambda}(z)\mathbf{A}(z^N)\mathbf{\Lambda}(z^{-1}),$$

where $\mathbf{\Lambda}(z)$ is the diagonal matrix $\text{diag}[1 \; z^{-1} \; \cdots \; z^{-N+1}]$. Show that $\mathbf{A}(z)$ is pseudocirculant if and only if $\mathbf{B}(z)$ is circulant. Furthermore, show that if $\mathbf{A}(z)$ is causal, then $\mathbf{B}(z)$ is also causal.

5.14 Using the results in Problems 5.11 and 5.13, show that the product of two $N \times N$ pseudocirculant matrices is also pseudocirculant.

5.15 Prove that any $N \times N$ pseudocirculant matrix $\mathbf{C}_{ps}(z)$ satisfies

$$\mathbf{C}_{ps}(z) = \mathbf{J}_N\mathbf{C}_{ps}^T(z)\mathbf{J}_N,$$

where $\mathbf{J}_N$ is the $N \times N$ reversal matrix (see Problem 5.8(b)). Suppose that the transceiver in Fig. 5.7 is ISI-free. Show that the new transceiver with the transmitting matrix $\mathbf{J}_N\mathbf{S}^T(z)$ and the receiving matrix $\mathbf{G}^T(z)\mathbf{J}_N$ is also ISI-free.

5.16 Show that in the CP scheme when $M = \nu$ the output vector $\widehat{\mathbf{r}}(n)$ is still related to $\mathbf{s}(n)$ by

$$\widehat{\mathbf{r}}(n) = \mathbf{C}_{circ}\mathbf{s}(n),$$

for some $M \times M$ circulant matrix $\mathbf{C}_{circ}$. Express $\mathbf{C}_{circ}$ in terms of the channel impulse response $c(0), \; c(1), \ldots, \; c(\nu)$.

5.17 Consider two sequences of length M, $c(n)$, and $s(n)$, for $0 \le n \le M-1$. Define the M-point sequence $x(n)$ as

$$x(n) = \sum_{k=0}^{M-1} s(k)c((n-k))_M.$$

Let $\mathbf{s} = [s(0) \; s(1) \; \cdots \; s(M-1)]^T$ and $\mathbf{x} = [x(0) \; x(1) \; \cdots \; x(M-1)]^T$. Show that there exists a circulant matrix $\mathbf{C}_{circ}$ such that $\mathbf{x} = \mathbf{C}_{circ}\mathbf{s}$. Express the entries of $\mathbf{C}_{circ}$ in terms of $c(n)$.

5.18 *CP system with $\nu > M$.* Suppose that in a CP system the input block size is $M = 3$ and the CP length is $\nu = 5$. Let $\mathbf{s} = [s(0) \; s(1) \; s(2)]$ be the input block. Then we can view the following sequence of length 8 as the result of adding ν cyclic-prefix samples to the input block:

$$[s(1) \; s(2) \; s(0) \; s(1) \; s(2) \; s(0) \; s(1) \; s(2)].$$

Suppose that the channel $c(n)$ has order 5. Show that the output vector $\widehat{\mathbf{r}}$ and the input block $\mathbf{s}(n)$ are related by $\widehat{\mathbf{r}}(n) = \mathbf{C}_{circ}\mathbf{s}(n)$ for some 3×3 circulant matrix $\mathbf{C}_{circ}$. Write $\mathbf{C}_{circ}$ in terms of $c(n)$.

5.19 *Cyclic-suffixed systems and cyclic-prefixed/suffixed systems.* Let $\mathbf{s}(n)$ be input blocks of size M. Let us copy the first ν samples at the beginning of each block into the ν locations at the end. The result is the vector

$$[\; \underbrace{s_0(n) \quad s_1(n) \quad \cdots \quad s_{M-1}(n)}_{\text{original block}} \; \vdots \; \underbrace{s_0(n) \quad \cdots \quad s_{\nu-1}(n)}_{\text{suffix}} \;]^T.$$

This operation is called **cyclic suffixing**. More generally we can combine cyclic prefixing with cyclic suffixing. The result is a hybrid system which will be called the **cyclic-prefixed/suffixed system**. For example, we can copy the last ν_0 samples at the end of each block into the ν_0 locations at the beginning and copy the first ν_1 samples at the beginning of each block into the ν_1 locations at the end, where $\nu_0 + \nu_1 = \nu$. The result is the vector $\tilde{\mathbf{x}}(n)$ given by

$$\left[\underbrace{s_{M-\nu_0} \quad \cdots \quad s_{M-1}}_{\text{prefix}} \vdots \underbrace{s_0 \quad s_1 \quad \cdots \quad s_{M-1}}_{\text{original block}} \vdots \underbrace{s_0 \quad s_1 \quad \cdots \quad s_{\nu_1-1}}_{\text{suffix}} \right]^T,$$

where we have dropped the dependency on n for simplicity. Suppose that the receiver of the cyclic-prefixed/suffixed system is the same as that of the cyclic-prefixed system in Fig. 5.15. Show that the retained block $\hat{\mathbf{r}}(n)$ is free from IBI. What is the relation between $\hat{\mathbf{r}}(n)$ and $\mathbf{s}(n)$?

5.20 Consider a minimum-redundancy block transceiver with $M = 2$. Suppose that the channel is $C(z) = 1 + z^{-1} + z^{-2}$. What are the minimum redundancy ρ_{min} and the Toeplitz matrix $\mathbf{C}_\rho$? Does there exist a zero-forcing equalizer?

5.21 Repeat the above problem for the channel $C(z) = 1 + z^{-1} + z^{-2} + z^{-3}$.

5.22 Consider an FSE system with oversampling factor $N = 2$. Suppose that $T = 1$ and the waveform $(p_1 * c_a * p_2)(t)$ is given by

$$(p_1 * c_a * p_2)(t) = \begin{cases} -0.5|t - 2| + 1, & \text{for } 0 \leq t \leq 4; \\ 0, & \text{otherwise.} \end{cases}$$

(a) Find the discrete-time equivalent channel $C(z)$.
(b) Find the Type 1 polyphase components $G_0(z)$ and $G_1(z)$ of the channel $C(z)$.
(c) Find a zero-forcing FIR filter $H(z)$.
(d) Give another zero-forcing FIR filter $H'(z)$ that is not a delayed or scaled version of the filter $H(z)$ in (c).

5.23 Suppose that we obtain an FSE system by increasing the sampling rate of the receiver of an SSE system and keeping the same receiving pulse. Show that if the FSE system does not have a causal stable zero-forcing equalizer, then neither does the SSE system.

5.24 *Maximum ratio combining (MRC).* Suppose that we transmit a symbol s with signal power $\mathcal{E}_s$ over a set of parallel subchannels, as shown in Fig. P5.24, where c_i are the subchannel gains (possibly complex) and q_i are the subchannel noise. Such a system of parallel subchannels may be the result of oversampling or employing multiple antennas at the receiver. Assume that q_i are zero-mean and uncorrelated with variances $\mathcal{N}_0$, and they are uncorrelated with the signal s. At the receiver, the received samples are $r_i = c_i s + q_i$. Suppose we linearly combine r_i to obtain $\hat{s}$ as follows:

$$\hat{s} = a_0 r_0 + a_1 r_1 + \cdots + a_{M-1} r_{M-1}.$$

We can write $\widehat{s} = \alpha s + \tau$, where the quantity τ depends only on the noise components q_i.

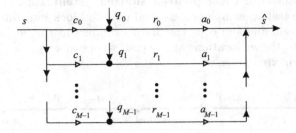

Figure P5.24. Maximal ratio combining transmission scheme.

(a) Define the vectors

$$\mathbf{c} = [c_0 \ c_1 \ \dots \ c_{M-1}]^T,$$
$$\mathbf{a} = [a_0 \ a_1 \ \dots \ a_{M-1}]^T,$$
$$\mathbf{q} = [q_0 \ q_1 \ \dots \ q_{M-1}]^T.$$

Express α, τ, and the signal to noise ratio

$$SNR = \frac{|\alpha|^2 \mathcal{E}_s}{E[|\tau|^2]}$$

in terms of the vectors $\mathbf{c}$, $\mathbf{a}$, and $\mathbf{q}$.

(b) Show that the SNR is maximized when $a_i = c_i^*$ for $i = 0, 1, \dots, M-1$, and the maximized SNR is given by

$$\frac{\left(|c_0|^2 + |c_1|^2 + \cdots + |c_{M-1}|^2\right) \mathcal{E}_s}{\mathcal{N}_0}.$$

Such an optimal receiver is known as the **maximal ratio combining (MRC)** receiver. The phase of the MRC receiver $a_i = c_i^*$ cancels the phase of c_i so that the signal component at the output of the receiver adds constructively. Moreover, the MRC receiver puts more weighting on the subchannels with larger received signal powers. The expression of the maximized SNR shows that each one of the M subchannels is useful, no matter how small the subchannel SNR is.

6

DFT-based transceivers

The most commonly used type of block transceiver is the so-called DFT-based transceiver, in which the polyphase matrices of the transmitter and receiver are related to low-cost DFT matrices in a simple way. The DFT-based transceiver has found applications in a wide range of transmission channels, wired [12, 154] or wireless [27]. It is typically called a DMT (discrete multi-tone) system for wired DSL (digital subscriber line) applications [7, 8] and an OFDM (orthogonal frequency division multiplexing) system for wireless local area networks [54] and broadcasting applications, e.g. digital audio broadcasting [39] and digital video broadcasting [40]. In an OFDM or DMT transceiver, the transmitter and receiver perform, respectively, IDFT and DFT computations. Another type of DFT-based transceiver, called a single-carrier system with cyclic prefix (SC-CP), has also been of great importance in wireless transmission [55, 128, 130]. The SC-CP system transmits a block of symbols directly after inserting a cyclic prefix while the receiver performs both DFT and IDFT computations.

For wireless transmission, the channel state information is usually not available to the transmitter. The transmitter is typically channel-independent and there is no bit or power allocation. Having a channel-independent transmitter is also a very useful feature for broadcasting applications, where there are many receivers with different transmission paths. In OFDM or SC-CP systems for wireless applications, there is usually no bit and power allocation. The transmitters have the desirable channel-independence property. The channel-dependent part of the transceiver is a set of M scalars at the receiver, where M is the number of subchannels. In DMT systems for wired DSL applications, signals are transmitted over copper lines. The channel does not vary rapidly. This allows time for the receiver to send back to the transmitter the channel state information, based on which bit and power allocation can be optimized. Using bit allocation, the disparity among the subchannel noise variances is exploited in the DMT system for bit rate maximization. The DMT system has been shown to be very efficient for high-speed transmission. In this chapter we will give a detailed analysis of DFT-based transceivers.

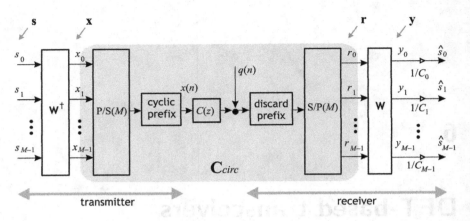

Figure 6.1. The OFDM system.

6.1 OFDM systems

Chapter 5 explains that with block transmission and proper insertion of redundant samples, there will be no interblock interference (IBI). Two common approaches of adding redundant samples are zero padding and cyclic prefixing.

The OFDM transceiver is an example of cyclic-prefixing systems. When the channel order is not larger than the prefix length ν, IBI can be easily removed by discarding the prefix at the receiver. Because there is no IBI, we will consider the transmission of only one block, i.e. *one-shot transmission*, in the discussion of system performance. Although the blocks are sent out consecutively in actual transmission, there is no loss of generality in considering the simple one-shot scenario. Figure 6.1 gives the block diagram of the OFDM system, where a single input block and a single output block are shown. The input of the transmitter $\mathbf{s}$ is an $M \times 1$ vector of modulation symbols. As there is usually no bit and power allocation in the OFDM system, the input symbols s_k have the same constellation and the same variance $\mathcal{E}_s$. The symbols are assumed to be zero-mean and uncorrelated, which is usually a reasonable assumption after proper interleaving of the input bit stream. The autocorrelation matrix of the input vector $\mathbf{s}$ is thus given by

$$\mathbf{R}_s = \mathcal{E}_s \mathbf{I}_M.$$

The channel $c(n)$ is an FIR filter with order $L \leq \nu$,

$$C(z) = \sum_{n=0}^{\nu} c(n) z^{-n},$$

where we have used index up to ν for convenience. (For $L < \nu$, $c(L+1) = c(L+2) = \cdots = c(\nu) = 0$.) For wireless applications, the channel noise $q(n)$ is usually modeled as a circularly symmetric complex Gaussian random process with zero mean and variance $\mathcal{N}_0$. The channel noise $q(n)$ is assumed to be uncorrelated with the symbols s_k.

At the transmitter IDFT is applied to the input symbols,

$$\mathbf{x} = \mathbf{W}^\dagger \mathbf{s},$$

where $\mathbf{W}$ is the normalized $M \times M$ DFT matrix given by

$$[\mathbf{W}]_{mn} = \frac{1}{\sqrt{M}} e^{-j2mn\pi/M}, \quad 0 \le m, n \le M - 1.$$

Then the vector $\mathbf{x}$ is unblocked and a cyclic prefix of length ν is inserted before the sequence is transmitted to the channel. At the receiver, the prefix is discarded and the DFT matrix is applied. When there is no channel noise, recognize that the system from $\mathbf{x}$ to $\mathbf{r}$ is the prefixed system discussed in great detail in Section 5.4. The transfer matrix is the $M \times M$ circulant matrix $\mathbf{C}_{circ}$ given in Definition 5.2. In the presence of channel noise, the received vector $\mathbf{r}$ is

$$\mathbf{r} = \mathbf{C}_{circ}\mathbf{x} + \mathbf{q}, \tag{6.1}$$

where $\mathbf{q}$ is the blocked channel noise vector of size M. The DFT matrix at the receiver has output vector $\mathbf{y}$ given by

$$\mathbf{y} = \mathbf{W}\mathbf{C}_{circ}\mathbf{x} + \boldsymbol{\tau}, \quad \text{where} \quad \boldsymbol{\tau} = \mathbf{W}\mathbf{q}.$$

We also know from Theorem 5.2 that circulant matrices can be diagonalized using DFT and IDFT matrices,

$$\mathbf{C}_{circ} = \mathbf{W}^\dagger \boldsymbol{\Gamma} \mathbf{W}. \tag{6.2}$$

The matrix $\boldsymbol{\Gamma}$ is a diagonal matrix with diagonal elements corresponding to the M-point DFT of $c(n)$, i.e.

$$\boldsymbol{\Gamma} = \text{diag}\begin{bmatrix} C_0 & C_1 & \cdots & C_{M-1} \end{bmatrix},$$

$$\text{where} \quad C_k = C(z)\Big|_{z=e^{j2\pi k/M}} = \sum_{n=0}^{\nu} c(n)e^{-j2\pi kn/M}.$$

The DFT decomposition of $\mathbf{C}_{circ}$ leads to the following expression for $\mathbf{y}$:

$$\mathbf{y} = \mathbf{W}\mathbf{C}_{circ}\underbrace{\mathbf{W}^\dagger \mathbf{s}}_{\mathbf{x}} + \boldsymbol{\tau} = \boldsymbol{\Gamma}\mathbf{s} + \boldsymbol{\tau},$$

or equivalently

$$y_k = C_k s_k + \tau_k. \tag{6.3}$$

Each y_k is simply s_k scaled by C_k plus noise τ_k. The channel is thus diagonalized using an IDFT matrix and a DFT matrix.

Equivalent parallel subchannels With the channel diagonalized, the scalar FIR channel $C(z)$ is converted to M parallel subchannels (Fig. 6.2(a)). The scalar C_k is called the kth **subchannel gain**. There is no inter subchannel ISI, and equalization can be easily done by multiplying y_k with the scalar $1/C_k$, which is usually referred to as a **frequency domain equalizer** (FEQ). It is so named because the equalization is carried out after DFT computations. The overall $M \times M$ transfer matrix $\mathbf{T}$ from the transmitter input vector $\mathbf{s}$ to the receiver output vector $\widehat{\mathbf{s}}$ is the $M \times M$ identity matrix and the receiver is **zero-forcing**. The overall transceiver can be viewed as a system of M parallel subchannels (Fig. 6.2(b)), where the only distortions are the additive noise sources.

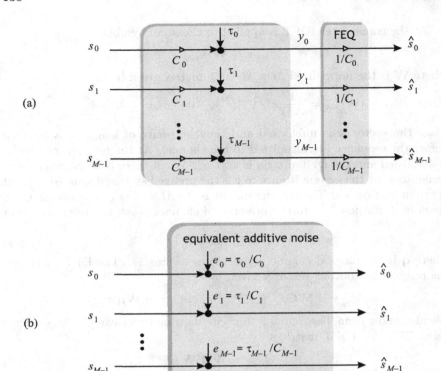

Figure 6.2. Equivalent parallel-subchannel model for the OFDM system.

Complexity In view of the block diagram of the OFDM system, we can see that the main computation of the transceiver comes from the IDFT and DFT matrices. For both matrices the Fast Fourier Transform (FFT) can be used in the computation when M is a power of 2, which is usually the choice of DFT size in practice. The complexity is in the order of $M \log_2 M$, i.e. in the order of $\log_2 M$ per input modulation symbol. The complexity of the transmitter is mostly that of an IDFT matrix. For the receiver, the complexity is that of a DFT matrix plus M multiplications for the FEQ part. In the context of system implementation, it is usually desirable to have as few channel-dependent components as possible. In particular, a channel-independent transmitter means there is no need to wait for the receiver to send back the channel information. This is especially important in wireless applications, where the channel can change rapidly and leaves little time for updating the transmitter. It is also of vital importance in broadcasting applications where there are one transmitter and many receiving ends. From the transmitter to each receiver, there is a corresponding channel. It would be very difficult to design the transmitter to accommodate different channels at the same time. The transmitter of the OFDM system has the much desired **channel-dependence** property. The whole OFDM system is channel-independent except for the FEQ coefficients. Only the FEQ coefficients need to be recomputed as the channel varies.

Cyclic prefix   The cyclic prefix acts as a buffer between consecutive blocks. When a signal is passed through the channel, what actually takes place is linear convolution. The insertion of a cyclic prefix allows us to fabricate **circular convolution**, which lends itself to channel diagonalization using simple, channel-independent DFT and IDFT matrices. However, prefixing also leads to bandwidth expansion by a factor of $\zeta = (M + \nu)/M$. The expansion will be minor if M is chosen to be significantly larger than ν, in which case ζ is close to unity. For example, the OFDM system in the application of wireless local area networks [54] has $M = 64$ and $\nu = 16$, so $\zeta = 1.25$, a 25% increase in bandwidth.

Peak to average power ratio   The peak to average power ratio (PAPR) of a signal $x(n)$ is defined as[1]

$$PAPR = \frac{\max_n |x(n)|^2}{E[|x(n)|^2]},$$

where $\max_n |x(n)|^2$ is used here to denote the maximum absolute of all possible values of the random variable $x(n)$. A transmitted signal with a large PAPR will translate to low efficiency for the power amplifier of the transmitter [149]. PAPR is a critical issue in the design of transmitters. For the OFDM system, the transmitter output signal $x(n)$ is obtained by unblocking the IDFT output vector and adding a cyclic prefix. Therefore the transmitter output has the same peak power and the same average power as the IDFT outputs, which are denoted by x_k in Fig. 6.1. We can consider the ratio

$$\frac{\max_k |x_k|^2}{E[|x_n|^2]}$$

instead. The output vector of the IDFT matrix $\mathbf{x} = \mathbf{W}^\dagger \mathbf{s}$ has autocorrelation matrix $\mathbf{R}_x = E[\mathbf{x}\mathbf{x}^\dagger]$ given by

$$\mathbf{R}_x = \mathbf{W}^\dagger \mathbf{R}_s \mathbf{W} = \mathcal{E}_s \mathbf{I}_M.$$

Therefore x_n are uncorrelated and their variances are the same, equal to $\mathcal{E}_s$. That is

$$E[|x_n|^2] = \mathcal{E}_s, \quad n = 0, 1, \ldots, M-1. \tag{6.4}$$

To compute the peak power, note that

$$x_n = \sum_{k=0}^{M-1} \frac{1}{\sqrt{M}} s_k e^{j\frac{2\pi}{M}kn} \tag{6.5}$$

---

[1] A more relevant quantity in practice is the continuous-time PAPR, which is defined as

$$PAPR_t = \frac{\max_t |x_a(t)|^2}{E[|x_a(t)|^2]},$$

where $x_a(t)$ is the continuous-time transmitted signal. In this definition, $PAPR_t$ depends on the transmitting pulse $p_1(t)$ at the transmitter. If a PAPR reduction method is not applied, $PAPR_t$ is usually only slightly larger than the PAPR defined over the discrete-time transmitted signal [149].

and
$$|x_n| = \left| \sum_{k=0}^{M-1} s_k \frac{1}{\sqrt{M}} e^{j\frac{2\pi}{M}kn} \right| \le \frac{1}{\sqrt{M}} \sum_{k=0}^{M-1} |s_k| \le \sqrt{M} \max_k |s_k|.$$

So we have
$$\max_n |x_n| = \sqrt{M} \max_k |s_k|. \tag{6.6}$$

The maximum value $\sqrt{M} \max_k |s_k|$ is attained when, for example, all input symbols have the same value and $|s_n| = \max_k |s_k|$. Combining (6.4) and (6.6), we arrive at the following expression of PAPR:

$$PAPR = M \frac{\max_k |s_k|^2}{E[|s_k|^2]}.$$

The ratio on the right-hand side is the PAPR of the input symbols. The PAPR for the OFDM system is M times the PAPR of the input modulation symbols. For some applications, e.g. fixed broadband wireless access systems [55], M can be as large as 2048, which leads to a 33 dB increase in PAPR. Many methods have been devised to reduce the PAPR of the OFDM system, e.g. coding [57] and tone reservation [148]. A more complete treatment of this topic and related references on PAPR reduction can be found in [48, 149].

Continuous-time transmitter output   The transmitter output $x(n)$ indicated in Fig. 6.1 is a discrete-time signal. The continuous-time output $x_a(t)$ is related to $x(n)$ by (Section 2.1)

$$x_a(t) = \sum x(k) p_1(t - kT),$$

where $p_1(t)$ is the transmitting pulse. The signal $x_a(t)$ can be further written in terms of the input symbols s_k using the relation in (6.5). In the literature, the output of the OFDM transmitter is commonly written as

$$x_a(t) = \sum s_k e^{-2\pi k f_0 t}$$

for some frequency f_0 or

$$x_a(t) = \sum s_k g(t) e^{-2\pi k f_0 t}$$

if a pulse shaping filter $g(t)$ is present. The expression is, in fact, *incorrect*. Such an expression is true only for an all-analog implementation of OFDM system. It can be shown that the output of an OFDM transmitter that is implemented using an IDFT matrix followed by a D/C converter (with or without cyclic prefixing) does not have the above expression [79].

6.1.1   Noise analysis

As the OFDM system in Fig. 6.1 is zero-forcing, the output error $e_k = \hat{s}_k - s_k$ comes solely from the channel noise. To analyze the effect of the channel noise, we draw in Fig. 6.3 the noise path at the receiver, where we consider only the channel noise and no signal. We assume the noise $q(n)$ is a circularly

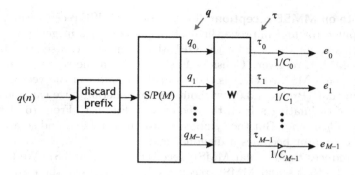

Figure 6.3. Noise path at the receiver of the OFDM system.

symmetric complex Gaussian random process with zero-mean and variance $\mathcal{N}_0$. The elements of the blocked noise vector $\mathbf{q}$ are uncorrelated and hence also independent because uncorrelated Gaussian random variables are independent. The autocorrelation matrix of $\mathbf{q}$ is $\mathbf{R}_q = \mathcal{N}_0\mathbf{I}_M$. The noise vector $\boldsymbol{\tau}$ at the output of the DFT matrix $\mathbf{W}$ has autocorrelation matrix

$$\mathbf{R}_\tau = \mathcal{N}_0\mathbf{W}\mathbf{W}^\dagger = \mathcal{N}_0\mathbf{I}_M.$$

Therefore τ_k are also uncorrelated Gaussian, with variance $\mathcal{N}_0$; the vector $\boldsymbol{\tau}$ has the same statistics as $\mathbf{q}$. The kth subchannel error is $e_k = \tau_k/C_k$. The errors continue to be uncorrelated Gaussian random variables, but with different variances. The variance of e_k is

$$\sigma_{e_k}^2 = \frac{\mathcal{N}_0}{|C_k|^2}. \tag{6.7}$$

The average mean squared error $\mathcal{E}_{rr} = 1/M \sum_{k=0}^{M-1} \sigma_{e_k}^2$ is hence

$$\mathcal{E}_{rr} = \frac{\mathcal{N}_0}{M} \sum_{i=0}^{M-1} \frac{1}{|C_i|^2}. \tag{6.8}$$

As the subchannel errors are independent, symbol detection can be performed for each subchannel separately without loss in performance.

The SNR of the kth subchannel $\beta(k) = \mathcal{E}_s/\sigma_{e_k}^2$ is

$$\beta_{ofdm}(k) = \gamma|C_k|^2, \quad \text{where} \quad \gamma = \mathcal{E}_s/\mathcal{N}_0. \tag{6.9}$$

We observe that the subchannel SNR $\beta_{ofdm}(k)$ is dependent on the channel response. For highly frequency-selective channels, $\beta_{ofdm}(k)$ can vary significantly with respect to k. If the channel profile is available to the transmitter, the disparity among subchannels can be exploited using bit and power allocation, which can improve the system performance significantly. This is routinely done in DMT systems for DSL applications, as we will explain in Section 6.6.

A note on MMSE reception   The use of an MMSE receiver can minimize the output error and improve the system performance in general. Unfortunately, this is not the case for the OFDM system. We can see this by examining its MMSE receiver. Consider Fig. 6.1. Given the received observation vector $\mathbf{r}$, the MMSE receiver is the optimal estimator of the vector $\mathbf{s}$. It can be shown that (Problem 3.9), without loss of generality, we can consider the problem of estimating $\mathbf{s}$ from the vector $\mathbf{y}$ in Fig. 6.1. From (6.3) we know that $y_k = C_k s_k + \tau_k$. But the symbols s_k are uncorrelated and so are τ_k. The optimal estimator becomes a diagonal matrix.   That is, the vector MMSE receiver reduces to M scalar MMSE receivers (Problem 3.8). We know from Example 3.6 that scalar MMSE receivers do not improve the error rate performance. Therefore for the OFDM system MMSE reception will not be used; we will only consider the zero-forcing receiver.

6.1.2   Bit error rate

For wireless applications, usually there is no bit allocation; all the symbols carry the same number of bits. In this case, the average bit error rate can be obtained by averaging the subchannel error rates. The calculation of bit error rate depends on the modulation scheme used. For QPSK symbols with power $\mathcal{E}_s$, the symbols have the form $\pm\sqrt{\mathcal{E}_s/2} \pm j\sqrt{\mathcal{E}_s/2}$. In this case the BER can be computed exactly (Section 2.3). The BER of the ith subchannel is

$$\mathcal{P}(i) = Q\left(\sqrt{\frac{\mathcal{E}_s}{\sigma_{e_i}^2}}\right) = Q\left(\sqrt{\beta_{ofdm}(i)}\right).$$

The function $Q(\cdot)$, as defined in Section 2.3 representing the area under the Gaussian tail, is reproduced below for convenience

$$Q(x) = \frac{1}{\sqrt{2\pi}} \int_x^\infty e^{-\tau^2/2} \, d\tau.$$

Using (6.9), the average BER of the OFDM system is given by

$$\mathcal{P}_{ofdm} = \frac{1}{M} \sum_{i=0}^{M-1} Q\left(\sqrt{\gamma |C_i|^2}\right). \tag{6.10}$$

For more general $2b$-bit QAM symbols, we can compute $\mathcal{P}(i)$ using the approximation in (2.18) and average them to obtain the average BER of the system.

Example 6.1 We assume that the noise is AWGN with variance $\mathcal{N}_0$. The modulation symbols are QPSK with values equal to $\pm\sqrt{\mathcal{E}_s/2} \pm j\sqrt{\mathcal{E}_s/2}$ and SNR $\gamma = \mathcal{E}_s/\mathcal{N}_0$. The number of subchannels M is 64 and the length of cyclic prefix ν is three. Two channels, as in Table 6.1, with four coefficients ($L = 3$) will be used. The magnitude responses (in dB) of the two channels $c_1(n)$ and $c_2(n)$ are shown in Fig. 6.4. The channel $c_1(n)$ has an almost flat magnitude response while $c_2(n)$ has a zero around $\omega = 1.1\pi$ and its magnitude response shows more variations. These two channels have the same energy. The BER performance will be obtained through Monte Carlo simulation (Section 2.3).

n	$c_1(n)$	$c_2(n)$
0	$0.3903 + j0.1049$	$0.3699 + j0.5782$
1	$0.6050 + j0.1422$	$0.4053 + j0.575$
2	$0.4402 + j0.0368$	$0.0834 + j0.0406$
3	$0.0714 + j0.5002$	$-0.1587 + j0.0156$

Table 6.1. Two channels with four coefficients.

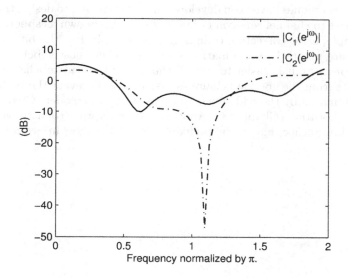

Figure 6.4. Magnitude responses of the two channels $c_1(n)$ and $c_2(n)$.

For channel $c_1(n)$, the subchannel SNRs for $\gamma = 16$ dB are shown in Fig. 6.5(a). Figure 6.5(b) shows the corresponding error rates for individual subchannels. The subchannels with error rates smaller than 10^{-6} are not shown in the figure. We can see that some subchannels have larger error rates and these correspond to where the channel has dips in the magnitude response. For $\gamma = 16$ dB, the good subchannels have error rates less than 10^{-6}. The error rates of the bad subchannels are as large as 10^{-2}, and these subchannels will be dominating when the average BER is computed. The average bit error rate as a function of γ is shown in Fig. 6.5(c). The average error rate for $\gamma = 16$ dB is around 10^{-3}. We perform the same set of experiments on channel $c_2(n)$ and show the results in Fig. 6.6. For $\gamma = 16$ dB, the subchannels near the channel null have a high error rate of around 0.5. As a result, the average error rate is dominated by these subchannels. We can see the average error rate stalls for $20 \lesssim \gamma \lesssim 50$ dB. This is because in this SNR range all the subchannels have error rates close to zero except for those close to the channel zero. Suppose one subchannel has an error rate of 0.5, the average error rate will be at least

$$0.5/M = 1/128.$$

The average error rate will stay around this value until γ is large enough to bring down the error rate of the worst subchannels. Note that the ranges of SNR in Fig. 6.5(c) and Fig. 6.6(c) are very different. For example, to obtain a BER of 10^{-4}, we need less than 20 dB for $c_1(n)$ but more than 40 dB for $c_2(n)$, although these two channels have the same energy. The difference is mainly due to the channel null. ∎

Many techniques have been developed in the literature to design transceivers more robust to channel zeros on the unit circle (also known as **spectral nulls**). For example, error-correcting code is generally applied to the bit stream before bits are mapped to modulation symbols. When most subchannels have small error rates, it is easier to correct the bits of the bad subchannels based on the bits from the good subchannels. We can also prevent the performance being dominated by the bad subchannels by using a precoder (Chapter 7), or using bit allocation (Chapter 8). Another way to improve robustness is to use zero padding rather than cyclic prefixing, as we will show in Section 6.2.

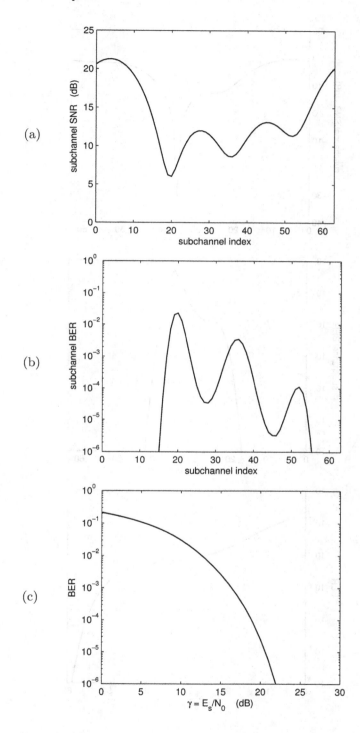

Figure 6.5. Example 6.1. Performance of the OFDM system over channel $c_1(n)$. (a) Subchannel SNRs for $\gamma = 16$ dB; (b) bit error rates of individual subchannels for $\gamma = 16$ dB; (c) average bit error rate.

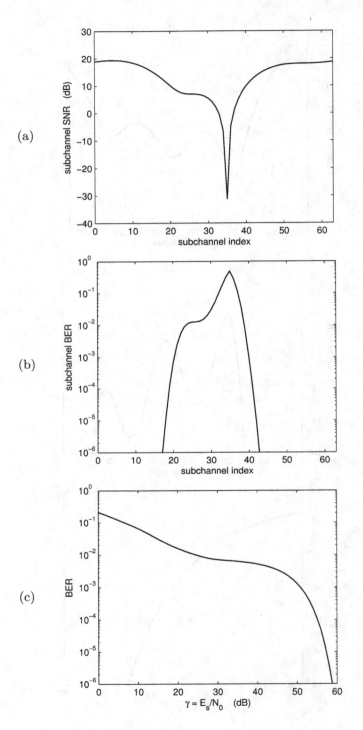

Figure 6.6. Example 6.1. Performance of the OFDM system over channel $c_2(n)$. (a) Subchannel SNRs for $\gamma = 16$ dB; (b) bit error rates of individual subchannels for $\gamma = 16$ dB; (c) average bit error rate.

6.2 Zero-padded OFDM systems

The insertion of guard intervals allows us to consider the simpler problem of one-shot block transmission. Cyclic prefixing is the most common type of guard interval. Another popular form of guard interval is zero padding. When the number of zeros padded to each block is not smaller than the order of the channel, there is no IBI and again we only need to consider the transmission of a single block. The transmitter of a zero-padded OFDM system (Fig. 6.7) is the same as that of the prefixed OFDM system, except that cyclic prefixing is replaced by zero padding. Note that in Fig. 6.7 the system from $\mathbf{x}$ to $\mathbf{r}$ as indicated by the gray box is the zero-padded system discussed in Section 5.4.1. The received vector $N \times 1$ vector $\mathbf{r}$, where $N = M + \nu$, is related to $\mathbf{x}$ by

$$\mathbf{r} = \mathbf{C}_{low}\mathbf{x} + \mathbf{q},$$

where $\mathbf{C}_{low}$ is the $N \times M$ lower triangular Toeplitz matrix defined in (5.16) and $\mathbf{q}$ is the $N \times 1$ blocked channel noise vector. When we substitute $\mathbf{x} = \mathbf{W}^{\dagger}\mathbf{s}$ into the above equation, we have

$$\mathbf{r} = \mathbf{C}_{low}\mathbf{W}^{\dagger}\mathbf{s} + \mathbf{q}. \tag{6.11}$$

The task of the receiver is to recover $\mathbf{s}$ from the received vector $\mathbf{r}$ by using an $M \times N$ receiving matrix $\mathbf{S}$, as shown in Fig. 6.7. Two types of receivers, **zero-forcing** and **MMSE**, will be derived below.

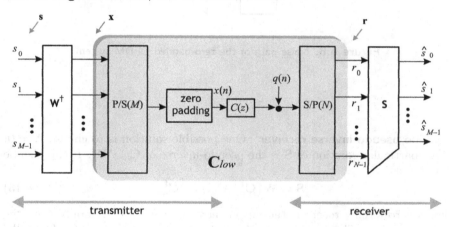

Figure 6.7. Zero-padded OFDM system.

6.2.1 Zero-forcing receivers

To obtain a zero-forcing solution, we observe from (6.11) that the receiving matrix $\mathbf{S}$ can be any left inverse of $\mathbf{C}_{low}\mathbf{W}^{\dagger}$. Let the singular value decomposition of $\mathbf{C}_{low}$ be

$$\mathbf{C}_{low} = \mathbf{U}\begin{bmatrix} \mathbf{\Lambda}_s \\ \mathbf{0} \end{bmatrix}\mathbf{V}^{\dagger},$$

where $\mathbf{U}$ and $\mathbf{V}$ are, respectively, $N \times N$ and $M \times M$ unitary matrices. The columns of $\mathbf{U}$ and $\mathbf{V}$ correspond, respectively, to the eigenvectors of $\mathbf{C}_{low}\mathbf{C}_{low}^{\dagger}$ and $\mathbf{C}_{low}^{\dagger}\mathbf{C}_{low}$. The matrix $\mathbf{\Lambda}_s$ is diagonal and the diagonal elements are the singular values of $\mathbf{C}_{low}$. We know $\mathbf{C}_{low}$ has full rank so long as $C(z) \neq 0$. Hence the diagonal elements of $\mathbf{\Lambda}_s$ are nonzero. We can choose $\mathbf{S}$ to be any matrix of the form

$$\mathbf{S} = \mathbf{W}\mathbf{V}\begin{bmatrix} \mathbf{\Lambda}_s^{-1} & \mathbf{A} \end{bmatrix}\mathbf{U}^{\dagger}, \tag{6.12}$$

where $\mathbf{A}$ is an arbitrary $M \times L$ matrix. We can freely choose $\mathbf{A}$ to obtain different solutions of $\mathbf{S}$. Although these are all zero-forcing receivers, they will behave differently when the channel noise comes into play. In what follows we consider two zero-forcing solutions.

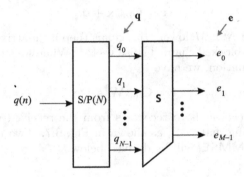

Figure 6.8. Noise path of the zero-padded OFDM system.

The pseudo-inverse receiver One possible solution is to choose $\mathbf{A} = \mathbf{0}$. This particular solution of $\mathbf{S}$ is the pseudo-inverse of $\mathbf{C}_{low}\mathbf{W}^{\dagger}$. It can also be written as

$$\mathbf{S} = \mathbf{W}\left(\mathbf{C}_{low}^{\dagger}\mathbf{C}_{low}\right)^{-1}\mathbf{C}_{low}^{\dagger}. \tag{6.13}$$

For a zero-forcing receiver, the output error $\mathbf{e} = \widehat{\mathbf{s}} - \mathbf{s}$ comes only from the channel noise. The overall system can be converted to a set of M parallel additive-noise subchannels, similar to that shown in Fig. 6.2 for the prefixed OFDM system. When the channel noise $q(n)$ is AWGN with zero-mean and variance $\mathcal{N}_0$, the autocorrelation matrix of the $N \times 1$ blocked noise vector $\mathbf{q}$ in Fig. 6.8 is $\mathcal{N}_0\mathbf{I}_N$. The error vector $\mathbf{e} = \widehat{\mathbf{s}} - \mathbf{s}$ has autocorrelation matrix

$$\mathbf{R}_e = \mathcal{N}_0\mathbf{S}\mathbf{S}^{\dagger} = \mathcal{N}_0\mathbf{W}\left(\mathbf{C}_{low}^{\dagger}\mathbf{C}_{low}\right)^{-1}\mathbf{W}^{\dagger}.$$

The kth subchannel error variance $\sigma_{e_k}^2$ is equal to the kth diagonal element of $\mathbf{R}_e$. In fact, the pseudo-inverse receiver is also the zero-forcing solution that yields the smallest total output noise when the channel noise is AWGN (Problem 6.8). Moreover, as the matrix $\mathbf{C}_{low}$ has full rank, the pseudo-inverse solution always exists, even in the presence of channel spectral nulls.

A computationally efficient receiver   The pseudo-inverse solution is no longer related to the DFT matrix as in the prefixed OFDM system. The receiver does not have as low-cost an implementation. Alternatively, we can use a zero-forcing receiver that is more computationally efficient. To derive such a receiver, we observe that the $N \times M$ matrix $\mathbf{C}_{low}$ has a lower triangular Toeplitz structure. It can be turned into the $M \times M$ circulant matrix $\mathbf{C}_{circ}$ by adding the bottom ν rows of $\mathbf{C}_{low}$ to the top ν rows. This can be done by using the following matrix manipulation:

$$\mathbf{\Upsilon}\mathbf{C}_{low} = \mathbf{C}_{circ}, \quad \text{where} \quad \mathbf{\Upsilon} = \begin{bmatrix} \mathbf{I}_\nu & \mathbf{0} & \mathbf{I}_\nu \\ \mathbf{0} & \mathbf{I}_{M-\nu} & \mathbf{0} \end{bmatrix}. \tag{6.14}$$

For example, when $M = 4$ and $\nu = 2$, the product $\mathbf{\Upsilon}\mathbf{C}_{low}$ is

$$\underbrace{\begin{bmatrix} 1 & 0 & 0 & 0 & | & 1 & 0 \\ 0 & 1 & 0 & 0 & | & 0 & 1 \\ 0 & 0 & 1 & 0 & | & 0 & 0 \\ 0 & 0 & 0 & 1 & | & 0 & 0 \end{bmatrix}}_{\mathbf{\Upsilon}} \underbrace{\begin{bmatrix} c(0) & 0 & 0 & 0 \\ c(1) & c(0) & 0 & 0 \\ c(2) & c(1) & c(0) & 0 \\ 0 & c(2) & c(1) & c(0) \\ \hline 0 & 0 & c(2) & c(1) \\ 0 & 0 & 0 & c(2) \end{bmatrix}}_{\mathbf{C}_{low}} = \underbrace{\begin{bmatrix} c(0) & 0 & c(2) & c(1) \\ c(1) & c(0) & 0 & c(2) \\ c(2) & c(1) & c(0) & 0 \\ 0 & c(2) & c(1) & c(0) \end{bmatrix}}_{\mathbf{C}_{circ}},$$

which is a 4×4 circulant matrix. Using the DFT decomposition $\mathbf{C}_{circ} = \mathbf{W}^\dagger\mathbf{\Gamma}\mathbf{W}$ in (6.2), we can obtain the following computationally efficient receiver:

$$\mathbf{S} = \mathbf{\Gamma}^{-1}\mathbf{W}\mathbf{\Upsilon}. \tag{6.15}$$

Compared to the receiver of the cyclic-prefixed OFDM system, for each received block there are only ν extra additions due to the matrix $\mathbf{\Upsilon}$. The complexity is roughly the same. With the efficient receiver, the subchannel noise variances can be verified to be (Problem 6.9)

$$\sigma_{e_i}^2 = \frac{N}{M}\frac{\mathcal{N}_0}{|C_i|^2}.$$

The factor N/M is also due to the matrix $\mathbf{\Upsilon}$. The subchannel SNR is $\beta(i) = \frac{M}{N}\gamma|C_i|^2$. The average mean squared error $\mathcal{E}_{rr}$ is given by

$$\mathcal{E}_{rr} = \frac{1}{M}\sum_{i=0}^{M-1}\sigma_{e_i}^2 = \frac{N}{M}\left[\frac{\mathcal{N}_0}{M}\sum_{i=0}^{M-1}\frac{1}{|C_i|^2}\right].$$

When we make a comparison with the cyclic-prefixed OFDM, we can observe the following properties.

- The subchannel SNRs of the zero-padded OFDM system are reduced by a factor of N/M when compared to that of the cyclic-prefixed OFDM system.

- We may also note that the average transmission power for the zero-padding case is slightly different due to the padding of zeros. When the input symbols s_k are uncorrelated, with zero-mean and variance $\mathcal{E}_s$, the outputs of the IDFT matrix also have zero-mean and variance $\mathcal{E}_s$. After zero padding, the transmitted output $x(n)$ has average power $\mathcal{E}_s M/N$, which is slightly less than the cyclic-prefixed OFDM system by a factor M/N.

- The decrease in subchannel SNRs will be a minor one when M is much larger than the prefix length ν and hence the factor is close to one. We can expect the performance of the zero-padded OFDM system with the efficient receiver to be very close to that of the prefixed system. If we fix the average transmission power to be the same as that in the prefixing case, then the error rates of these two systems will be just the same.

6.2.2   The MMSE receiver

The receiver output error can be minimized by using the MMSE receiver. Unlike the prefixed OFDM system, whose MMSE receiver reduces to M scalar MMSE receivers, the performance of the zero-padded OFDM system can be improved using MMSE reception, as we will see. We assume the input symbols s_k are uncorrelated, with zero-mean and variance $\mathcal{E}_s$. Also assume the channel $q(n)$ is AWGN with zero-mean and variance $\mathcal{N}_0$. By the orthogonality principle in Section 3.2, the mean squared error $E[||\widehat{\mathbf{s}} - \mathbf{s}||^2]$ is at its smallest when the error $\mathbf{e} = \widehat{\mathbf{s}} - \mathbf{s}$ is orthogonal to the observation vector $\mathbf{r}$. That is,

$$E[\mathbf{e}\mathbf{r}^\dagger] = \mathbf{0}.$$

The receiving matrix that satisfies this condition is $\mathbf{S} = E[\mathbf{s}\mathbf{r}^\dagger]\left(E[\mathbf{r}\mathbf{r}^\dagger]\right)^{-1}$. As the input symbols and noise are uncorrelated, i.e. $E[\mathbf{s}\mathbf{q}^\dagger] = \mathbf{0}$, we have

$$E[\mathbf{s}\mathbf{r}^\dagger] = \mathcal{E}_s \mathbf{W}\mathbf{C}_{low}^\dagger \quad \text{and} \quad E[\mathbf{r}\mathbf{r}^\dagger] = \mathcal{E}_s \mathbf{C}_{low}\mathbf{C}_{low}^\dagger + \mathcal{N}_0\mathbf{I}_N.$$

Thus,

$$\mathbf{S} = \gamma\mathbf{W}\mathbf{C}_{low}^\dagger\left(\gamma\mathbf{C}_{low}\mathbf{C}_{low}^\dagger + \mathbf{I}_N\right)^{-1}. \tag{6.16}$$

The overall transfer matrix $\mathbf{T}$ from the transmitter input $\mathbf{s}$ to the receiver output $\widehat{\mathbf{s}}$ is

$$\mathbf{T} = \mathbf{S}\mathbf{C}_{low}\mathbf{W}^\dagger.$$

It is now no longer the identity matrix, and the diagonal elements of $\mathbf{T}$ are not unity. The error $e_i = \hat{s}_i - s_i$ does not come from channel noise alone. It is a combination of s_i, channel noise, and other modulation symbols. Because e_i contains the term s_i, the receiver output $\hat{s}_i$ is a **biased estimate** of s_i (Section 3.2.1) and the ratio

$$\beta_{biased}(i) = \mathcal{E}_s/\sigma_{e_i}^2$$

is a biased SNR quantity. The unbiased subchannel SNR $\beta(i)$, as explained in Section 3.2.2, is given by

$$\beta(i) = \beta_{biased}(i) - 1.$$

We can then use $\beta(i)$ to obtain an accurate estimate of the actual error rate.

Remarks

- The efficient receiver in (6.15) has very low channel dependence. The channel-dependent part is a set of M scalars, like the cyclic-prefixed system. The pseudo-inverse receiver and MMSE receiver, however, depend heavily on the channel. Also, matrix inversions are needed to compute the coefficients in the receiving matrices. For the MMSE receiver, the SNR quantity γ is also required in computing the receiver.

- The implementation cost of the efficient receiver is comparable to that of the prefixed system. For the pseudo-inverse and MMSE receivers, there is no corresponding DFT structure. Although the implementation cannot be carried out using a low-cost DFT matrix, it is possible to reduce the complexity for these two receivers [26].

- We know in general the MMSE receiver will become a zero-forcing one when the SNR is large. For the zero-padded OFDM system, the solution of zero-forcing receivers is not unique. Which zero-forcing solution will the MMSE receiver reduce to as the SNR approaches infinity? To answer this question, we rewrite the MMSE receiver in (6.16) as (Problem 6.7)

$$\mathbf{S} = \mathbf{W} \left(\mathbf{C}_{low}^{\dagger} \mathbf{C}_{low} + \gamma^{-1} \mathbf{I}_M \right)^{-1} \mathbf{C}_{low}^{\dagger}. \qquad (6.17)$$

When the SNR γ is large, the MMSE receiver reduces to the pseudo-inverse solution in (6.13)! As a result, the performance of the pseudo-inverse receiver comes close to that of the MMSE receiver for a sufficiently large SNR.

Example 6.2 Zero-padded OFDM systems. The simulation environment and parameters used in this example are the same as those in Example 6.1. For the channel $c_1(n)$, Fig. 6.9(a) shows the BER performances of the zero-padded OFDM system with the three types of receivers discussed in this section: the efficient receiver in (6.12), the pseudo-inverse receiver in (6.13), and the MMSE receiver in (6.16). The BERs are denoted, respectively, by $\mathcal{P}_{ofdm-zp}$ (efficient), $\mathcal{P}_{ofdm-zp}$ (pseudo), and $\mathcal{P}_{ofdm-zp,mmse}$ in the figure. For comparison, the error rate of the prefixed OFDM system is also shown in the figure. The BERs are plotted as a function of $\gamma = \mathcal{E}_s/\mathcal{N}_0$. For the same γ, the average transmission power of the prefixed system is N/M times larger than the zero-padded case. The performances for the channel $c_2(n)$ are shown in Fig. 6.9(b).

For both channels the error rate of the efficient receiver is slightly larger than that of the prefixed system. This is because the ratio $N/M = 17/16$ is very close to one. (We are comparing two systems with the same γ. If the comparison is done for the same transmission power, the performances will be exactly the same.) In the case of channel $c_2(n)$, these two systems are badly affected by the zero close to the unit circle (two curves indistinguishable in the figure). For the zero-padded OFDM systems, the performances of the pseudo-inverse and the MMSE receivers are much less affected by the channel null. The curves of these two receivers are very close for both channels. For high SNR, one is indistinguishable from the other in both figures. We also note the difference in the SNR ranges between the top and bottom plots. A much higher SNR is needed to achieve the same error rate for $c_2(n)$. The zero of $c_2(n)$ close to the unit circle makes the recovery of input symbols at the receiver a much more difficult task, even though $c_2(n)$ has the same energy as $c_1(n)$. ∎

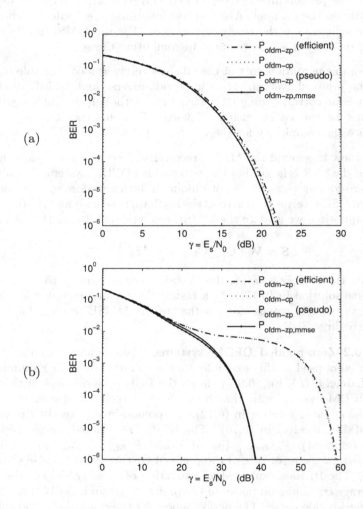

Figure 6.9. **Example 6.2.** Zero-padded OFDM system: BER performances for (a) the channel $c_1(n)$ and (b) the channel $c_2(n)$.

6.3 Single-carrier systems with cyclic prefix (SC-CP)

In the OFDM system, modulation symbols are sent after IDFT operation and adding a prefix. If the symbols are sent directly after redundant samples are added, we call it a block-based single-carrier system. Redundant samples can take the form of cyclic prefixing or zero padding. Correspondingly, the system is called the single-carrier system with cyclic prefix (SC-CP) or the single-carrier system with zero padding (SC-ZP). It is called a single carrier system because the modulation symbols are directly transmitted after the insertion of redundant samples, as opposed to the multicarrier OFDM system, where the channel is divided into multiple subchannels via the DFT matrix.

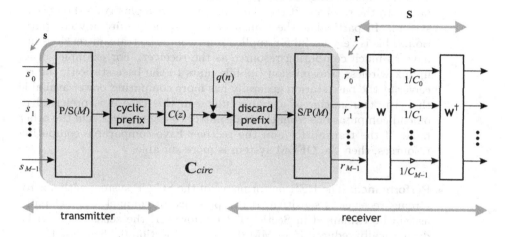

Figure 6.10. SC-CP system with a zero-forcing receiver.

The system block diagram of the SC-CP system is shown in Fig. 6.10. There is no processing of the modulation symbols at the transmitter. We can see that the system from the vector **s** at the transmitter to the vector **r** at the receiver is the familiar cyclic-prefixed system shown in Fig. 5.16 (Section 5.4). The corresponding transfer matrix is the $M \times M$ circulant matrix $\mathbf{C}_{circ}$. We can remove ISI by inverting $\mathbf{C}_{circ}$. The resulting zero-forcing receiver is

$$\mathbf{S} = \mathbf{C}_{circ}^{-1} = \mathbf{W}^{\dagger}\mathbf{\Gamma}^{-1}\mathbf{W}, \tag{6.18}$$

which is a circulant matrix as it is a diagonal matrix sandwiched between an IDFT and a DFT matrix (Theorem 5.2). The block diagram of the zero-forcing receiver is as shown in Fig. 6.10. A few observations on the SC-CP system are in order.

- **Channel independence.** Like the OFDM system, the SC-CP system is also a DFT-based transceiver with a channel-independent transmitter. A cyclic prefix is inserted as in the OFDM system. The only channel-dependent part is a set of M scalars $1/C_k$ at the receiver. These scalars are called FEQs in the OFDM receiver. Therefore the SC-CP system is also known as a single-carrier system with frequency domain equalization (**SC-FDE**) [129].

- **PAPR.** The SC-CP system has the advantage that the PAPR of the transmitted signal is very low. This is because the modulation symbols are transmitted directly, without further processing, and the PAPR of the transmitted signal is the same as the PAPR of the input modulation symbols. The PAPR is much lower than that in the OFDM system, especially for large M.

- **Implementation.** If we compare the SC-CP system to the OFDM system, it is like moving the computation of the IDFT matrix at the transmitter to the receiver. It has the same overall complexity as the OFDM system. In particular, the transmitter does not require any computation. This is very useful for applications where the transmitter does not have as much computing resource as the receiver. For example, in an uplink wireless transmission (mobile units to the base station), the receiver at the base station generally has more computing power available than the transmitter of a mobile unit. In this case it is preferred to offload computations from the transmitter to the receiver. On the other hand, if the transmitter and the receiver have comparable computing resources, then the OFDM system is more suitable.

- **Performance.** The BER performance of the SC-CP system with a zero-forcing receiver is sensitive to the presence of channel spectral nulls, as will be explained in Section 6.3.1. However, the sensitivity can be dramatically reduced if an MMSE receiver (Section 6.3.3) is used.

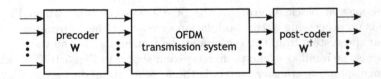

Figure 6.11. SC-CP system viewed as a precoded OFDM system.

We can also look at the SC-CP from an alternative viewpoint, which allows us to derive the SC-CP system from the OFDM system. For an OFDM system, if we cascade a DFT matrix before the transmitter and an IDFT matrix at the end of the receiver (Fig. 6.11), then the IDFT and DFT matrices at the transmitter cancel each other out. What we get is an SC-CP system! Thus the SC-CP system can be viewed as an OFDM system with a DFT precoder and IDFT post-coder. Having a precoder can significantly alter the characteristics of the system, for example PAPR. Another example is the behavior of subchannel noise, which will be explained in the following noise analysis. (More detailed discussion on precoded OFDM systems will be given in Chapter 7.)

6.3.1 Noise analysis: zero-forcing case

The noise path at the receiver is shown in Fig. 6.12. A comparison with the noise path of the OFDM system in Fig. 6.3 shows that the noise vector $\boldsymbol{\mu}$ has the same statistics as the output noise vector in the OFDM system. Thus μ_k are uncorrelated Gaussian, with variances given by $\mathcal{N}_0/|C_k|^2$. The autocorrelation matrix $\mathbf{R}_\mu$ is a diagonal matrix,

$$\mathbf{R}_\mu = \mathrm{diag}\left[\frac{\mathcal{N}_0}{|C_0|^2} \quad \frac{\mathcal{N}_0}{|C_1|^2} \quad \cdots \quad \frac{\mathcal{N}_0}{|C_{M-1}|^2}\right].$$

The autocorrelation matrix of the output noise vector $\mathbf{e}$ is given by

$$\mathbf{R}_e = \mathbf{W}^\dagger \mathbf{R}_\mu \mathbf{W}.$$

It is a circulant matrix as $\mathbf{R}_\mu$ is diagonal (Theorem 5.2). The diagonal elements of $\mathbf{R}_e$ are the same, equal to the average of $\sigma_{\mu_i}^2$. Therefore the subchannel noise variances, corresponding to the diagonal elements of $\mathbf{R}_e$, are the same for all subchannels. We have

$$\sigma_{e_i}^2 = \mathcal{E}_{rr} = \frac{\mathcal{N}_0}{M} \sum_{\ell=0}^{M-1} \frac{1}{|C_\ell|^2}. \tag{6.19}$$

The average error is the same as that in the OFDM system.

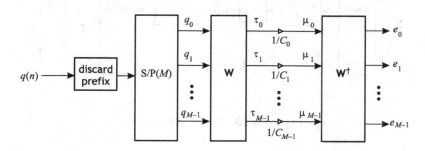

Figure 6.12. Noise path at the receiver of the zero-forcing SC-CP system.

Identical subchannel noise variance means identical subchannel SNRs. The subchannel SNR $\beta_{sc\text{-}cp} = \mathcal{E}_s/\mathcal{E}_{rr}$ is given by

$$\beta_{sc\text{-}cp} = \frac{1}{\frac{1}{M}\sum_{i=0}^{M-1}\frac{1}{\gamma|C_i|^2}}.$$

When QPSK modulation is used, the bit error rate of the SC-CP system is

$$\mathcal{P}_{sc\text{-}cp} = Q\left(\sqrt{\beta_{sc\text{-}cp}}\right). \tag{6.20}$$

In view of the subchannel SNRs, the SC-CP system is very different from the OFDM system. The subchannel SNRs are the same even when the channel is not flat. We will see in Chapter 8 that there is little gain when bit and power allocation are allowed.

Channels with spectral nulls   When we examine the expression in (6.19), we observe that all the subchannel errors will become very large when the channel has a spectral null. Suppose one DFT coefficient, say C_{ℓ_0}, is equal to zero. All the subchannels are equally affected; the subchannel noise variances will go to infinity and the bit error rate is 0.5 for all the subchannels. Such a disaster can be avoided by using an MMSE receiver, to be discussed next.

6.3.2   The MMSE receiver

For the prefixed OFDM system, the MMSE receiver degenerates to M scalar MMSE receivers. There is no performance improvement. In the case of the SC-CP system, there is no such degeneration. The use of MMSE reception does improve the performance in terms of BER. It turns out that the MMSE receiver can be easily obtained from the zero-forcing receiver by replacing the channel-dependent scalars with another set of coefficients.

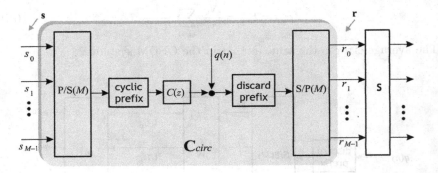

Figure 6.13. The SC-CP system.

Suppose we are given the transmitter of the SC-CP system in Fig. 6.13, which does nothing more than insert a cyclic prefix to each input block of modulation symbols. What would be the corresponding MMSE receiving matrix $\mathbf{S}$ that minimizes the subchannel errors? To answer this question, recall that the received vector can be written as $\mathbf{r} = \mathbf{C}_{circ}\mathbf{s} + \mathbf{q}$. The receiver output error is

$$\mathbf{e} = \widehat{\mathbf{s}} - \mathbf{s} = \mathbf{Sr} - \mathbf{s}.$$

By the orthogonality principle in Section 3.2, the mean squared error $E[\|\mathbf{e}\|^2]$ is minimized if $\mathbf{S}$ is such that the error $\mathbf{e}$ is orthogonal to the observation vector $\mathbf{r}$. That is,

$$E[\mathbf{er}^\dagger] = \mathbf{0}.$$

We can verify that the optimal $\mathbf{S}$ satisfying the above condition can be expressed in the form (Problem 6.10)

$$\mathbf{S} = \mathbf{W}^\dagger \mathbf{\Lambda} \mathbf{W}, \qquad (6.21)$$

where $\boldsymbol{\Lambda}$ is an $M \times M$ diagonal matrix with the ℓth diagonal entry given by

$$\lambda_\ell = \frac{\gamma C_\ell^*}{1 + \gamma |C_\ell|^2}, \quad \ell = 0, 1, \ldots, M - 1.$$

When γ approaches infinity, the ℓth diagonal entry reduces to $\lambda_\ell = 1/C_\ell$ and the MMSE receiver reduces to the zero-forcing case. The MMSE receiving matrix has the interesting property that it is also a circulant matrix. It is a channel-dependent diagonal matrix $\boldsymbol{\Lambda}$ sandwiched between an IDFT and a DFT matrix (Fig. 6.14). It has the same structure as the zero-forcing receiver. The implementation complexity is also the same as the zero-forcing receiver: one M-point DFT, one M-point IDFT, and M channel-dependent scalars. In terms of channel dependence, the SNR quantity $\gamma = \mathcal{E}_s/\mathcal{N}_0$ is also needed for the MMSE receiver, while the zero-forcing receiver requires only the channel response.

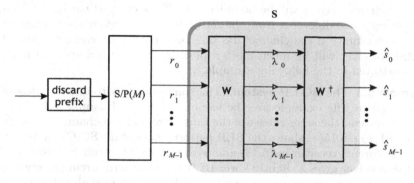

Figure 6.14. The MMSE receiver for the SC-CP system.

6.3.3 Error analysis: MMSE case

When the MMSE receiver is used, we replace the receiving matrix $\mathbf{S}$ in Fig. 6.10 by the MMSE solution in (6.21). The overall transfer matrix $\mathbf{T}$ becomes

$$\mathbf{T} = \mathbf{SC}_{circ} = \mathbf{W}^\dagger \boldsymbol{\Lambda\Gamma W}, \tag{6.22}$$

which is no longer the identity matrix. The ith output error $e_i = \hat{s}_i - s_i$ does not come from channel noise alone; it is a mixture of channel noise and interference. The receiver output vector is $\hat{\mathbf{s}} = \mathbf{Ts} + \mathbf{Sq}$. The output error vector $\mathbf{e} = \hat{\mathbf{s}} - \mathbf{s}$ can be expressed as

$$\mathbf{e} = (\mathbf{T} - \mathbf{I})\mathbf{s} + \mathbf{Sq} = \mathbf{W}^\dagger \underbrace{(\boldsymbol{\Lambda\Gamma} - \mathbf{I})}_{\mathbf{D}} \mathbf{Ws} + \mathbf{Sq}.$$

When the signal $\mathbf{s}$ and noise $\mathbf{q}$ are uncorrelated, the autocorrelation matrix of the output error $\mathbf{R}_e = E[\mathbf{ee}^\dagger]$ is

$$\mathbf{R}_e = \mathbf{W}^\dagger (\mathcal{E}_s \mathbf{DD}^\dagger + \mathcal{N}_0 \boldsymbol{\Lambda\Lambda}^\dagger) \mathbf{W},$$

which is a circulant matrix as the matrix sandwiched between $\mathbf{W}^\dagger$ and $\mathbf{W}$ is diagonal. This implies that all the subchannel errors have the same variance, just as in the zero-forcing case. The subchannel error variances are equal to their average $\mathcal{E}_{rr}$. Using the definitions of $\mathbf{D}$ and $\mathbf{\Lambda}$, we can verify that the average error $\mathcal{E}_{rr}$ is given by

$$\mathcal{E}_{rr} = \sigma_{e_i}^2 = \frac{\mathcal{E}_s}{M} \sum_{\ell=0}^{M-1} \frac{1}{1+\gamma|C_\ell|^2}, \quad i = 0, 1, \ldots, M-1. \quad (6.23)$$

We can use the subchannel error variance to compute the biased SNR, $\beta_{biased} = \mathcal{E}_s/\mathcal{E}_{rr}$. The unbiased SNR β can be obtained using $\beta = \beta_{biased} - 1$.

Comparing with the zero-forcing receiver, we can see from (6.23) that the MMSE receiver always has a smaller error due to the extra unity in the denominator. We may also observe that each of the terms in the summation in (6.23) is less than one. The average error will be upper bounded by $\mathcal{E}_s$. The subchannel errors will be bounded even if the channel has spectral nulls. This is a marked difference from the zero-forcing case, where the average error can go to infinity if the channel has one DFT coefficient equal to zero. The MMSE receiver will be much more robust against channel spectral nulls as demonstrated in the following example.

Example 6.3 The SC-CP system. The simulation environment and parameters used in this example are the same as those in Example 6.1. For the SC-CP system, the error rates are the same across all subchannels. For channel $c_1(n)$, Fig. 6.15(a) shows the BER performances of the SC-CP system with the zero-forcing receiver in (6.18) and with the MMSE receiver in (6.21). The MMSE receiver gives a slightly lower BER than the zero-forcing receiver. For comparison, the error rate of the prefixed OFDM system is also shown. Comparing the OFDM and the SC-CP systems, both with zero-forcing receivers, the total output errors are the same but the SC-CP system outperforms the OFDM system for a moderate value of BER. The difference is significant when SNR is large. The performances for channel $c_2(n)$ are shown in Fig. 6.15(b). Although there is no significant difference in terms of performance between zero-forcing and MMSE receivers for an almost-flat channel like $c_1(n)$, for channels with spectral nulls the gap between the two is much larger. The MMSE receiver considerably improves the error rate, as we can see from Fig. 6.15(b). ∎

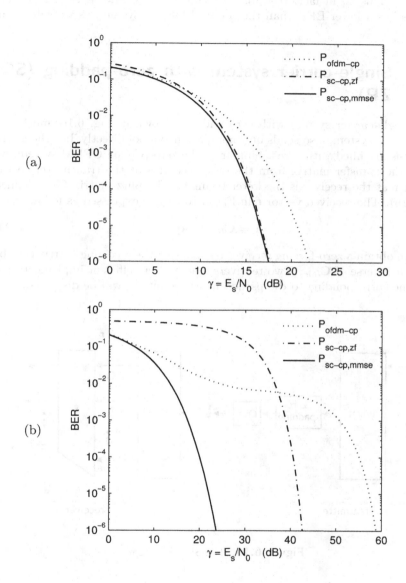

Figure 6.15. Example 6.3. The SC-CP system: BER performances for (a) channel $c_1(n)$ and (b) channel $c_2(n)$.

We observe from Fig. 6.15 that, for both channels, the BER curves of the MMSE SC-CP system lie under those of the prefixed OFDM system for all SNR γ. This is, in fact, true for all channels. The MMSE SC-CP system always has a lower BER than the prefixed OFDM system, as we will see in Chapter 7.

6.4 Single-carrier system with zero-padding (SC-ZP)

The single-carrier system with zero padding shown in Fig. 6.16, much like the SC-CP system, also sends out modulation symbol directly, but the guard intervals are filled with zeros. Similar to the zero-padded OFDM system, we know the transfer matrix from the input vector $\mathbf{s}$ at the transmitter to the vector $\mathbf{r}$ at the receiver is the lower triangular Toeplitz matrix $\mathbf{C}_{low}$ defined in (5.16). The received vector $\mathbf{r}$ in Fig. 6.16 can be expressed as follows:

$$\mathbf{r} = \mathbf{C}_{low}\mathbf{s} + \mathbf{q}. \tag{6.24}$$

We can obtain a zero-forcing receiver by choosing the receiving matrix $\mathbf{S}$ to be any left inverse of $\mathbf{C}_{low}$. Two receivers, one with an efficient implementation and one corresponding to the pseudo-inverse solution, will be discussed.

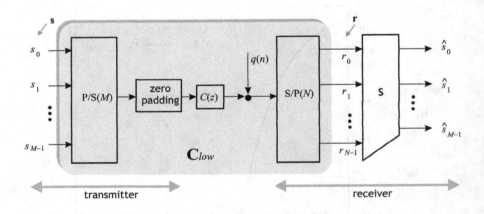

Figure 6.16. The SC-ZP system.

One zero-forcing choice of $\mathbf{S}$ can be obtained by first turning $\mathbf{C}_{low}$ into the $M \times M$ circulant $\mathbf{C}_{circ}$ using (6.14). We can then take the inverse of $\mathbf{C}_{circ}$. The resulting solution is the following receiver:

$$\mathbf{S} = \mathbf{W}^{\dagger}\mathbf{\Gamma}^{-1}\mathbf{W}\mathbf{\Upsilon}, \tag{6.25}$$

where $\mathbf{\Upsilon}$ is the $M \times N$ matrix defined in (6.14). The above receiver differs from the zero-forcing SC-CP receiver only in $\mathbf{\Upsilon}$, which requires only ν additions per block. Such a receiver has the advantage that it can be implemented efficiently. Note that the matrix $\mathbf{\Upsilon}$ has little effect when M is much larger than ν. As a result, the performance of the SC-ZP system with the efficient receiver will be similar to the SC-CP system with the zero-forcing receiver. Both receivers contain the term $\mathbf{\Gamma}^{-1}$, i.e. the inverses of the channel DFT coefficients. When the channel has one DFT coefficient equal to zero, the performance can be seriously affected.

Another possible zero-forcing choice of $\mathbf{S}$ is the pseudo-inverse of $\mathbf{C}_{low}$,

$$\mathbf{S} = (\mathbf{C}_{low}^{\dagger}\mathbf{C}_{low})^{-1}\mathbf{C}_{low}^{\dagger}. \tag{6.26}$$

This does not have the simple DFT–IDFT structure and cannot be implemented as efficiently as the receiver in (6.25) in general. In terms of noise analysis, this is also the zero-forcing receiver that leads to the smallest total output noise (similar to the zero-padded OFDM case). Like the zero-padded OFDM system, the pseudo-inverse solution always exists irrespective of channel spectral nulls. This stands in contrast to the efficient receiver given in (6.25), which cannot recover the transmitted symbols when the channel has one or more DFT coefficients equal to zero.

MMSE receiver With the received vector $\mathbf{r}$ given in (6.24), we can directly obtain the MMSE receiver using the orthogonality principle. It is given by

$$\mathbf{S} = \gamma\mathbf{C}_{low}^{\dagger}(\gamma\mathbf{C}_{low}\mathbf{C}_{low}^{\dagger} + \mathbf{I}_N)^{-1}. \tag{6.27}$$

The overall transfer matrix for the MMSE receiver is

$$\mathbf{T} = \gamma\mathbf{C}_{low}^{\dagger}(\gamma\mathbf{C}_{low}\mathbf{C}_{low}^{\dagger} + \mathbf{I}_N)^{-1}\mathbf{C}_{low}.$$

The subchannel SNR $\beta(i)$ can be computed using an approach similar to that for the zero-padded OFDM system. Unlike the MMSE receiver of the SC-CP system in (6.21), the MMSE receiver of the SC-ZP is not a circulant matrix. It cannot be implemented efficiently using DFT and IDFT matrices.

Example 6.4 The SC-ZP system. The simulation environment and parameters used in this example are the same as those in Example 6.1. Figure 6.17 shows the BER performances of the SC-ZP system with the efficient receiver in (6.25), with the pseudo-inverse receiver in (6.26), and with the MMSE receiver in (6.27). The BERs of the three receivers for the SC-ZP system are denoted, respectively, by $\mathcal{P}_{sc-zp}$ (efficient), $\mathcal{P}_{sc-zp}$ (pseudo), and $\mathcal{P}_{sc-zp,mmse}$ in the figure. For comparison, the error rates of the SC-CP system with the zero-forcing and the MMSE receivers are also shown. The error rate of the zero-padded system with the efficient receiver is very close to the SC-CP system for both channels, as we had expected. For the channel $c_1(n)$, all the curves are very close. The pseudo-inverse and MMSE receivers do not have much gain over the efficient receiver. For the second channel, the MMSE and pseudo-inverse receivers are less affected by the zero that is close to the unit circle. The pseudo-inverse receiver can achieve a gain of around 15 dB over the efficient receiver. The MMSE receiver has an even more significant gain. ∎

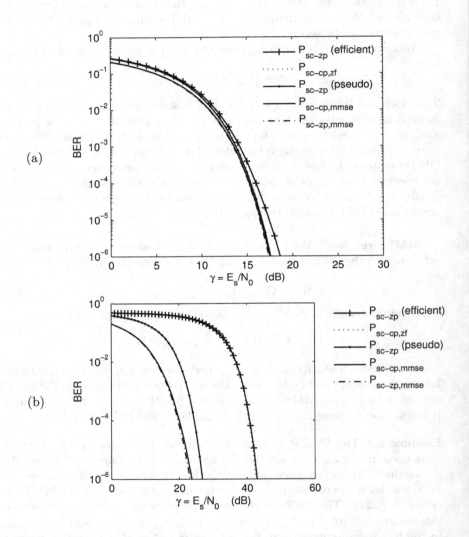

Figure 6.17. Example 6.4. The SC-ZP system: BER performances for (a) channel $c_1(n)$ and (b) channel $c_2(n)$.

6.5 Filter bank representation of OFDM systems

In this section, we will look at the OFDM system from the viewpoint of filter banks. This viewpoint allows us to see how the transmitted power spectrum is related to the input modulation symbols, which will give us more insight into how the symbols are transmitted in the frequency domain.

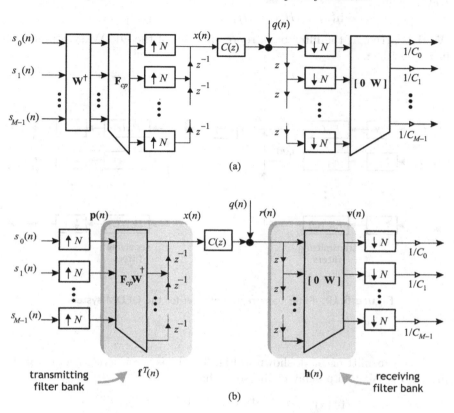

(a)

(b)

Figure 6.18. Derivations of the filter bank representation for the OFDM system.

To derive the equivalent filter bank structure, let us redraw the OFDM system using the matrix representation in Fig. 6.18(a), where the details of blocking and unblocking mechanisms, and the insertion and removal of prefix, are shown explicitly. The $N \times M$ matrix $\mathbf{F}_{cp}$ is the prefix insertion matrix defined in (5.14). As $\mathbf{F}_{cp}$ and $\mathbf{W}^\dagger$ are constant matrices, we can exchange these two matrices and the expanders. The resulting transmitter is as shown in Fig. 6.18(b), where we have lumped $\mathbf{F}_{cp}$ and $\mathbf{W}^\dagger$ together. Similarly, we can exchange the decimators and the matrix $\begin{bmatrix} \mathbf{0} & \mathbf{W} \end{bmatrix}$ at the receiver to yield the receiver shown in Fig. 6.18(b). We note that the M-input one-output system from $\mathbf{p}(n)$ to $x(n)$ at the transmitter is an LTI system, which we call the transmitting filter bank. Denoting the system by $\mathbf{f}^T(z)$ and its kth filter by $F_k(z)$, we have

$$\mathbf{f}^T(z) = \begin{bmatrix} F_0(z) & F_1(z) & \cdots & F_{M-1}(z) \end{bmatrix}.$$

Similarly, at the receiving end, the one-input M-output system from $r(n)$ to $\mathbf{v}(n)$ is also LTI. We call the $M \times 1$ system $\mathbf{h}(z)$ and denote its kth filter as $H_k(z)$. Thus

$$\mathbf{h}(z) = \begin{bmatrix} H_0(z) & H_1(z) & \cdots H_{M-1}(z) \end{bmatrix}^T.$$

We thus arrive at the filter bank representation of the OFDM system shown in Fig. 6.19.

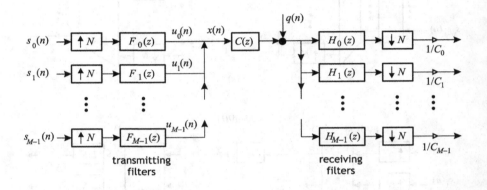

Figure 6.19. Filter bank representation for the OFDM system.

The transmitting bank shown in Fig. 6.18(b) is the interconnection of the matrix $\mathbf{F}_{cp}\mathbf{W}^{\dagger}$ and a delay chain, so we have

$$\mathbf{f}^T(z) = \begin{bmatrix} 1 & z^{-1} & \cdots & z^{-(N-1)} \end{bmatrix} \mathbf{F}_{cp}\mathbf{W}^{\dagger}.$$

The kth transmitting filter is determined by $F_k(z) = \sum_{i=0}^{N-1} z^{-i}[\mathbf{F}_{cp}\mathbf{W}^{\dagger}]_{ik}$. Its impulse response is

$$f_k(n) = \frac{1}{\sqrt{M}}W^{-(n-\nu)k}, \quad n = 0, 1, \ldots, N-1, \qquad (6.28)$$

where ν is the length of the cyclic prefix. The transmitting filters are all DFT filters. The filter $f_0(n)$ is a rectangular window of length N. All the other transmitting filters are scaled and frequency-shifted versions of the **prototype** $F_0(z)$,

$$F_k(z) = W^{\nu k}F_0(zW^k) \quad \text{or} \quad F_k(e^{j\omega}) = W^{\nu k}F_0(e^{j(\omega - 2k\pi/M)}). \qquad (6.29)$$

The transmitting filters form a *DFT bank*. Figure 6.20(a) shows the magnitude response of $F_0(z)$. As $f_0(n)$ is a rectangular window of length N, $F_0(e^{j\omega})$ has zeros at integer multiples of $2\pi/N$ (except $\omega = 0$). The magnitude responses of the other filters are shifted versions of $|F_0(e^{j\omega})|$, the shifts being integer multiples of $2\pi/M$. Figure 6.20(b) gives a schematic plot of the magnitude responses of the transmitting filters. The kth filter $F_k(z)$ is a bandpass filter with the passband centered around $2k\pi/M$. The stopband attenuation of the filters is around 13 dB and the stopband of $F_0(e^{j\omega})$ decays slowly with frequency [105].

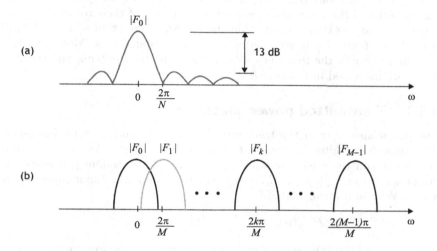

Figure 6.20. (a) The magnitude response of the prototype $F_0(z)$; (b) schematic plot of the magnitude responses of the transmitting filters in the OFDM system.

Similarly, the receiving bank in Fig. 6.18(b) is the interconnection of an advance chain and $\begin{bmatrix} 0 & \mathbf{W} \end{bmatrix}$, so we have

$$\mathbf{h}(z) = \begin{bmatrix} 0 & \mathbf{W} \end{bmatrix} \begin{bmatrix} 1 \\ z \\ \vdots \\ z^{N-1} \end{bmatrix}.$$

The impulse response of the kth filter is given by

$$h_k(n) = \frac{1}{\sqrt{M}} W^{-(n+\nu)k}, \quad n = -N+1, -N+2, \ldots, -\nu. \tag{6.30}$$

The receiving filters are also DFT filters. They are noncausal due to the advance chain but can be made causal easily by introducing enough delays in the receiver. The filter $H_0(z)$ is also a rectangular window, but of length M. All the other receiving filters are scaled and frequency-shifted versions of the prototype $H_0(z)$,

$$H_k(z) = W^{-\nu k} H_0(zW^k) \quad \text{or} \quad H_k(e^{j\omega}) = W^{-\nu k} H_0(e^{j(\omega - 2k\pi/M)}).$$

A schematic plot of the magnitude responses of the receiving filters would look very similar to that of transmitting filters in Fig. 6.20. A slight difference is that the first zero of $H_0(e^{j\omega})$ is at $2\pi/M$ (instead of $2\pi/N$) as its length is M. The kth filter is bandpass with the passband centered at $2k\pi/M$.

   A numerical example of the transmitting and receiving filters for $M = 8$ and $\nu = 2$ is shown in Fig. 6.21. The magnitude response of the two prototype filters $F_0(z)$ and $H_0(z)$ are drawn with a thicker line. The magnitude responses of all the other filters are shifted versions of their respective prototypes. We can see that the first sidelobe has an attenuation of around 13 dB only. The attenuation is not adequate in many applications. More details on the importance of the frequency selectivity of the transmitting and receiving filters are discussed in Chapter 9.

6.5.1   Transmitted power spectrum

The power spectrum of the transmitted signal $x(n)$ in Fig. 6.19 depends on the transmitting filters as well as the transmitter inputs. Assume the inputs of the transmitter $s_k(n)$ are white, zero-mean WSS random processes with power spectrum $S_{s_k}(e^{j\omega}) = \mathcal{E}_s$. Furthermore, assume the input processes are jointly WSS and uncorrelated,[2]

$$E[s_k(n)s_j^*(i)] = 0, \quad \forall n, i, \text{ and } k \neq j.$$

The output of the kth transmitting filter, $u_k(n)$, has a power spectrum given by (Appendix B)

$$S_{u_k}(e^{j\omega}) = \frac{\mathcal{E}_s}{N} |F_k(e^{j\omega})|^2.$$

When the inputs of the transmitter are jointly WSS and uncorrelated, the transmitted signal $x(n)$ is CWSS(N) with power spectrum

$$S_x(e^{j\omega}) = \frac{\mathcal{E}_s}{N} \sum_{k \in \mathcal{A}} |F_k(e^{j\omega})|^2, \qquad (6.31)$$

where $\mathcal{A}$ is the set of subchannels that are actually used for transmission.[3] The contribution from the k-subchannel occupies the frequency bin corresponding to the passband of $F_k(z)$, i.e. the bin centered at $2k\pi/M$. Thus the data from each subchannel occupy a separate frequency band and the subchannels are separated via the frequency-selective property of the transmitting filters.

   When we pass the discrete-time transmitted signal $x(n)$ through the D/C converter, we obtain the continuous-time transmitted signal $x_a(t)$. We can use $S_x(e^{j\omega})$ to obtain the power spectrum of $x_a(t)$. When $x(n)$ is CWSS(N)

---

[2]The transmitted power spectrum may also be derived with a more relaxed condition using the result in [131], which uses only the assumption that the inputs form a WSS vector process.

   [3]In some applications of the OFDM system, for example wireless local area networks [54], certain subchannels are reserved and only a subset of the subchannels are actually used for data transmission.

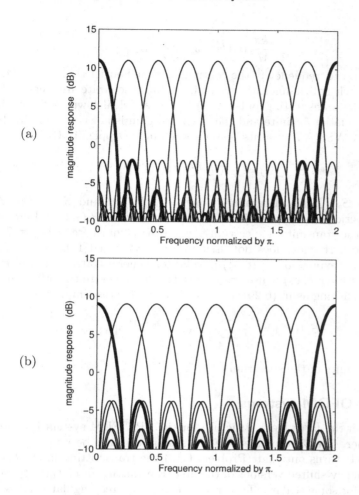

Figure 6.21. Magnitude responses of the transmitting and receiving filters in an OFDM system for $M = 8$ and $\nu = 2$: (a) the transmitting filters; (b) the receiving filters.

and the sampling period of the D/C converter is T, the transmitted signal $x_a(t)$ is a continuous-time CWSS process with period NT, i.e. CWSS(NT) (see Appendix B.4 for details). Assume the transmitting pulse $p_1(t)$ at the transmitter is an ideal lowpass filter with cutoff frequency π/T and a constant unity gain in the passband, where T is the underlying sample spacing. Then the power spectrum of $x_a(t)$ is (Appendix B.4)

$$S_{x_a}(j\Omega) = \begin{cases} \frac{1}{T} S_x(e^{j\Omega T}), & |\Omega| < \pi/T, \\ 0, & \text{otherwise.} \end{cases}$$

Recall that the kth transmitting filter is a bandpass filter centered at $\omega = 2\pi k/M$. As a result, the kth subchannel corresponds to the frequency bin centered at

$$\frac{1}{T}\frac{2k\pi}{M} \text{ rad/s, or } \frac{k}{MT} \text{ Hz}$$

with respect to the frequency of the continuous-time Fourier transform. Thus the data of each subchannel are transmitted in a separate frequency bin. In practice, the transmitting pulse $p_1(t)$ is not an ideal lowpass filter. Its stopband attenuation is finite and the passband gain is not constant. In this case, $x_a(t)$ is CWSS(NT) and its power spectrum is (Appendix B)

$$S_{x_a}(j\Omega) = \frac{1}{T}S_x(e^{j\Omega T})|P_1(j\Omega)|^2. \tag{6.32}$$

The shape of $S_{x_a}(j\Omega)$ is determined jointly by $P_1(j\Omega)$ and $S_x(e^{j\Omega T})$. As $S_x(e^{j\Omega T})$ is periodic with period $2\pi/T$, the stopband of $P_1(j\Omega)$ will dictate the rolloff of the transmitted spectrum beyond the Nyquist frequency π/T.

The above derivation does not assume $F_k(z)$ to be DFT filters. The power spectrum expression in (6.31) is valid for general transmitting filters. When the first filter $F_0(z)$ is used as a prototype to generate the other filters by frequency shifting as in (6.29), the spectrum $S_x(e^{j\omega})$ is simply the sum

$$S_x(e^{j\omega}) = \frac{\mathcal{E}_s}{N}\sum_{k\in\mathcal{A}}|F_0(e^{j(\omega-2k\pi/M)})|^2.$$

It consists of shifts of the spectrum of $F_0(z)$.

6.5.2   ZP-OFDM systems

We can obtain the filter bank structure for the ZP-OFDM system in a very similar manner. The zero-padded OFDM system can also be represented as in Fig. 6.19. It turns out that (Problem 6.16) the transmitting filters $F_k(z)$ are also frequency-shifted versions of the first transmitting filter $F_0(z)$ (prototype) except for some scalars. The prototype is now a rectangular window of length M. When the receiver is chosen to be the one that has the efficient implementation in (6.15), the first receiving filter $H_0(z)$ is a rectangular window of a longer length N. All the other receiving filters are frequency-shifted versions of $H_0(z)$, except possibly for some scalars. For other types of receivers, both pseudo-inverse and MMSE receivers, the frequency-shifting property no longer holds in general. As the derivation of the transmitted power spectrum in (6.31) does not require assumptions on the length or coefficients of the transmitting filters, the transmitted power spectrum for the ZP-OFDM system also has the form in (6.31).

6.6   DMT systems

The discrete multitone (DMT) system, often considered for wired applications, is very similar to the OFDM system. The DMT system has been successfully applied to high-speed data transmission over digital subscriber lines (DSL) such as ADSL (asymmetric digital subscriber lines) [7] and VDSL (very-high-Speed digital subscriber lines) [8]. In DSL applications the channel can have a very long impulse response. To avoid IBI, we need to have

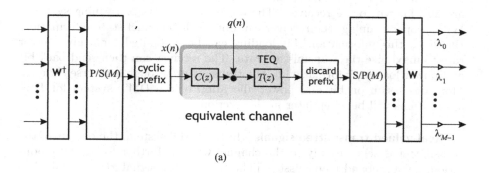

(a)

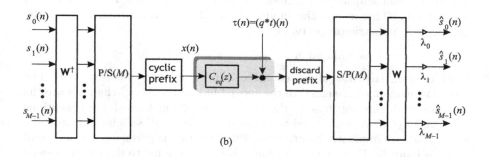

(b)

Figure 6.22. (a) The DMT system with a time domain equalizer; (b) the DMT system with the equivalent shortened channel.

a cyclic prefix that is as long as the longest impulse response. A long prefix can seriously reduce the transmission rate. In these applications, a **time domain equalizer** (TEQ) $t(n)$ is usually included at the receiver to shorten the effective channel impulse response, as shown in Fig. 6.22(a). TEQ plays an important role in improving the transmission bit rate of the DMT system. There has been extensive research on the design of TEQ [1, 6, 10, 90]. See Section 3.5 for TEQ designs.

Upon time domain equalization, the impulse response of the shortened channel is $c_{eq}(n) = t(n) * c(n)$ and the equivalent channel noise is $\tau(n) = q(n) * t(n)$ (Fig. 6.22(b)). The original channel noise $q(n)$ in DSL applications is often **colored** due to crosstalk (to be explained later). This is different from the OFDM system for wireless applications, where the noise can usually be assumed to be white. Even if the original channel noise is white, the equivalent noise $\tau(n)$ is in general colored due to time domain equalization. With the equivalent system in Fig. 6.22(b), we can obtain the filter bank representation for the DMT system just like in the OFDM case. The filter bank representation will be the same as that in Fig. 6.19 except that the channel $C(z)$ is replaced by the shortened channel $C_{eq}(z)$. The subchannels in DMT systems are often referred to as **tones**.

For wired transmission, the channel does not change as rapidly as in wireless applications. This allows the transmitter to send out symbols first that are also known to the receiver. These are called pilots or training symbols. With proper training, the receiver can obtain the channel state information. Based on this information, bit loading can be computed for a given error rate to maximize the transmission rate. The receiver can then send back bit and power loading information to the transmitter through a reverse channel. More discussion on bit and power allocation of the DMT system for DSL applications will be given later in this section.

Real-valued transmitted signals In the DMT system, the output of the transmitter is sent directly to the channel without further frequency modulation, i.e. baseband transmission. The transmitted signal $x(n)$ is real. The inputs of the transmitter s_k are typically complex QAM modulation symbols. To have a real transmitted signal, the outputs of the IDFT matrix should be real too. This means the inputs of the IDFT matrix need to have the conjugate symmetric property, i.e.

$$s_0 \text{ is real and } s_k = s_{M-k}^*, \quad k = 1, 2, \ldots, M - 1. \tag{6.33}$$

If M is even, then the above equation implies that $s_{M/2}$ is also real. The modulation symbols assigned to the second half of the M subchannels are the complex conjugates of those in the first half subchannels, with the exceptions of s_0 and $s_{M/2}$. We only need to consider the first half subchannels and the second half are implicitly determined. Bit allocation is performed for only half the subchannels. Usually the subchannels corresponding to the DC frequency and Nyquist frequency, i.e. zeroth and $(M/2)$th subchannels, are reserved for other purposes. In this case, only tones 1 to $(M/2-1)$ will be considered. Due to the conjugate symmetric property of the inputs, it is possible to compute the transmitter outputs using only real additions and real multiplications instead of complex computations. It can be shown that the IDFT operation for conjugate symmetric inputs can be computed using DCT (discrete cosine transform) matrices and DST (discrete sine transform) matrices using only real computations [174] (Problem 6.19).

The transmitted power spectrum In the formulation of the transmitted power spectrum for the OFDM system, we have assumed that the inputs are uncorrelated WSS random processes. For the DMT system, we can no longer make such an assumption as the inputs are in conjugate pairs and hence strongly correlated. However, with a slight modification of the assumption, the transmitted power spectrum of DMT system has a form similar to that of the OFDM system in (6.31). To be more specific, let $E[|s_k(n)|^2] = \mathcal{E}_{s,k}$. We have added the subscript k to indicate that the subchannels can have different symbol powers. For those tones with complex symbols, we express $s_k(n)$ in terms of its real part $a_k(n)$ and imaginary part $b_k(n)$ as

$$s_k(n) = a_k(n) + jb_k(n).$$

We assume, reasonably, that these random processes $a_k(n)$ and $b_k(n)$ are uncorrelated, jointly WSS, with power spectrum

$$S_{a_k}(e^{j\omega}) = S_{b_k}(e^{j\omega}) = \frac{1}{2}\mathcal{E}_{s,k}, \quad k = 1, 2, \ldots, M/2 - 1.$$

The scalar $1/2$ is included so that $E[|s_k(n)|^2] = \mathcal{E}_{s,k}$.

Consider the kth and the $(M-k)$th subchannels. The inputs are a conjugate pair and so are the filter coefficients,

$$f_{M-k}(n) = f_k^*(n).$$

The corresponding outputs $u_k(n)$ and $u_{M-k}(n)$ (indicated in Fig. 6.19) are thus also a conjugate pair. The sum of these two subchannels is given by

$$u_k'(n) = u_k(n) + u_{M-k}(n) = 2\text{Real}\{u_k(n)\}.$$

This can be written as

$$u_k'(n) = 2\sum_{\ell=\infty}^{\infty}\left(a_k(\ell)f_{k,r}(n-N\ell) - b_k(\ell)f_{k,i}(n-N\ell)\right), \qquad (6.34)$$

where $f_{k,r}(n)$ and $f_{k,i}(n)$ are, respectively, the real and imaginary parts of $f_k(n)$. As $a_k(n)$ and $b_k(n)$ are uncorrelated, we have

$$S_{u_k'}(e^{j\omega}) = \frac{\mathcal{E}_{s,k}}{N}\left(2|F_{k,r}(e^{j\omega})|^2 + 2|F_{k,i}(e^{j\omega})|^2\right).$$

Using the fact that $f_{M-k}(n)$ and $f_k(n)$ form a conjugate pair, we can verify that the sum in the above parenthesis is equal to $|F_k(e^{j\omega})|^2 + |F_{M-k}(e^{j\omega})|^2$. As a result we have

$$S_{u_k'}(e^{j\omega}) = \frac{\mathcal{E}_{s,k}}{N}\left(|F_k(e^{j\omega})|^2 + |F_{M-k}(e^{j\omega})|^2\right).$$

Summing up the contributions from individual subchannels, we can obtain the transmitted power spectrum. When frequency division multiplexing (to be explained in the following) is used, not all the subchannels are used. In this case,

$$S_x(e^{j\omega}) = \frac{1}{N}\sum_{k\in\mathcal{A}}\mathcal{E}_{s,k}|F_k(e^{j\omega})|^2, \qquad (6.35)$$

where $\mathcal{A}$ is the set of subchannels that are actually used for transmission. (We can also arrive at the same result using properties from [131].) Like the OFDM case (given in (6.32)), the power spectrum of the transmitted continuous-time signal depends on the sampling period and the transmitting pulse.

Remarks

- Note that the transmitted power spectrum of the DMT system is of the same form as that of the OFDM system if an equal power allocation is used, i.e. $\mathcal{E}_{s,k} = \mathcal{E}_s$ for all k.

- In the above derivation of the transmitted power spectrum, the transmitting filters only need to be in conjugate pairs $(f_{M-k}(n) = f_k^*(n))$. The transmitting filters are not necessarily DFT filters for the expression in (6.35) to hold. They can be arbitrary filters that have the conjugate-pair property.

Frequency division multiplexing In a DMT system, signals are usually sent in both directions simultaneously over the same transmission medium such as the twisted pairs in ADSL systems, from the customer to the network (**upstream** direction) and from the network to the customer (**downstream** direction). The downstream and upstream signals can be separated by using frequency division multiplexing (FDM), in which different frequency bands are allocated for each direction. The DMT system lends itself nicely to FDM, since the subchannels are divided in a frequency based manner. To do this, we can set aside some tones for upstream transmission and some for downstream transmission. For example, in the ADSL system, the total number of subchannels M is 512. Tones 1–31 are used for upstream transmission and the rest of the higher frequency tones (32–255) for downstream transmission as illustrated in Fig. 6.23. (Usually a few close-to-DC tones are reserved for voice and guard band.) In downstream transmission, only the downstream tones are used; the inputs of the upstream tones are zero. Similarly, the inputs of the downstream tones are zero in upstream transmission. By having a much wider bandwidth, the downstream direction can have a much higher transmission rate than the upstream direction. That is why ADSL, where "A" stands for *asymmetric*, is so named. The narrower upstream band is placed in a lower frequency region than the downstream band. This way upstream transmission is still possible even when the channel is much attenuated in high frequency, which is usually the case with long DSL lines.

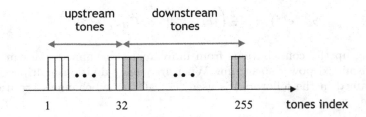

Figure 6.23. Upstream and downstream tones in the ADSL system.

In the standards for ADSL and VDSL systems [7, 8] the power spectrum for each transmitting direction is constrained through a **spectrum mask**. The spectrum mask sets an upper bound for the spectrum of the transmitted signal as a function of frequency. Figure 6.24 shows an example of spectrum mask for the downstream ADSL transmission. The mask is given in dBm/Hz, where

$$1 \text{ dBm} = 10 \log(10^{-3} \text{ watts}).$$

Also shown is an example of the downstream signal's spectrum, which is below the spectrum mask for all frequency.

Noise impairments In a DSL environment, a number of noise sources contribute to the channel noise $q(n)$. Usually DSL lines are bundled together

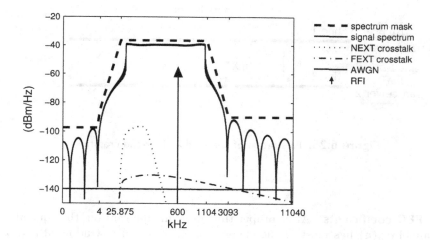

Figure 6.24. Example of spectrum mask for downstream ADSL transmission and the spectrum of a downstream signal. Also shown in the figure are noise impairments in a DSL environment, including NEXT crosstalk, FEXT crosstalk, AWGN, and RFI.

in multiples of 25 in cables and this makes the channel prone to **crosstalk** from other lines in the same cable. The crosstalk may be near-end crosstalk (NEXT) or far-end crosstalk (FEXT). NEXT comes in when the disturbing signal is traveling in the direction opposite to the received signal. For example, in Fig. 6.25, the upstream signal of line 1 couples into line 2 and becomes the source of NEXT in downstream transmission on line 2. FEXT refers to the case when the disturbing signal is traveling in the same direction as the received signal. Figure 6.25 shows the downstream signal of line 1 as a FEXT source for the downstream direction on line 2. FEXT is usually the dominating noise when FDM is used because the spectrum of the signal traveling in the opposite direction occupies a different frequency band. In fact, it is possible to eliminate interference from the other direction and NEXT by using a longer cyclic prefix and carefully synchronizing downstream and upstream transmission [139]. Both types of crosstalk can be determined from the disturbers' power spectrum and crosstalk transfer functions [7, 8] and they can be modeled as colored Gaussian noises [122, 144].

In addition, the channel noise will contain AWGN due to thermal noise. Also, radio frequency signals such as amateur radio (HAM) and AM radio may interfere and result in radio frequency interference (RFI). These RFI signals are of a narrowband nature but have a large amplitude (in the frequency domain). Many tones in the vicinity of the RFI carrier frequency may be affected due to the poor frequency separation of DFT filters of the DMT receiver. Figure 6.24 shows the various noise impairments in a DSL environment, including FEXT and NEXT crosstalk, AWGN, and RFI. The AWGN noise in the figure is −140 dBm/Hz and the RFI is around 600 kHz with an amplitude of around −55 dBm.

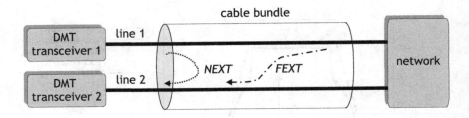

Figure 6.25. Illustration of near-end and far-end crosstalks.

FEQ coefficients   After proper time domain equalization, the equivalent channel $c_{eq}(n)$ has most of the energy concentrated in a window of ν taps. Say the window spans the range from n_0 to $n_0 + \nu$. We can obtain the FEQ coefficients λ_k by computing the reciprocals of the M-point DFT of the channel coefficients. That is,

$$\lambda_k = 1/C_{eq,k}, \quad \text{where} \quad C_{eq,k} = \sum_{n=n_0}^{n_0+\nu} c_{eq}(n)e^{-2\pi kn/M}. \tag{6.36}$$

A precise calculation of FEQ coefficients involves the coefficients of the channel outside the window too [81]. But (6.36) gives a very good estimation, especially when the TEQ does a good job of shortening the channel.

Bit allocation   Although time domain equalization can shorten the channel, the equivalent channel $c_{eq}(n)$ will still have nonzero coefficients outside the window. These nonzero coefficients result in residual interference from adjacent blocks and also from other subchannels of the same block. In particular, if we compute the error $e_k(n) = \hat{s}_k(n) - s_k(n)$, it will contain channel noise as well as interference from other symbols. We can obtain the SNR of the kth tone by computing

$$SNR_k = \frac{E[|s_k(n)|^2]}{E[|e_k(n)|^2]}, \tag{6.37}$$

where the symbol power $E[|s_k(n)|^2]$ is constrained by the spectrum mask in DSL applications (Problem 6.20). Although e_k is a mixture of interference and noise, the Gaussian distribution renders a nice approximation as we have explained in Chapter 3. We can use (6.37) to compute the number of bits that can be loaded to the kth tone,

$$b_k = \log_2\left(1 + \frac{SNR_k}{\Gamma}\right), \tag{6.38}$$

where Γ is the SNR gap for a given error rate as described in (2.21). For those tones that are not used for transmission, no bits are allocated, $b_i = 0$. The total number of bits transmitted in each block is $\sum_{i=0}^{M-1} b_i$. Each block takes up a transmission time of NT seconds, where T is the underlying sample spacing of the D/C converter. The transmission rate R_b is given by

$$R_b = \frac{1}{NT} \sum_{i=0}^{M-1} b_i.$$

The bit assignments computed in (6.38) are not integers in general. We can use, for example, rounding, to have an integer bit allocation. If rounding is used, the QAM symbol of the kth tone will carry

$$\hat{b}_k = \text{round} \left[\log_2 \left(1 + \frac{SNR_k}{\Gamma} \right) \right] \text{ bits.} \qquad (6.39)$$

The value $\hat{b}_k$ can be even or odd. After rounding, the error rate will be slightly different from the target error rate. A minor adjustment of the subchannel powers can be carried out to compensate for the rounding error (Problem 21). In ADSL and VDSL applications [7, 8], the receiver computes bit allocation and power adjustment, and sends the information back to the transmitter.

Example 6.5 A DMT system for ADSL application. The DFT size $M = 512$ and the cyclic-prefix length $\nu = 40$. The number of blocks to be transmitted per second is 4000 [7]. The sampling rate is

$$(512 + 40) \times 4000 = 2.208 \times 10^6 \text{ samples/s.}$$

We will consider downstream transmission in this example. The tones used are 33–255. The transmission power used is -4 dB. Figure 6.26(a) shows a typical ADSL channel – loop 6 [7]. The length of the loop is 18,000 feet. The magnitude response of the channel is shown in Fig. 6.26(b). The channel has a typical lowpass characteristic, more attenuated in high-frequency band. The low-frequency tones can carry more bits than high-frequency tones. Figure 6.26(c) shows the power spectrum of the channel noise, which comprises NEXT and FEXT as well as AWGN noise. The channel noise is more prominent in the frequency band used for upstream transmission due to NEXT.

After time domain equalization, the resulting equivalent channel, shown in Fig. 6.26(a), has energy much more concentrated and hence much "shorter" than the original channel. The TEQ has six taps and the MSSNR method in Section 3.5 [90] has been used to design the TEQ. Equations (6.37) and (6.39) are used to compute the subchannel SNRs and the bit loading of QAM symbols, shown, respectively, in Fig. 6.26(d) and (e). The target symbol error rate in this case is 10^{-7}. The total number of bits transmitted in each block is 1742 and the transmission rate is 7.0 Mbits/sec.   ∎

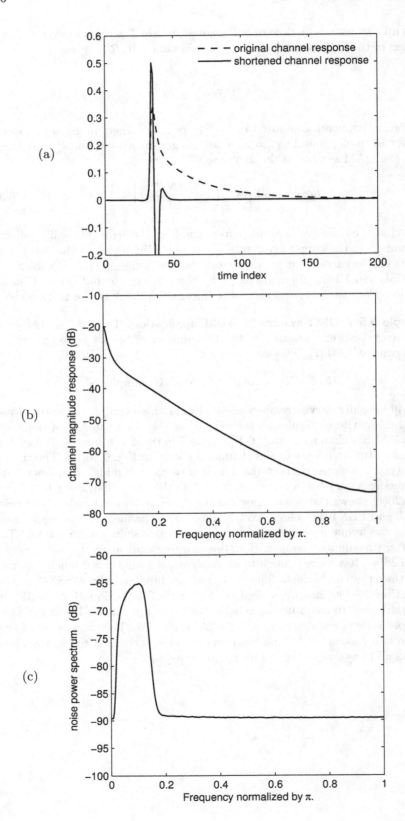

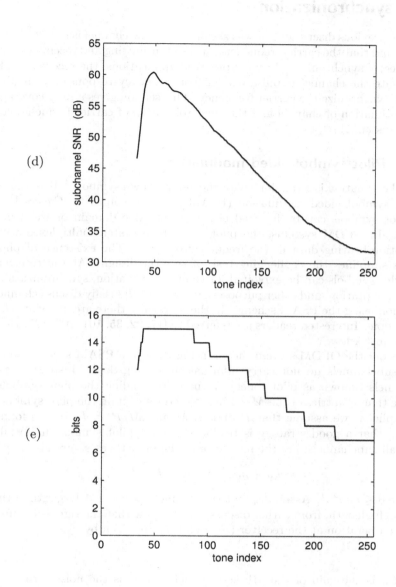

Figure 6.26. Example 6.5. A DMT system. (a) The impulse responses of the channel and the equivalent shortened channel; (b) the magnitude response of the channel; (c) the channel noise spectrum; (d) the subchannel SNRs; (e) the subchannel bit loading.

6.7 Channel estimation and carrier frequency synchronization

In all the previous discussions, it was assumed that the channel is known at the receiver and that the carrier frequencies at the transmitting and receiving sides are perfectly synchronized. In most practical applications, the receiver needs to estimate the channel impulse response or frequency response and it also needs to synchronize the carrier frequency. This section gives a brief coverage of the estimation of channel and the synchronization of carrier frequency for OFDM systems.

6.7.1 Pilot symbol aided modulation

Channel estimation is often done with the aid of known symbols.[4] It is known as pilot symbol aided modulation (PSAM), a term coined by Cavers [17]. The pilot symbols can be inserted either in the time domain or frequency domain. In OFDM systems, the pilot symbols are often multiplexed with information-bearing data in the frequency domain. The existence of pilot symbols simplifies the challenging task of receiver design. At the receiver, these pilot symbols can be exploited for channel estimation, synchronization, receiver adaptation, and other purposes. Below we will briefly discuss channel estimation using the PSAM scheme. In the literature, there are many articles on this topic. Interested readers are referred to [17, 22, 35, 101, 103, 125, 153], to name just a few.

Consider the OFDM system shown in Fig. 6.1. In a PSAM scheme, some of the subchannels do not carry information-bearing data. Instead, these subchannels (known as pilot tones) are used for sending the pilot symbols. Suppose that P entries of the $M \times 1$ input vector $\mathbf{s}$ contain the pilot symbols. For simplicity, we assume that P divides M and $M/P = J$. It was found [35, 153] that a good strategy is to distribute the pilot symbols uniformly among all subchannels; i.e. the indexes n of the pilot tones s_n are given by

$$ n = kJ + n_0, \quad \text{for } 0 \le k < P, $$

for some $0 \le n_0 < J$. Assuming that the channel order is not larger than the cyclic-prefix length, from earlier discussions we know that the signals y_n after the DFT operation at the receiver (see Fig. 6.1) are given by

$$ y_n = C_n s_n + q_n, $$

where C_n is the gain of the nth subchannel and q_n is the noise term. As the pilot symbols s_{kJ+n_0} are known, the receiver can estimate the subchannel gains for these pilot tones as

$$ \widehat{C}_{kJ+n_0} = \frac{y_{kJ+n_0}}{s_{kJ+n_0}}. $$

To obtain the subchannel gain C_n for $n \ne kJ + n_0$, one can use the interpolation scheme [36]. When the channel has a smooth frequency response, an

---

[4]It is also possible to estimate the channel blindly. However, blind channel estimation is seldom employed in practice due to its high complexity and lower estimation accuracy. Readers interested in this topic are referred to [21, 45, 147, 183].

accurate estimate of $\widehat{C}_n$ can be obtained by interpolation. Another way of getting all C_n is by the following time domain approach. Using the fact that C_n are the DFT coefficients of $c(\ell)$, we can write

$$\widehat{C}_{kJ+n_0} = \sum_{\ell=0}^{\nu} W^{(kJ+n_0)\ell}\widehat{c}(\ell),$$

for $0 \leq k < P$. Let $\mathbf{v}$ be the $P \times 1$ vector containing $\widehat{C}_{kJ+n_0}$ for $0 \leq k < P$ and let $\mathbf{u}$ be the $(\nu+1) \times 1$ vector containing $\widehat{c}(\ell)$. Then the above system of linear equations can be re-expressed as $\mathbf{v} = \widetilde{\mathbf{W}}\mathbf{u}$, where $\widetilde{\mathbf{W}}$ is a $P \times (\nu+1)$ matrix with (k,l)th entry $W^{(kJ+n_0)l}$. When $P > \nu$, we can solve the matrix equation for $\widehat{c}(\ell)$ and obtain

$$\mathbf{u} = (\widetilde{\mathbf{W}}^{\dagger}\widetilde{\mathbf{W}})^{-1}\widetilde{\mathbf{W}}^{\dagger}\mathbf{v}.$$

Taking the M-point DFT of $\widehat{c}(\ell)$, we get $\widehat{C}_n$ for all $0 \leq n < M$. The complexity of the time domain approach is usually higher than the interpolation scheme, but it gives a more accurate estimate of C_n, especially when P is much larger than ν.

6.7.2 Synchronization of carrier frequency

For passband communications, the transmitted signal is modulated to a carrier frequency at the transmitter. At the receiver, we need to perform carrier demodulation to get the baseband signal. The frequency of the oscillator at the receiver is usually slightly different from that of the oscillator at the transmitter. This results in carrier frequency offset (CFO). The OFDM system is very sensitive to CFO [118], which destroys the orthogonality of subchannels and degrades the performance of OFDM systems significantly [118]. Thus, at the receiver there is a need to perform an accurate synchronization of carrier frequency. Many approaches have been developed in the past. For example, see [9, 53, 71, 85, 91, 93, 95, 136, 166]. In many practical applications, a periodic training sequence known to the receiver is sent at the beginning of transmission. By exploiting the periodic nature of the training sequence, we can use a simple method for correcting the CFO, as we shall explain below.

Consider Fig. 6.1. Suppose that two identical vectors $\mathbf{s}$ are sent consecutively at the transmitter. When there is no CFO, the corresponding two received vectors, denoted by $\mathbf{r}_1$ and $\mathbf{r}_2$, respectively, will have the following form (see (6.1)):

$$\mathbf{r}_1 = \mathbf{C}_{circ}\mathbf{x} + \mathbf{q}_1, \qquad \mathbf{r}_2 = \mathbf{C}_{circ}\mathbf{x} + \mathbf{q}_2, \tag{6.40}$$

where $\mathbf{x} = \mathbf{W}^{\dagger}\mathbf{s}$. They are different only because of the noise vectors. However, when there is CFO, the two received vectors are no longer given by (6.40). Let f_c be the oscillator frequency at the transmitter and let $f_c - \delta_f$ be the oscillator frequency at the receiver. Then due to the mismatch of the oscillator frequencies, the received baseband signal will be modulated by an extra exponential term $e^{j\epsilon n}$. The quantity ϵ is the normalized CFO parameter and it is related to δ_f by

$$\epsilon = 2\pi\delta_f T,$$

where T is the sample spacing. As a result, the two received vectors become [53, 71, 85]

$$\tilde{\mathbf{r}}_1 = \mathbf{D}\mathbf{r}_1, \quad \tilde{\mathbf{r}}_2 = e^{jN\epsilon}\mathbf{D}\mathbf{r}_2, \tag{6.41}$$

where $\mathbf{D}$ is a matrix diagonal, given by

$$\mathbf{D} = \mathrm{diag}[e^{\nu\epsilon} \quad e^{(\nu+1)\epsilon} \quad \cdots \quad e^{(N-1)\epsilon}],$$

$N = M + \nu$, and ν is the length of cyclic prefix. From (6.41), it is clear from (6.40) and (6.41) that we can obtain an estimate of the CFO ϵ as

$$\hat{\epsilon} = \frac{1}{N} \angle \left(\tilde{\mathbf{r}}_1^\dagger \tilde{\mathbf{r}}_2 \right),$$

where $\angle z$ denotes the angle of the complex number z. The above method, though simple, gives a very accurate estimate of the CFO [95]. For more advanced approaches to carrier frequency synchronization, see [9, 53, 71, 85, 91, 93, 95, 136, 166].

6.8   A historical note and further reading

The use of frequency division multiplexing (FDM) to divide the channel into narrower subchannels and send data in parallel dates back to the 1950s. For each subchannel, the bandwidth is narrower. The transmitted signal "sees" a flatter channel, and equalization becomes easier. The system is potentially less sensitive to channel distortion and wideband impulsive noise. The first study on designing continuous-time transmitting filters for maintaining orthogonality among the subchannels at the receiver was made in [18], and later in [127] and [19]. With the advances of digital signal processing, digital implementation was made more practical at a lower cost. An efficient implementation of FDM systems using Discrete Fourier Transform (DFT) was developed in [178]. This eliminated the need for an expensive array of oscillators and coherent demodulators used in earlier works. Bit loading optimization for multichannel (multitone) systems was derived in [59, 60]. In earlier FDM systems there was no prefix. The orthogonality among the subchannels was usually destroyed after channel filtering, and the receiver could not remove ISI as easily. The task of equalizing the FDM system was greatly simplified when cyclic prefixing was introduced at the transmitter in [116]. The insertion of a cyclic prefix allowed the channel to be equalized using low-cost IDFT and DFT matrices. Bits were assigned according to subchannel SNRs to achieve a higher transmission rate [116].

The first practical application of cyclic-prefixed DFT-based transceivers was proposed in [154] for DSL applications. The cyclic-prefixed DFT-based transceiver was termed the discrete multitone (DMT) system in [24]. Performance evaluations of the DMT system were given in [24] for HDSL (high-bit-rate digital subscriber lines) application, and in [25] for ADSL and VDSL applications. Subsequently the DMT system has been repeatedly demonstrated to outperform other competing single-carrier solutions significantly. It was later adopted in the standards for ADSL [7] and VDSL [8] transmission. Multichannel FDM transmission was proposed for wireless environment by Cimini

[27], and the system is termed OFDM (orthogonal frequency division multiplexing). The OFDM system is now one of the most popular techniques for wireless transmission. Cyclic-prefixed OFDM systems have been adopted in standards such as digital audio broadcasting [39], digital video broadcasting [40], wireless local area networks [54], and broadband wireless access [55]. Comparisons of cyclic-prefixed and zero-padded OFDM can be found in [96].

Frequency domain equalization for single-carrier systems was first proposed in [173]. There was no redundancy in the transmitted signals. The introduction of cyclic prefix to single-carrier systems first appeared in [129]. As equalization was performed at the receiver after DFT operation, it was also called a single-carrier system with frequency domain equalization (SC-FDE). Extensive performance evaluation given in [41, 175] showed that the SC-CP system compares favorably with the OFDM system. The SC-CP system enjoys a smaller PAPR, especially for a large number of subchannels. It is also less sensitive to the so-called carrier frequency offset, which is an important issue in applying OFDM to wireless channels [118, 175]. An analytic comparison of the SC-CP and the OFDM systems was given in [77, 78] to show that the SC-CP always has a smaller uncoded BER. The SC-CP system is now part of the broadband wireless access standard [55]. For zero-padded single-carrier systems, an extensive study was given in [104], and an analytic BER comparison with SC-CP system was also given therein.

6.9 Problems

6.1 Let $M = 4$ and let the cyclic-prefix length $\nu = 2$. Suppose the channel is $C(z) = 1 + z^{-1}$, the channel noise is AWGN, the SNR quantity $\gamma = \mathcal{E}_s/\mathcal{N}_0$ is 10 dB, and the modulation symbols are QPSK. Compute the subchannel SNRs $\beta(i)$, the average mean squared error $\mathcal{E}_{rr}$, and the average bit error rate $\mathcal{P}$ of (a) the OFDM system, (b) the SC-CP system with a zero-forcing receiver, and (c) the SC-CP system with an MMSE receiver.

6.2 *Channel estimation error.* In the OFDM system, the receiver needs to estimate the channel to compute the FEQ coefficients. When there is channel noise, there will be estimation error. Suppose the estimated channel is $\hat{c}(n) = 0.9\delta(n)$, while the actual channel is $c(n) = \delta(n)$. We use the channel estimate $\hat{c}(n)$ to compute the FEQ coefficients and implement the receiver. Assume the rest of the setting is the same as in Problem 6.1. What are the bit error rates for the three systems in Problem 6.1?

6.3 *Modified OFDM system with rotated cyclic prefix.* In this problem, we consider an OFDM system with a different type of prefix – a rotated cyclic prefix. When we copy the last ν samples of each block and place them at the beginning of the block, we apply an extra rotation $e^{jM\theta}$ to the prefix. The rotated prefix samples are

$$e^{-jM\theta}x_{M-\nu} \quad e^{-jM\theta}x_{M-\nu+1} \quad \cdots \quad e^{-jM\theta}x_{M-1}.$$

Assume the channel $C(z)$ is an FIR filter of order $\leq \nu$. The new prefixed system as shown in Fig. P6.3 is the same as that in Fig. 5.16 except for the rotated prefix.

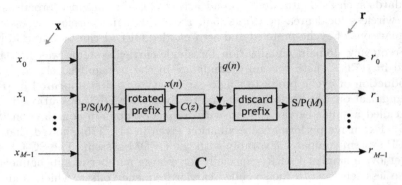

Figure P6.3. A prefixed system with rotation.

(a) Find the transfer matrix $\mathbf{C}$ from $\mathbf{x}$ to $\mathbf{r}$.

(b) Show that $\mathbf{\Lambda}(e^{j\theta})\mathbf{C}\mathbf{\Lambda}(e^{-j\theta})$ is a circulant matrix, where $\mathbf{\Lambda}(z)$ is the diagonal matrix

$$\mathbf{\Lambda}(z) = \text{diag}\begin{bmatrix} 1 & z^{-1} & \cdots & z^{-(M-1)} \end{bmatrix}.$$

(c) Design a transmitting matrix $\mathbf{G}$ and a receiving matrix $\mathbf{S}$ for the new prefixed system so that the overall system is ISI-free by modifying the receiver of the OFDM system. In such a modified OFDM system, the ISI-free property continues to hold for any FIR channel of order ν. The transmitter is still channel-independent and the only channel-dependent element is a set of scalars at the receiver.

(d) Suppose we already know that the channel has a spectral null at $2\pi/M$. Choose θ so that the receiver can recover all the data in the absence of channel noise. In this way, the spectral null can be avoided by choosing θ carefully.

6.4 *Differential OFDM* [39]. In the OFDM system shown in Fig. 6.1, only one block of inputs is shown. In consecutive transmission, we denote the kth subchannel input of the nth block by $s_k(n)$. Suppose the QPSK symbols to be transmitted are $a_k(n) = \pm 1/\sqrt{2} \pm j/\sqrt{2}$, for $0 \leq k < M$ and $n \geq 0$. We set

$$s_k(-1) = \sqrt{\mathcal{E}_s/2}(1+j), \quad s_k(n) = s_k(n-1)a_k(n), \quad n \geq 0.$$

Assume the channel does not change over two consecutive blocks. In the absence of channel noise, show how the receiver can retrieve $a_k(n)$ without knowing the channel impulse response.

6.5 *OFDM with cyclic prefix and suffix.* In the transmitter of the OFDM system in Fig. 6.1, a cyclic prefix of length ν is added and the transmitted signal $x(n)$ has $N = M + \nu$ samples in each block. In particular, if we consider one block, the samples are

$$\underbrace{x_{M-\nu} \quad \cdots \quad x_{M-1}}_{\text{prefix}} \quad \underbrace{x_0 \quad x_1 \quad \cdots \quad x_{M-1}}_{\text{IDFT outputs}}$$

Now suppose, instead of cyclic prefix, the transmitter adds a combination of cyclic prefix of length ν_0 and cyclic suffix of length $\nu - \nu_0$ as in Problem 5.19. The resulting transmitted signal is,

$$\underbrace{x_{M-\nu_0} \quad \cdots \quad x_{M-1}}_{\text{prefix}} \quad \underbrace{x_0 \quad x_1 \quad \cdots \quad x_{M-1}}_{\text{IDFT outputs}} \quad \underbrace{x_0 \quad x_1 \quad \cdots \quad x_{\nu-\nu_0-1}}_{\text{suffix}}$$

Modify the receiver in Fig. 6.1 so that zero-forcing is achieved.

6.6 Consider the ZP-OFDM system in Fig. 6.7.

(a) Show that the receiver is zero-forcing if and only if the receiving matrix $\mathbf{S}$ is given by (6.12).

(b) Show that when $\mathbf{A} = \mathbf{0}$, the receiver becomes the pseudo-inverse solution in (6.13).

6.7 Show that the MMSE receiver of the ZP-OFDM system in (6.16) can be rewritten as the matrix in (6.17). (Hint. The identity $\mathbf{B}(\mathbf{B}^\dagger\mathbf{B} + \mathbf{I}_n) = (\mathbf{B}\mathbf{B}^\dagger + \mathbf{I}_m)\mathbf{B}$ holds for an arbitrary $m \times n$ matrix $\mathbf{B}$.)

6.8 Suppose the channel is a MIMO system with a constant $m \times n$ ($m > n$) transfer matrix $\mathbf{C}$. The transmitted vector $\mathbf{s}$ is a vector of size n. The output of the channel is $\mathbf{r} = \mathbf{Cs} + \mathbf{q}$. The channel noise $\mathbf{q}$ has autocorrelation matrix $N_0\mathbf{I}_m$. The receiver $\mathbf{S}$ is a zero-forcing receiver with $\mathbf{SC} = \mathbf{I}_n$. The output noise is $\mathbf{e} = \mathbf{Sr} - \mathbf{s}$.

(a) Find the zero-forcing receiver that leads to the smallest output noise $E[\mathbf{e}^\dagger\mathbf{e}]$. (Hint. Use the singular value decomposition of $\mathbf{C}$.)

(b) Show that the receiver in (a) also minimizes the noise in each subchannel.

6.9 When the ZP-OFDM system has the DFT-based efficient receiver as given in (6.15), show that the ith subchannel noise variance is $\sigma_{e_i}^2 = \frac{N}{M}\frac{N_0}{|C_i|^2}$.

6.10 For the SC-CP system, derive the MMSE receiver in (6.21) using the orthogonality principle.

6.11 In the error analysis of MMSE SC-CP system (Section 6.3.3), we derived the autocorrelation matrix of the output error and used it to obtain the subchannel error variances and subchannel unbiased SNRs. Alternatively we can obtain these values using diagonal elements t_{ii} of the overall transfer matrix $\mathbf{T}$ (6.22) using (3.20) and (3.18).

(a) Express t_{ii} in terms of the channel DFT coefficients C_ℓ.

(b) Find $\beta(i)$ and $\beta_{biased}(i)$.

(c) Find $\sigma^2_{e_i}$.

6.12 It is known that the magnitude response of an LTI filter $C(z)$ stays the same when we replace one zero α of $C(z)$ by its conjugate reciprocal $1/\alpha^*$. In this case only the phase response is altered. Show that for the cyclic-prefixed OFDM and SC-CP systems, the performance is affected only by the magnitude response, but not the phase response. This means that the system performance is the same if the channel is replaced by another that has the same magnitude response. Therefore it does not matter if the channel has zeros outside the unit circle. The performance of the system remains the same.

6.13 Consider the SC-ZP system. Suppose the channel noise is AWGN with variance $\mathcal{N}_0$ and the receiver is the one with an efficient implementation as in (6.25). Compute the subchannel noise variances $\sigma^2_{e_i}$ at the output of the receiver.

6.14 The SC-ZP transmitter pads ν zeros for every block of size M. Suppose we have a receiver that has a computation mechanism that can perform N-point DFT and IDFT, where $N = M + \nu$. For such a receiver can we use N-point DFT and IDFT instead of M-point DFT and IDFT to achieve zero-forcing? If so, find a block diagram of the receiving end that can recover the transmitted symbols in the absence of channel noise. Find the solutions of all zero-forcing receivers in this case.

6.15 *DFT-based MMSE SC-ZP system.* Consider the SC-ZP system in Fig. P6.15, where the receiver has a DFT-based structure. The matrix Υ is as defined in (6.14).

(a) Let $\hat{\mathbf{r}}$ be the output vector of the matrix Υ as indicated in the figure. We can write $\hat{\mathbf{r}}$ as $\mathbf{C}_{circ}\mathbf{s} + \boldsymbol{\tau}$, where $\boldsymbol{\tau}$ comes entirely from the channel noise. Determine the autocorrelation matrix $\mathbf{R}_\tau$.

(b) Let $\mathbf{y} = \mathbf{W}\mathbf{s}$. Show that

$$E[||\hat{\mathbf{y}} - \mathbf{y}||^2] = E[||\hat{\mathbf{s}} - \mathbf{s}||^2],$$

where $\hat{\mathbf{y}}$ is as indicated in the figure.

(c) Use the orthogonality principle to find the scalars $\lambda_0, \lambda_1, \ldots, \lambda_{M-1}$ such that $E[||\hat{\mathbf{s}} - \mathbf{s}||^2]$ (or equivalently $E[||\hat{\mathbf{y}} - \mathbf{y}||^2]$) is minimized.

The MMSE receiver for the SC-ZP system in Section 6.4 does not have an efficient implementation like the SC-CP system. In this problem the receiver is optimized to minimize MSE subject to an efficient DFT-based receiver. The readers can verify by simulations that the performance comes very close to the unconstrained MMSE receiver.

6.16 *Zero-padded OFDM systems.* Consider the ZP-OFDM system with the efficient zero-forcing receiver in (6.15).

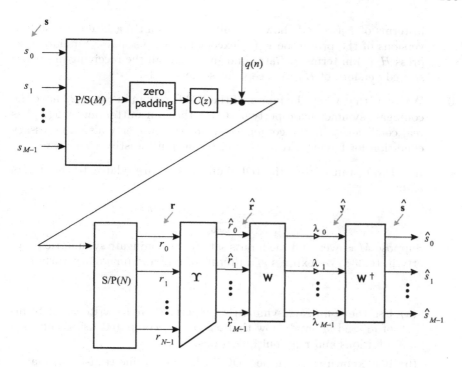

Figure P6.15. DFT-based MMSE SC-ZP system

(a) Derive the transmitting and receiving filters $F_k(z)$ and $H_k(z)$ in the corresponding filter bank representation. Show that $F_0(z)$ and $H_0(z)$ are both rectangular windows of length M and N, respectively. Express the other transmitting (receiving) filters as frequency-shifted versions of the first transmitting (receiving) filter except for a scalar.

(b) Using the filter bank representation, the kth noise path at the receiver is as shown in Fig. P6.16. Suppose the channel noise $q(n)$ is AWGN with variance $\mathcal{N}_0$. Use the receiving filter $H_k(z)$ in (a) to obtain the subchannel noise variance $\sigma_{e_k}^2$.

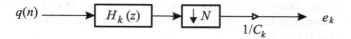

Figure P6.16. Noise path for the kth subchannel.

6.17 Derive the filter bank representation for the SC-CP system by finding the transmitting filters $F_k(z)$ and receiving filters $H_k(z)$. Express $F_k(z)$

in terms of $F_0(z)$ and show that all the transmitting filters are shifted versions of the prototype $F_0(z)$ except for some scalars. Similarly, express $H_k(z)$ in terms of $H_0(z)$ and show that all the receiving filters are shifted versions of $H_0(z)$ except for some scalars.

6.18 We know that when the inputs of the DMT system in Fig. 6.22 have the conjugate symmetric property in (6.33), the transmitter output $x(n)$ has real coefficients. Is the conjugate symmetric property also a necessary condition for having a real transmitter output? Justify your answer.

6.19 In a DMT transmitter, the IDFT outputs x_n are related to the inputs s_k by

$$x_n = \frac{1}{\sqrt{M}} \sum_{k=0}^{M-1} s_k e^{j2\pi kn/M}.$$

Suppose M is even and the inputs satisfy the conjugate symmetric property in (6.33). We express s_k in terms of real and imaginary parts by

$$s_k = a_k + jb_k.$$

(a) Use the conjugate symmetric property of s_k to write x_n in terms of a_k and b_k, and show that x_n can be computed using only real additions and real multiplications.

(b) It is known that a type I DCT (discrete cosine transform) matrix [187] $\mathbf{C}_I$ of dimensions $(K+1) \times (K+1)$ is given by

$$[\mathbf{C}_I]_{nm} = \alpha_n \alpha_m \sqrt{\frac{2}{K}} \cos\left(\frac{\pi}{K}nm\right), \quad 0 \le k, m \le K,$$

where

$$\alpha_n = \begin{cases} 1/\sqrt{2}, & n = 0, K, \\ 1, & \text{otherwise.} \end{cases}$$

A type I DST (discrete sine transform) matrix $\mathbf{S}_I$ of dimensions $(K-1) \times (K-1)$ is given by

$$[\mathbf{S}_I]_{nm} = \sqrt{\frac{2}{K}} \sin\left(\frac{\pi}{K}(m+1)(n+1)\right), \quad 0 \le m, n < K-1.$$

Use (a) to show that x_n can be computed using $\mathbf{C}_I$ and $\mathbf{S}_I$ matrices, which are known to have fast algorithms.

If we are to compute x_n using the IDFT formula directly, complex additions and complex multiplications are required. Exploiting the conjugate symmetric property of the inputs, we can use DCT and DST with only real computations.

6.20 For the DMT system, the power spectrum of the transmitted discrete time signal $x(n)$ is as given in (6.35). When $F_0(z)$ is the rectangular window and all the other transmitting filters are frequency-shifted versions of $F_0(z)$, we can approximate $S_x(e^{j2k\pi/M})$ by

$$S_x(e^{j2k\pi/M}) \approx \frac{\mathcal{E}_{s,k}}{N}|F_0(e^{j0})|^2,$$

where $k \in \mathcal{A}$. Use this expression to obtain an approximation for $S_{x_a}(jk\frac{2\pi}{MT})$. Suppose it is required that $S_{x_a}(jk\frac{2\pi}{MT}) \leq A_k$. Find the maximum allowable value $\mathcal{E}_{s,k}$.

6.21 In (6.39) we have used rounding to obtain integer bit assignments for a given set of subchannel SNR_k. The resulting subchannel error rates will be slightly smaller or higher than the target error rate.

   (a) Show that we can maintain the same error rate by adjusting the kth symbol power by a factor of

$$\Gamma(2^{\hat{b}_k} - 1)/SNR_k = (2^{\hat{b}_k} - 1)/(2^{b_k} - 1).$$

   (b) Show that the maximum amount of power compensation is around ± 1.5 dB.

6.22 In the DMT system shown in Fig. 6.22, a time domain equalizer (TEQ) is included to shorten the channel. This affects the subchannel signal to noise ratios. Assume the TEQ has shortened the channel so that IBI is negligible. Denote the M-point DFT coefficients of $T(z)$ by T_k, $k = 0, 1, \ldots, M - 1$.

   (a) Find the subchannel gains.

   (b) Find an expression for subchannel noise variances in terms of $T(e^{j\omega})$ and receiving filters $H_k(e^{j\omega})$.

   (c) Suppose the TEQ has one DFT coefficient equal to zero, say $T_{k_0} = 0$. Show that the k_0 subchannel noise variance is not zero in general.

This demonstrates that including a TEQ will change the subchannel SNRs.

6.23 *Zipper–avoidance of NEXT crosstalk* [139]. Consider a DMT system with transmitter output signal $x(n)$. The channel $c(n)$ has order L. Due to NEXT crosstalk, the received signal is

$$r(n) = x(n - \Delta) * c(n) + x'(n) * c_{next}(n),$$

where Δ is the propagation delay. The signal $x'(n)$ is the transmitted signal of another DMT system that comes into the received signal because of coupling, and $c_{next}(n)$ is the impulse response of the crosstalk transfer function (order less than L). Suppose the cyclic-prefix length ν satisfies $\nu \geq L + \Delta$. The receiver retains the following M samples:

$$\mathbf{r} = \begin{bmatrix} r(\nu) & r(\nu + 1) & \cdots & r(N - 1) \end{bmatrix}^T$$

and applies DFT on $\mathbf{r}$ to obtain $\mathbf{y} = \mathbf{W}\mathbf{r}$.

   (a) Show that $\mathbf{r}$ depends on samples of $x(n)$ that are from the same transmitter input block and on samples of $x'(n)$ that are also from the same transmitter input block. This means that we can consider one-shot transmission in this case.

(b) Let the input symbols of the two DMT transmitters be, respectively, s_k and s'_k. How is y_k related to s_k and s'_k, where y_k is the kth element of $\mathbf{y}$?

(c) When FDM is used, the downstream and upstream signals use different tones. The tones are divided into two sets, $\mathcal{A}$ and $\mathcal{A}'$. The tones in $\mathcal{A}$ are used by $x(n)$ and the tones in $\mathcal{A}'$ are used by $x'(n)$. How is y_k related to s_k and s'_k, for $k \in \mathcal{A}$ and for $k \in \mathcal{A}'$?

(d) Suppose a combination of cyclic prefix and suffix (Problem 5.19) is used at the transmitter, i.e. a suffix of length Δ and a prefix of length $(\nu - \Delta)$. How is y_k related to s_k and s'_k now? Such a DMT system is known as a *zipper*.

In (c) and (d) the NEXT crosstalk can be completely avoided by using extra redundant samples (extra Δ samples) and by carefully synchronizing the downstream and upstream signals.

6.24 *Oversampled OFDM system* [150]. In Section 5.5, we considered oversampling receivers. Figure P6.24(a) shows an OFDM system with an oversampling receiver. The transmitter sends out a sample every T seconds, whereas the C/D of the receiver takes samples every T/Q seconds, where Q (a positive integer) is the oversampling factor. At the receiver, the number of prefix samples discarded now is $Q\nu$ and the DFT size is MQ due to oversampling. From Section 5.5, we know the system from $x(n)$ to $r(n)$ can be replaced by a discrete-time equivalent channel model, as shown in Fig. P6.24(b). The discrete-time channel in this case is

$$c(n) = (p_1 * c_a * p_2)(t)|_{t=nT/Q},$$

where $p_1(t)$ is the transmitting pulse and $p_2(t)$ is the receiving pulse. Assume $c(n)$ has order $\leq Q\nu$. The equivalent block diagram of the oversampled OFDM system is shown in Fig. P6.24(c).

(a) Show that the transmitter in Fig. P6.24(c) has the alternative representation given in Fig. P6.24(d). The inputs a_k shown in Fig. P6.24(c) are related to s_k by

$$a_k = \frac{1}{\sqrt{Q}} s_{((k))_M}, \quad k = 0, 1, \ldots, MQ - 1,$$

where the $((\cdot))_M$ denotes modulo M operation. That is, the inputs of the new transmitter are the original inputs repeated Q times. These two transmitters produce the same output $x_e(n)$.

(b) Show that, in the absence of channel noise, the receiver outputs y_k are given by

$$y_k = C_k a_k = \frac{1}{\sqrt{Q}} C_k s_{((k))_M}, \quad k = 0, 1, \ldots, MQ - 1,$$

where C_k are the MQ-point DFT of $c(n)$. Therefore the outputs $y_k, y_{k+M}, \ldots, y_{k+(Q-1)M}$ are all from the same transmitter input s_k, but scaled by different DFT coefficients of the channel.

(c) Suppose the channel noise $q(n)$ is AWGN with variance $\mathcal{N}_0$. Given $y_k, y_{k+M}, \ldots, y_{k+(Q-1)M}$, find the optimal estimate of s_k, for $0 \le k < M$. What is the kth subchannel SNR using the optimal estimate of s_k? Compare with the case without oversampling, i.e. $Q = 1$, and determine the gain of oversampling in terms of subchannel SNRs.

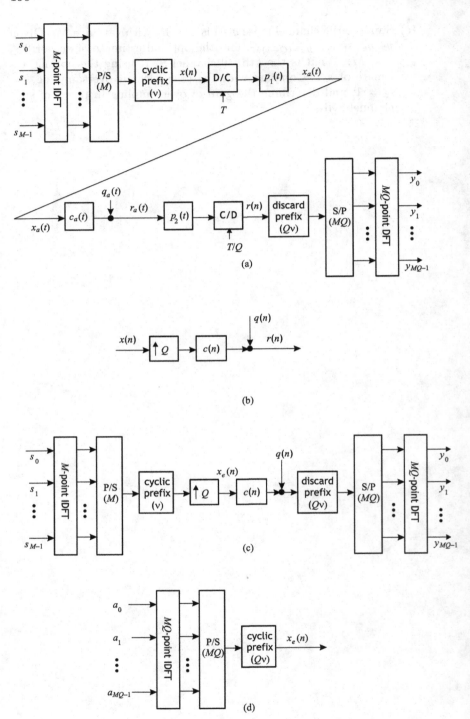

Figure P6.24. (a) Oversampled OFDM system; (b) discrete-time equivalent channel; (c) equivalent block diagram of (a); (d) alternative representation of the transmitter in (c).

6.25 The OFDM system can also be used in data transmission for more than one user. In this case, it is known as OFDM-FDMA (frequency division multiple access) or OFDMA. Each user is allocated a subset of the M subchannels. As an example, suppose there are K users and each user is assigned $P = M/K$ subchannels for transmission, assuming K divides M. For the kth user, subchannels $kP, kP+1, \ldots, kP+P-1$ are used and the channel from the kth user to the base station is $C_k(z)$, with order less than the length of cyclic prefix, as shown in Fig. P6.25. What the receiver has is the sum of signals from all the users plus channel noise.

(a) Determine the outputs of the receiver for the first P subchannels.

(b) How can we recover the data of the kth user in the absence of channel noise?

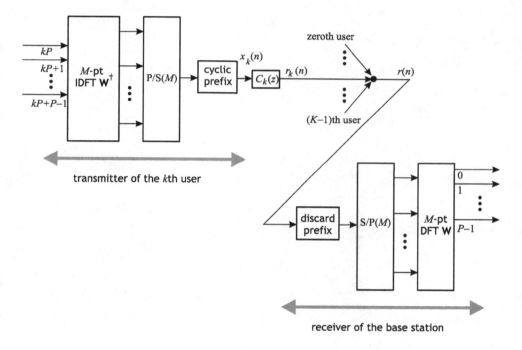

Figure P6.25. OFDMA system.

7

Precoded OFDM systems

In practical communication systems, the transmitter usually sends out training symbols to the receiver, based on which the channel can be estimated at the receiver. It is therefore reasonable to assume that the channel is known to the receiver. When the channel state information is also known to the transmitter, we can optimize the transmitter to better the system performance. Having this knowledge available to the transmitter requires the receiver to send back the information, which takes time. For wireless transmission, the channel varies rapidly. By the time the transmitter receives the channel profile, the channel may have changed. Therefore, for wireless applications it is often desirable to have a transmitter that is channel-independent. In such a channel-independent transmitter, there is no bit/power allocation. Having a channel-independent transmitter is also of vital importance for broadcasting applications, where there are many receivers with different transmission paths. The OFDM system has the much desired feature that the transmitter is channel-independent and furthermore the channel-dependent part of the transceiver is only a set of M scalars at the receiver. Moreover, the main processing at the transmitter (receiver) is M-point IDFT (DFT), which can be implemented efficiently using fast algorithms, and the complexity is in the order of $M \log_2 M$ instead of M^2.

The discussion in Chapter 6 suggests that the OFDM system can be severely affected by channel spectral nulls. For high SNR the error rate is usually limited by those subchannels that have low SNRs. One method that prevents the performance being dominated by a few bad subchannels is to have a precoder. Figure 7.1 shows an OFDM transmission system with a precoder at the transmitter and a post-coder at the receiver. We have seen earlier that when the precoder is the DFT matrix, the precoder DFT matrix will cancel out the IDFT at the transmitter and the precoded system becomes the SC-CP system shown in Fig. 6.10. The SC-CP system illustrates that having a precoder can alter the BER behavior of the system, although the total output mean squared error is unchanged in this case. In a conventional single-band transmission system, BER is directly tied to mean squared error. For multi-subchannel systems like OFDM, SC-CP, or the more general precoded OFDM systems, this is no longer true. Transceivers with the same mean squared error can have very different performances of average BER (over all M subchannels).

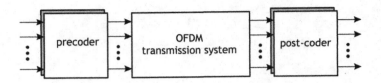

Figure 7.1. Precoded OFDM system.

In this chapter, we will consider optimal unitary precoders in OFDM systems for minimum BER. The objective will be BER, rather than mean squared error. To maintain the channel independence property at the transmitter, there is no bit/power allocation. Transceiver designs with bit and power allocation are considered in Chapter 8. The presentation of this chapter is mostly based on [78]. See Section 7.6 for other precoded systems and a further reading.

7.1 Zero-forcing precoded OFDM systems

Let us choose the precoder to be an $M \times M$ unitary matrix $\mathbf{P}$ with $\mathbf{P}^\dagger \mathbf{P} = \mathbf{I}_M$ and correspondingly the post-coder at the receiver $\mathbf{P}^\dagger$. The resulting precoded OFDM system becomes the one shown in Fig. 7.2. For the OFDM transmission system alone, we know from the previous chapter that the overall transfer matrix is the identity matrix $\mathbf{I}_M$. As the precoder and post-coder satisfy $\mathbf{P}^\dagger \mathbf{P} = \mathbf{I}_M$, the precoded system remains zero-forcing. The special case $\mathbf{P} = \mathbf{I}$ corresponds to the conventional OFDM system, while $\mathbf{P} = \mathbf{W}$ corresponds to the SC-CP system. Unless it is mentioned otherwise, an OFDM system in this chapter refers to the cyclic-prefixing case.

The new transmitting and receiving matrices $\mathbf{G}$ and $\mathbf{S}$ (shown in Fig. 7.2) are, respectively,

$$\mathbf{G} = \mathbf{W}^\dagger \mathbf{P}, \quad \mathbf{S} = \mathbf{P}^\dagger \mathbf{\Gamma}^{-1} \mathbf{W},$$

where $\mathbf{\Gamma} = \mathrm{diag} \begin{bmatrix} C_0 & C_1 & \cdots & C_{M-1} \end{bmatrix}$ and C_k are the DFT coefficients of the channel impulse response. The resulting transmitting matrix is a general unitary matrix. As in the previous chapter, we assume that the transmitter input symbols s_k are uncorrelated and of the same variance $\mathcal{E}_s$. Due to the unitary property of $\mathbf{G}$, the outputs of the transmitting matrix also have variance $\mathcal{E}_s$. So the transmission power $\mathcal{E}_s$ is the same for all unitary precoders. As $\mathbf{G}$ is an arbitrary unitary matrix, the problem of finding the optimal precoder is equivalent to one of designing an optimal block transceiver with a unitary transmitting matrix. Let us first analyze the effect of the precoder on the subchannel SNRs. Optimal precoders that minimize the error rate will be derived in subsequent sections.

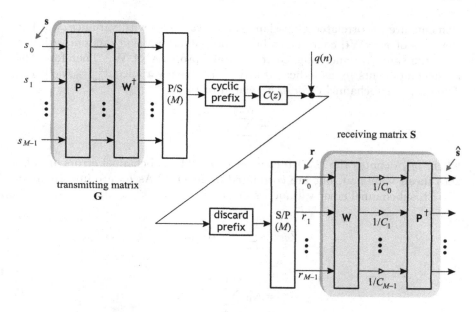

Figure 7.2. Zero-forcing precoded OFDM system with a precoder $\mathbf{P}$ and post-coder $\mathbf{P}^{\dagger}$.

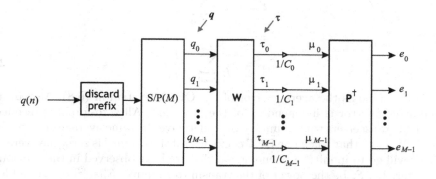

Figure 7.3. Illustration of the noise path of a zero-forcing receiver.

Noise analysis

Assume the channel noise $q(n)$ is complex circularly symmetric AWGN with zero-mean and variance $\mathcal{N}_0$. Denote the receiver output vector by $\widehat{\mathbf{s}}$ as indicated in Fig. 7.2. Then the output error vector $\mathbf{e} = \mathbf{s} - \widehat{\mathbf{s}}$ comes entirely from the channel noise as the receiver is zero-forcing. The error vector $\mathbf{e}$ can be analyzed by considering the receiver block diagram in Fig. 7.3, where only the channel noise is shown. The vector $\mathbf{q}$ is the blocked channel noise. Its elements are uncorrelated Gaussian random variables with variance $\mathcal{N}_0$. The elements of $\boldsymbol{\tau} = \mathbf{Wq}$ continue to be uncorrelated Gaussian random variables with the same variance, due to the unitary property of $\mathbf{W}$. Therefore, the noise components μ_k as indicated in Fig. 7.3 have variance $\sigma_{\mu_k}^2 = \mathcal{N}_0/|C_k|^2$. The output subchannel error e_i is related to μ_k by

$$e_i = \sum_{k=0}^{M-1} p_{k,i}^* \mu_k,$$

where $p_{k,i}$ denotes the (k, i)th element of $\mathbf{P}$. The subchannel errors are also zero mean Gaussian, but are correlated in general. As μ_k are uncorrelated, the ith subchannel error variance is given by

$$\sigma_{e_i}^2 = \sum_{k=0}^{M-1} |p_{k,i}|^2 \sigma_{\mu_k}^2.$$

That is,

$$\sigma_{e_i}^2 = \mathcal{N}_0 \sum_{k=0}^{M-1} \frac{|p_{k,i}|^2}{|C_k|^2}. \tag{7.1}$$

The precoder $\mathbf{P}$ is unitary and thus each of its rows has unit energy. That is, $\sum_{i=0}^{M-1} |p_{k,i}|^2 = 1$, for $k = 0, 1, \ldots, M-1$. Using this fact, we can write the average mean squared error

$$\mathcal{E}_{rr} = \frac{1}{M} \sum_{i=0}^{M-1} \sigma_{e_i}^2$$

as

$$\mathcal{E}_{rr} = \frac{\mathcal{N}_0}{M} \sum_{k=0}^{M-1} \frac{1}{|C_k|^2}. \tag{7.2}$$

It is the same as the average error of the OFDM system in (6.8). Moreover, the average error is independent of the precoder. All zero-forcing precoded OFDM transceivers with a unitary precoder have the same average error given in (7.2). Note that when one DFT coefficient of the channel is zero, the average error will go to infinity, a phenomenon that we have observed in the previous chapter. Let $\mathcal{E}_s$ be the power of the transmitter inputs ($E[|s_k|^2] = \mathcal{E}_s$) and let $\beta(i) = \mathcal{E}_s/\sigma_{e_i}^2$ be the SNR of the ith subchannel, then we have

$$\beta(i) = \frac{\gamma}{\sum_{k=0}^{M-1} \frac{|p_{k,i}|^2}{|C_k|^2}}, \quad \text{where} \quad \gamma = \mathcal{E}_s/\mathcal{N}_0. \tag{7.3}$$

The subchannel SNRs depend on the precoder $\mathbf{P}$. Although the mean squared error is the same regardless of $\mathbf{P}$, the precoder affects how the same amount of noise is distributed among the subchannels. For convenience, we repeat the results of the OFDM and SC-CP systems from Chapter 6 in the following.

- **The OFDM system** When $\mathbf{P} = \mathbf{I}_M$, the precoded system reduces to the conventional OFDM system. The ith subchannel has error variance

$$\sigma_{e_i, ofdm}^2 = \mathcal{N}_0/|C_i|^2. \tag{7.4}$$

The corresponding subchannel SNR is

$$\beta_{ofdm}(i) = \gamma|C_i|^2. \tag{7.5}$$

For the ith subchannel, both the error variance and SNR depend only on the ith subchannel gain C_i.

- **The SC-CP system** When the unitary precoder $\mathbf{P}$ is the normalized DFT matrix $\mathbf{W}$, the transmitting matrix becomes $\mathbf{G} = \mathbf{I}_M$. The postcoder $\mathbf{P}^\dagger$ is the IDFT matrix $\mathbf{W}^\dagger$. The resulting precoded system becomes the SC-CP system in Fig. 6.10. All the subchannels have the same error variance and the same SNR,

$$\sigma_{sc\text{-}cp}^2 = \mathcal{E}_{rr}, \quad \beta_{sc\text{-}cp} = \mathcal{E}_s/\mathcal{E}_{rr}.$$

Bounds on subchannel errors The BER performance is determined by the subchannel SNRs. The unitary property of $\mathbf{P}$ allows us to establish upper and lower bounds on the subchannel error variances. First we observe one interesting property of the error variances that ties the precoded system nicely to the OFDM system. With an inspection of the error variances given in (7.1), we can see that they are related to those of the OFDM system in (7.4) by

$$\sigma_{e_i}^2 = \sum_{k=0}^{M-1} |p_{k,i}|^2 \sigma_{e_k,ofdm}^2.$$

Each is a linear combination of subchannel error variances of the OFDM system. And the linear combination coefficients are $|p_{k,i}|^2$, which add up to unity all for i. The linear combination property means

$$\min_k \sigma_{e_k,ofdm}^2 \le \sigma_{e_i}^2 \le \max_k \sigma_{e_k,ofdm}^2.$$

The subchannel error variances of the precoded system are always bounded between those of the best and worst subchannels of the OFDM system. The upper and lower bounds hold for any unitary precoder $\mathbf{P}$. For different choice of $\mathbf{P}$, the noise variances are distributed differently.

Best and worst case SNRs The bounds on $\sigma_{e_i}^2$ lead to the following bounds on the subchannel SNRs $\beta(i)$:

$$\min_k \beta_{ofdm}(k) \le \beta(i) \le \max_k \beta_{ofdm}(k), \tag{7.6}$$

where $\beta_{ofdm}(k) = \gamma|C_k|^2$ is the kth subchannel SNR of the OFDM system. For any unitary precoder, the SNR of the best subchannel is no better than the best subchannel of the OFDM system and the SNR of the worst subchannel is no worse than the worst subchannel of the OFDM system.

The inverse of the subchannel SNR, $1/\beta(i)$, is the noise to signal ratio (**NSR**) of the ith subchannel. The subchannel NSRs have the same linear combination property that holds for the subchannel error variances, i.e.

$$\frac{1}{\beta(i)} = \sum_{k=0}^{M-1} \frac{|p_{k,i}|^2}{\gamma|C_k|^2} = \sum_{k=0}^{M-1} |p_{k,i}|^2 \frac{1}{\beta_{ofdm}(k)}.$$

The subchannel NSR $1/\beta(i)$ is a linear combination of those of the OFDM system. The NSR is a quantity that rarely arises in the discussion of communication systems, but we will find it more convenient to work with NSR instead of SNR in some discussions of this chapter.

7.2 Optimal precoders for QPSK modulation

In this section, we optimize the precoder $\mathbf{P}$ to minimize BER for the zero-forcing precoded OFDM system given in Section 7.1. The computation of error rate depends on the modulation scheme used. We will consider QPSK modulation in this section. The results will be extended to QAM of larger constellations and also to other modulation schemes in Section 7.3. For QPSK modulation with symbol power $\mathcal{E}_s$, $s_k = \pm\sqrt{\mathcal{E}_s/2} \pm j\sqrt{\mathcal{E}_s/2}$. The BER was derived in Section 2.3. The BER of the ith subchannel is given by $\mathcal{P}(i) = Q(\sqrt{\beta(i)})$, where the function $Q(\cdot)$ is as defined in (2.11). The average BER $\mathcal{P} = \frac{1}{M}\sum_{i=0}^{M-1}\mathcal{P}(i)$ is

$$\mathcal{P} = \frac{1}{M}\sum_{i=0}^{M-1} Q\left(\sqrt{\beta(i)}\right). \tag{7.7}$$

The choice of precoder determines the subchannel SNRs and hence the error rate. We would like to find the optimal $\mathbf{P}$ such that the average BER is minimized. For convenience, we introduce the function

$$f(y) \triangleq Q(1/\sqrt{y}), \quad y > 0. \tag{7.8}$$

We have the following alternative expression for the subchannel BERs:

$$\mathcal{P}(i) = Q\left(\sqrt{\beta(i)}\right) = f\left(\frac{1}{\beta(i)}\right) = f\left(\sum_{k=0}^{M-1} \frac{|p_{k,i}|^2}{\gamma|C_k|^2}\right). \tag{7.9}$$

The argument of the function $f(\cdot)$ in the above equation is $1/\beta(i)$, i.e. the ith subchannel NSR. Now we rewrite the subchannel error rates of the OFDM and SC-CP systems, respectively, as

$$\mathcal{P}_{ofdm}(i) = f\left(\frac{1}{\gamma|C_i|^2}\right), \quad \mathcal{P}_{sc\text{-}cp} = f\left(\frac{1}{M}\sum_{k=0}^{M-1}\frac{1}{\gamma|C_k|^2}\right).$$

The bit error rate performance is directly related to the behavior of the function $f(y)$. The following lemma gives some important properties of $f(y)$.

Lemma 7.1 The function $f(y) = Q(1/\sqrt{y})$ is monotone increasing. It is convex[1] when $y \le 1/3$ and concave when $y > 1/3$. ∎

Proof. Let $u(y) \triangleq 1/\sqrt{y}$ for $y > 0$, then $f(y) = Q(u(y))$. The lemma can be proved by computing the first and second derivative of $f(y)$.

---

[1] See Appendix A for definitions of convex and concave functions.

The function $u(y) = 1/\sqrt{y}$ for $y > 0$ is convex with the first derivative $u'(y) = -\frac{1}{2}y^{-3/2}$ and the second derivative $u''(y) = \frac{3}{4}y^{-5/2}$. The function $Q(x)$ for $x \geq 0$ is also convex with $Q'(x) = -\frac{1}{\sqrt{2\pi}}e^{-x^2/2}$ and

$$Q''(x) = \frac{x}{\sqrt{2\pi}}e^{-x^2/2}.$$

We have $f'(y) = Q'(u(y))u'(y)$ equal to $f'(y) = \frac{1}{2\sqrt{2\pi}}e^{-\frac{1}{2y}}y^{-3/2}$, which is positive for $y > 0$. This means $f(y)$ is monotone increasing. We can verify that $f''(y) = Q''(u(y))[u'(y)]^2 + Q'(u(y))u''(y)$ can be expressed as

$$f''(y) = \frac{1}{4\sqrt{2\pi}}e^{-\frac{1}{2y}}y^{-7/2}(1 - 3y).$$

Therefore, $f''(y) \geq 0$ for $y \leq 1/3$ and $f''(y) < 0$ for $y > 1/3$, which proves the lemma. ∎

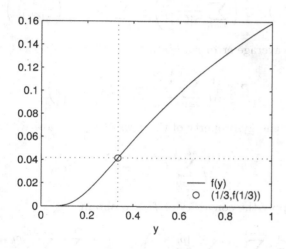

Figure 7.4. Plot of $f(y) = Q(1/\sqrt{y})$.

A plot of $f(y)$ is shown in Fig. 7.4. Each subchannel is operating in the convex or the concave region of the function $f(\cdot)$ depending on the subchannel SNR $\beta(i)$. In particular the case $\beta(i) > 3$ (NSR $1/\beta(i) < 1/3$) corresponds to the convex region of $f(\cdot)$. The case $\beta(i) < 3$ corresponds to the concave region.

Convex and concave regions of $f(\cdot)$ For a sufficiently large γ, all the subchannels in the OFDM system will be operating in the convex region of $f(\cdot)$. That is, $\beta_{ofdm}(i) = \gamma|C_i|^2 \geq 3$ for all i. This requires

$$\gamma \geq \max_i \frac{3}{|C_i|^2} \overset{\triangle}{=} \gamma_1. \tag{7.10}$$

When this happens all the subchannels in the precoded system will also be operating in the convex region due the SNR bounds in (7.6). In a similar manner, for a sufficiently small γ, all the subchannels in the OFDM system will be operating in the concave region of $f(\cdot)$. This is true if $\beta_{ofdm}(i) = \gamma|C_i|^2 < 3$ for all i, which requires

$$\gamma \leq \min_i \frac{3}{|C_i|^2} \stackrel{\triangle}{=} \gamma_0. \tag{7.11}$$

For convenience, we define three SNR regions:

$$\mathcal{R}_{low} = \{\gamma|\gamma \leq \gamma_0\}, \quad \mathcal{R}_{mid} = \{\gamma|\gamma_0 < \gamma < \gamma_1\}, \quad \mathcal{R}_{high} = \{\gamma|\gamma \geq \gamma_1\}.$$

We are now ready to establish a more precise connection among the BER performances of the three systems, OFDM, SC-CP, and a precoded OFDM system for the three SNR regions. Let us first consider the case $\mathcal{R}_{high}$. For this SNR range we can use the convexity of $f(\cdot)$ (Appendix A) to obtain an upper bound for the subchannel BERs given in (7.9):

$$\mathcal{P}(i) = f\left(\sum_{k=0}^{M-1} \frac{|p_{k,i}|^2}{\gamma|C_k|^2}\right) \leq \sum_{k=0}^{M-1} |p_{k,i}|^2 f\left(\frac{1}{\gamma|C_k|^2}\right).$$

Therefore the average error rate satisfies

$$\mathcal{P} \leq \frac{1}{M} \sum_{i=0}^{M-1} \sum_{k=0}^{M-1} |p_{k,i}|^2 f\left(\frac{1}{\gamma|C_k|^2}\right).$$

Using the unit energy property of the rows of $\mathbf{P}$, we have

$$\mathcal{P} \leq \frac{1}{M} \sum_{k=0}^{M-1} f\left(\frac{1}{\gamma|C_k|^2}\right) = \mathcal{P}_{ofdm}. \tag{7.12}$$

We can also apply the convexity of $f(\cdot)$ to obtain

$$\mathcal{P} = \frac{1}{M} \sum_{i=0}^{M-1} f\left(\underbrace{\sum_{k=0}^{M-1} \frac{|p_{k,i}|^2}{\gamma|C_k|^2}}_{1/\beta(i)}\right) \geq f\left(\frac{1}{M} \sum_{i=0}^{M-1} \sum_{k=0}^{M-1} \frac{|p_{k,i}|^2}{\gamma|C_k|^2}\right).$$

The inequality becomes an equality if and only if the quantity $\sum_{k=0}^{M-1} \frac{|p_{k,i}|^2}{\gamma|C_k|^2}$, which is the subchannel NSR $1/\beta(i)$, is the same for all i; or equivalently all the subchannel noise variances are the same. Again, using the unit energy property of the rows of $\mathbf{P}$, we have

$$\mathcal{P} \geq f\left(\frac{1}{M} \sum_{k=0}^{M-1} \frac{1}{\gamma|C_k|^2}\right) = \mathcal{P}_{sc-cp}. \tag{7.13}$$

Combining the two inequalities in (7.12) and (7.13), we obtain $\mathcal{P}_{ofdm} \geq \mathcal{P} \geq \mathcal{P}_{sc-cp}$ for $\gamma \in \mathcal{R}_{high}$. Similarly, when $\gamma \in \mathcal{R}_{low}$, we can use the concavity of $f(\cdot)$ to show that $\mathcal{P}_{ofdm} \leq \mathcal{P} \leq \mathcal{P}_{sc-cp}$. Summarizing, we have the following theorem.

Theorem 7.1 Zero-forcing case. The average BER $\mathcal{P}$ of the precoded OFDM system in Fig. 7.2 with QPSK modulation satisfies

$$\mathcal{P}_{ofdm} \leq \mathcal{P} \leq \mathcal{P}_{sc\text{-}cp}, \quad \text{for } \gamma \in \mathcal{R}_{low},$$

$$\mathcal{P}_{ofdm} \geq \mathcal{P} \geq \mathcal{P}_{sc\text{-}cp}, \quad \text{for } \gamma \in \mathcal{R}_{high}.$$

Each of the two inequalities relating $\mathcal{P}$ and $\mathcal{P}_{sc\text{-}cp}$ becomes an equality if and only if the subchannel error variances of the precoded system are the same, i.e. $\sigma_{e_i}^2 = \mathcal{E}_{rr}$, where $\mathcal{E}_{rr}$ is as given in (7.2). ∎

For the low SNR region $\mathcal{R}_{low}$, $\mathbf{P} = \mathbf{I}_M$ is the optimal solution; the OFDM system has the smallest error rate. But for $\gamma \in \mathcal{R}_{high}$, it is the worst solution and its error rate is the largest. However, as we will see next, the SNR region $\mathcal{R}_{low}$ corresponds to a high error rate, while $\mathcal{R}_{high}$ corresponds to a more useful range of BER.

BER behavior in $\mathcal{R}_{low}$, $\mathcal{R}_{mid}$, and $\mathcal{R}_{high}$

- $\gamma \in \mathcal{R}_{low}$: For the OFDM system, all the subchannels have $\beta_{ofdm}(k) \leq 3$ and hence

$$\mathcal{P}_{ofdm} \geq Q(\sqrt{3}) = 0.0416, \quad \text{for } \gamma \in \mathcal{R}_{low}.$$

  In this range of SNR the error rate $\mathcal{P}_{ofdm}$ is at least 0.0416, a relatively large BER. The error rate of the precoded system is lower bounded by that of the OFDM system. So the BERs of precoded systems, including the SC-CP system, will be even higher.

- $\gamma \in \mathcal{R}_{high}$: For this range, all the subchannels in the precoded system have $\beta(i) \geq 3$ and subchannel error rate less than $Q(\sqrt{3})$; the average error rate $\mathcal{P} \leq 0.0416$. Note that when $\gamma = \gamma_1$, the worst subchannel of the OFDM system has an error rate $Q(\sqrt{3}) = 0.0416$, and the average BER will be at least $Q(\sqrt{3})/M$. In other words, γ_1 is the minimum SNR for the OFDM system to achieve an error rate lower than $Q(\sqrt{3})/M$. As an example, when $\gamma = \gamma_1$, the error rate is at least $Q(\sqrt{3})/16 = 0.0026$ for $M = 16$ and at least $Q(\sqrt{3})/64 = 6.4 \times 10^{-4}$ for $M = 64$. Therefore the SNR region $\mathcal{R}_{high}$ corresponds to a more useful range of BER.

- $\gamma \in \mathcal{R}_{mid}$: If we plot $\mathcal{P}_{sc\text{-}cp}$ and $\mathcal{P}_{ofdm}$ as functions of γ, the curves of $\mathcal{P}_{ofdm}$ and $\mathcal{P}_{sc\text{-}cp}$ will cross each other in this range. This is because $\mathcal{P}_{ofdm}$ is smaller than $\mathcal{P}_{sc\text{-}cp}$ for $\gamma \leq \gamma_0$ and larger than $\mathcal{P}_{sc\text{-}cp}$ for $\gamma \geq \gamma_1$.

Error-equalizing precoders

The above theorem states that the error rate of a precoded OFDM is the same as that of the SC-CP system if and only if the subchannel noise variances $\sigma_{e_i}^2$ are equalized to the average value $\mathcal{E}_{rr}$. That is,

$$\sigma_{e_i}^2 = N_0 \sum_{k=0}^{M-1} \frac{|p_{k,i}|^2}{|C_k|^2} = \mathcal{E}_{rr} = \frac{N_0}{M} \sum_{k=0}^{M-1} \frac{1}{|C_k|^2}.$$

The above error-equalizing property can be achieved using a unitary precoder that satisfies

$$|p_{m,n}| = \frac{1}{\sqrt{M}}, \quad 0 \le m, n \le M - 1. \tag{7.14}$$

In this case all the subchannel BERs are the same, $\mathcal{P}(i) = \mathcal{P} = \mathcal{P}_{sc\text{-}cp}$. Two well-known unitary matrices satisfying the equal magnitude property in (7.14) are the DFT matrix and the Hadamard matrix [2]. When $\mathbf{P} = \mathbf{W}$, the transmitting matrix $\mathbf{G} = \mathbf{I}_M$, and the transceiver in Fig. 7.2 becomes the SC-CP system. The Hadamard matrices can be generated recursively when M is a power of 2. The 2×2 Hadamard matrix is given by

$$\mathbf{H}_2 = \frac{1}{\sqrt{2}} \begin{bmatrix} 1 & 1 \\ 1 & -1 \end{bmatrix}.$$

The $2n \times 2n$ Hadamard matrix can be given in terms of the $n \times n$ Hadamard matrix by

$$\mathbf{H}_{2n} = \frac{1}{\sqrt{2}} \begin{bmatrix} \mathbf{H}_n & \mathbf{H}_n \\ \mathbf{H}_n & -\mathbf{H}_n \end{bmatrix}.$$

Except for a scalar, the Hadamard matrix is real with elements equal to ± 1. The resulting transmitting matrix $\mathbf{G} = \mathbf{W}^\dagger \mathbf{H}$ will be complex. The implementation of Hadamard matrices requires only additions. The overall complexity of the transceiver will be more than that of the OFDM system (or the SC-CP system) due to the two extra Hadamard matrices, one at the transmitter and one at the receiver. With the Hadamard precoder, the complexity of the transmitter is higher compared to the OFDM or SC-CP systems, but the complexity of the receiver is between those of the OFDM and SC-CP systems. Having a Hadamard precoder also has the advantage of having a PAPR smaller than that of the OFDM system [112], though larger than that of the SC-CP system. When we have a unitary precoder that has the equal magnitude property in (7.14), we can use it to generate other unitary matrices satisfying the equal magnitude property. An example is given in Problem 7.8. Both the DFT and Hadamard matrices are **channel-independent**, so the resulting transmitter is also channel-independent.

7.3 Optimal precoders: other modulations

The computation of error rate in Section 7.2 is carried out for QPSK modulation. We can extend the results to PAM, QAM, and PSK (phase shift keying) with slight modifications. Let us take QAM modulation as an example. Suppose the inputs are $2b$-bit QAM symbols with variance $\mathcal{E}_s$. The ith subchannel BER $\mathcal{P}(i)$ can be approximated by (Section 2.3),

$$\mathcal{P}(i) \approx aQ(\sqrt{\zeta\beta(i)}) = af\left(\frac{1}{\zeta\beta(i)}\right), \tag{7.15}$$

$$\text{where} \quad a = \frac{2}{b}\left(1 - \frac{1}{2^b}\right) \text{ and } \zeta = 3/(2^{2b} - 1). \tag{7.16}$$

As every subchannel carries the same number of bits, the average error rate can also be obtained by averaging the subchannel error rates,

$$P = \frac{1}{M} \sum_{i=0}^{M-1} P(i) = \frac{a}{M} \sum_{i=0}^{M-1} f\left(\frac{1}{\zeta\beta(i)}\right). \tag{7.17}$$

The subchannel SNRs are independent of the modulation scheme used and they observe the same upper and lower bounds given in (7.6), $\min_k \beta_{ofdm}(k) \leq \beta(i) \leq \max_k \beta_{ofdm}(k)$. When γ is large enough such that all the subchannels are operating in the convex region of $f(\cdot)$, equalizing the subchannel noise variances will minimize the approximated error rate given in (7.17). The same error-equalizing precoders given in Section 7.2 will be optimal in this case too. One sufficient condition for this is $\zeta\beta_{ofdm}(i) \geq 3$, which is equivalent to $\gamma\zeta|C_i|^2 \geq 3$, for all i. This means

$$\gamma \geq \gamma_1, \quad \text{where} \quad \gamma_1 = \max_k \frac{(2^{2b} - 1)}{|C_k|^2}. \tag{7.18}$$

On the other hand, when $\zeta\beta_{ofdm}(i) \leq 3$ for all i, the conventional OFDM system is the optimal transceiver. The condition for this is

$$\gamma \leq \gamma_0, \quad \text{where} \quad \gamma_0 = \min_k \frac{(2^{2b} - 1)}{|C_k|^2}.$$

The conditions now depend on the QAM constellation. For a larger constellation, i.e. larger b, both γ_0 and γ_1 also become larger. As the constellation size increases, the SC-CP system will become optimal at a higher SNR.

The above derivation is valid for any modulation scheme in which the subchannel error probability can be either approximated or expressed as

$$aQ\left(\sqrt{\zeta\beta(i)}\right) = af\left(\frac{1}{\zeta\beta(i)}\right),$$

for some constants a and ζ that are independent of subchannels. Examples of such a case include PAM, QAM, and PSK modulation schemes. Once the error probability is in such a form, we can invoke the convexity and concavity of $f(y)$ to obtain the SNR ranges for which the OFDM system or the SC-CP system is optimal. Similar to the QPSK case, we can conclude that the SC-CP system is optimal for the high SNR case while the OFDM system is optimal for the low SNR case.

7.4 MMSE precoded OFDM systems

In this section we consider the case where the receiver has minimum mean squared error. Similar to the SC-CP system, the precoded system benefits considerably from MMSE reception. Also similar to the SC-CP system, the precoded system with an MMSE receiver is more robust to channel spectral nulls while having the same implementation cost as the zero-forcing receiver.

7.4.1   MMSE receivers

Let us start with a receiving matrix $\mathbf{S}$ that can be any $M \times M$ matrix (Figure 7.5). Let $\widehat{\mathbf{s}}$ be the receiver output and let the error vector $\mathbf{e} = \widehat{\mathbf{s}} - \mathbf{s}$. For a given received vector $\mathbf{r}$, we would like to find the optimal receiving matrix $\mathbf{S}$ that minimizes the mean squared error $E[\mathbf{e}^\dagger \mathbf{e}]$.

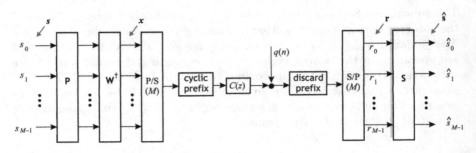

Figure 7.5. Precoded OFDM system with an MMSE receiver.

By the orthogonality principle, the optimal $\mathbf{S}$ is such that the error is orthogonal to the observation $\mathbf{r}$, $E\left[(\mathbf{Sr} - \mathbf{s})\mathbf{r}^\dagger\right] = 0$, or equivalently

$$\mathbf{S} = E\left[\mathbf{s}\mathbf{r}^\dagger\right]\left(E\left[\mathbf{r}\mathbf{r}^\dagger\right]\right)^{-1}.$$

We know that the transfer matrix from $\mathbf{x}$ to $\mathbf{r}$ is $\mathbf{C}_{circ}$, where $\mathbf{C}_{circ}$ is the circulant matrix defined in Definition 5.2. So

$$\mathbf{r} = \mathbf{C}_{circ}\mathbf{W}^\dagger\mathbf{Ps} + \mathbf{q} = \mathbf{W}^\dagger\mathbf{\Gamma Ps} + \mathbf{q},$$

where we have used the DFT decomposition $\mathbf{C}_{circ} = \mathbf{W}^\dagger\mathbf{\Gamma W}$ (Theorem 5.2). Assume the transmitter inputs are uncorrelated, i.e. $E[\mathbf{ss}^\dagger] = \mathcal{E}_s\mathbf{I}$, and uncorrelated with the noise, i.e. $E[\mathbf{sq}^\dagger] = \mathbf{0}$. Using the above expression for $\mathbf{r}$, we have

$$E\left[\mathbf{s}\mathbf{r}^\dagger\right] = \mathcal{E}_s\mathbf{P}^\dagger\mathbf{\Gamma}^\dagger\mathbf{W} \quad \text{and} \quad E\left[\mathbf{r}\mathbf{r}^\dagger\right] = \mathcal{E}_s\mathbf{W}^\dagger\mathbf{\Gamma}\mathbf{\Gamma}^\dagger\mathbf{W} + \mathcal{N}_0\mathbf{I}_M.$$

By direct substitution, we can obtain the optimal receiving matrix $\mathbf{S}$ that minimizes $E[\mathbf{e}^\dagger \mathbf{e}]$ as

$$\mathbf{S} = \mathbf{P}^\dagger\mathbf{\Lambda W}, \tag{7.19}$$

where $\mathbf{\Lambda}$ is a diagonal matrix with the kth diagonal elements λ_k given by

$$\lambda_k = \frac{\gamma C_k^*}{1 + \gamma|C_k|^2}.$$

The optimal $\mathbf{S}$ has the structure of a diagonal matrix sandwiched between $\mathbf{P}^\dagger$ and $\mathbf{W}$. To obtain the MMSE receiver we can simply replace the M channel-dependent scalars $1/C_k$ in the zero-forcing receiver by λ_k.

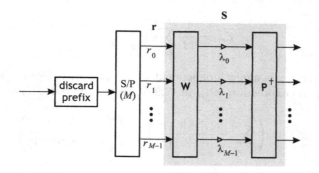

Figure 7.6. MMSE receiver for the precoded OFDM system.

With the MMSE receiver, the overall transfer matrix $\mathbf{T}$ from the input $\mathbf{s}$ of the transmitter to the output $\hat{\mathbf{s}}$ of the receiver is

$$\mathbf{T} = \mathbf{P}^\dagger \mathbf{\Lambda} \mathbf{\Gamma} \mathbf{P}. \qquad (7.20)$$

Let the (i,j)th element of $\mathbf{T}$ be t_{ij}. We can write the ith output as

$$\hat{s}_i = t_{ii}s_i + \tau_i, \qquad (7.21)$$

where τ_i consists of ISI from other symbols and channel noise and is given by

$$\tau_i = \sum_{j \neq i} t_{ij}s_j + [\mathbf{Sq}]_i.$$

Comparing the expression $\hat{s}_i = t_{ii}s_i + \tau_i$ with that in (3.13), we realize the two have the same form. From the discussion in Section 3.2, we know that the mean squared error and the unbiased SNR of an MMSE receiver are, respectively, given by (3.20) and (3.18). Therefore we get the ith subchannel error variance and SNR in terms of t_{ii} as

$$\sigma_{e_i}^2 = (1 - t_{ii})\mathcal{E}_s, \quad \beta(i) = t_{ii}/(1 - t_{ii}).$$

Note that $e_i = (t_{ii} - 1)s_i + \tau_i$ has mean $E[e_i] = 0$, although the conditional mean $E[e_i|s_i] = (t_{ii}-1)s_i$ is not zero. By directly multiplying out the matrices in (7.20), we get

$$t_{ii} = \sum_{k=0}^{M-1} |p_{k,i}|^2 \frac{\gamma|C_k|^2}{1 + \gamma|C_k|^2}. \qquad (7.22)$$

As each row of the unitary precoder $\mathbf{P}$ has unit energy, the subchannel error variance $\sigma_{e_i}^2$ has the form

$$\sigma_{e_i}^2 = \sum_{k=0}^{M-1} |p_{k,i}|^2 \frac{\mathcal{E}_s}{1 + \gamma|C_k|^2}. \qquad (7.23)$$

It follows that the average mean squared error $\mathcal{E}_{rr} = \frac{1}{M} E[\mathbf{e}^\dagger \mathbf{e}]$ is

$$\mathcal{E}_{rr} = \frac{1}{M} \sum_{i=0}^{M-1} \frac{\mathcal{E}_s}{1 + \gamma |C_i|^2}.$$

The average error is again independent of the precoder, like the zero-forcing case. Using the expression of t_{ii} in (7.22), we can express $\beta(i)$ as

$$\beta(i) = \left(\sum_{k=0}^{M-1} \frac{|p_{k,i}|^2 \gamma |C_k|^2}{1 + \gamma |C_k|^2} \right) \Big/ \left(\sum_{k=0}^{M-1} \frac{|p_{k,i}|^2}{1 + \gamma |C_k|^2} \right). \qquad (7.24)$$

Recall that for the zero-forcing receiver, the output noise can become infinitely large in the presence of a channel spectral null. In the MMSE case, even if the channel has some DFT coefficients equal to zero, the subchannel SNRs are not zero. The system will be more robust to channel spectral nulls. Recall the following two special cases.

- **The OFDM system** The precoder $\mathbf{P}$ is the identity matrix. The ith subchannel error variance is

$$\sigma_{e_i, ofdm}^2 = \frac{\mathcal{E}_s}{1 + \gamma |C_i|^2}. \qquad (7.25)$$

It is smaller than the error variance in (7.4) of a zero-forcing receiver. However, the subchannel SNR computed from $\mathcal{E}_s / \sigma_{e_i, ofdm}^2$ will be biased,

$$\beta_{ofdm, biased}(i) = 1 + \gamma |C_i|^2.$$

From Lemma 3.1, we know that the biased and unbiased SNRs are related by $\beta(i) = \beta_{biased}(i) - 1$. The unbiased SNR of the ith subchannel is therefore $\gamma |C_i|^2$, the same as in the zero-forcing case (7.5).

- **The SC-CP system** The precoder $\mathbf{P}$ is the normalized DFT matrix with $|p_{k,i}| = 1/\sqrt{M}$. In this case, the ith subchannel error variance becomes

$$\sigma_{e_i, sc-cp}^2 = \frac{1}{M} \sum_{k=0}^{M-1} \frac{\mathcal{E}_s}{1 + \gamma |C_k|^2},$$

which is the same as we had earlier in (6.23).

Subchannel error variances In the zero-forcing precoded system, the subchannel error variances can be expressed as linear combinations of those in the OFDM system. Here we can observe a similar property in the MMSE case. The expression in (7.23) is, in fact, a linear combination of the error variances in (7.25). Note that the ratio $\sigma_{e_i}^2 / \mathcal{E}_s$ is the biased noise to signal ratio. We get

$$\frac{1}{\beta_{biased}(i)} = \frac{1}{1 + \beta(i)} = \sum_{k=0}^{M-1} |p_{k,i}|^2 \frac{1}{1 + \gamma |C_i|^2}. \qquad (7.26)$$

We observe that $\frac{1}{\beta_{biased}(i)}$ can be expressed as the same linear combinations of the biased subchannel NSRs in the OFDM system, just like the subchannel error variances.

When an MMSE receiver is used, the system is not ISI-free, and the output errors do not come from channel noise alone. We can use the Gaussian approximation to compute the bit error rate as in Chapter 6. Such an approximation allows us to have a nice closed-form expression of BER that can be used for finding the optimal precoder later. The computation of bit error rate depends on the modulation scheme used. We look into the QPSK case next. Other modulation schemes are considered in Section 7.4.3.

7.4.2   Optimal precoders for QPSK modulation

For QPSK modulation, the bit error rate of the ith subchannel has the simple expression $Q(\sqrt{\beta(i)})$. In Section 7.2, we exploited the convexity and concavity of the function $f(\cdot)$ to derive the optimal precoder. Unfortunately we cannot repeat the same trick here because the unbiased NSR $1/\beta(i)$ is no longer a linear combination of $1/\beta_{ofdm}(i)$ like the zero-forcing case. To simplify derivations, let us define a new function

$$h(y) \triangleq Q(\sqrt{y^{-1} - 1}), \quad \text{for} \quad 0 < y < 1. \tag{7.27}$$

The introduction of $h(y)$ gives the following error rate expression:

$$Q(\sqrt{\beta(i)}) = h\left(\frac{1}{1 + \beta(i)}\right) = h\left(\frac{1}{\beta_{biased}(i)}\right).$$

Note the argument of the function $h(\cdot)$ is the biased ith subchannel NSR. Using the expression of biased subchannel NSRs in (7.26), the average bit error rate is given by

$$\mathcal{P}_{mmse} = \frac{1}{M} \sum_{i=0}^{M-1} h\left(\sum_{k=0}^{M-1} \frac{|p_{k,i}|^2}{1 + \gamma|C_k|^2}\right). \tag{7.28}$$

Substituting $\mathbf{P} = \mathbf{I}_M$ and $\mathbf{P} = \mathbf{W}$ into the above equation, we get, respectively, the BERs of MMSE OFDM and SC-CP systems:

$$\mathcal{P}_{ofdm,mmse} = \mathcal{P}_{ofdm} = \frac{1}{M} \sum_{i=0}^{M-1} h\left(\frac{1}{1 + \gamma|C_i|^2}\right), \tag{7.29}$$

$$\mathcal{P}_{sc\text{-}cp,mmse} = h\left(\frac{1}{M} \sum_{i=0}^{M-1} \frac{1}{1 + \gamma|C_i|^2}\right). \tag{7.30}$$

To study the behavior of $\mathcal{P}_{mmse}$, let us first look into some properties of the function $h(y)$. A plot of $h(y)$ is given in Fig. 7.7. It is a monotone increasing function defined for the interval $0 < y < 1$. Note that the argument of $h(\cdot)$ in (7.28) is always between 0 and 1. It can be shown that (Problem 7.11) $h(y)$ is strictly convex with

$$h'(y) > 0 \quad \text{and} \quad h''(y) \geq 0, \quad 0 < y < 1.$$

Using the convexity of $h(\cdot)$, we can show the following theorem.

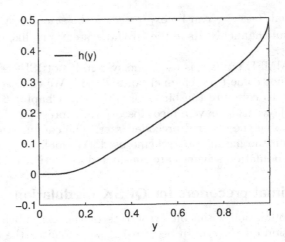

Figure 7.7. Plot of $h(y) = Q(\sqrt{y^{-1} - 1})$.

Theorem 7.2 MMSE case. Consider the precoded OFDM systems with an MMSE receiver. For QPSK modulation, the bit error rate $\mathcal{P}_{mmse}$ satisfies

$$\mathcal{P}_{sc\text{-}cp,mmse} \le \mathcal{P}_{mmse} \le \mathcal{P}_{ofdm}.$$

The first inequality becomes an equality if and only if the subchannel error variances $\sigma_{e_i}^2$ are the same.   ∎

Unlike the zero-forcing case in Theorem 7.1, the inequalities in the above theorem hold for any value of $\gamma = \mathcal{E}_s/\mathcal{N}_0$. With an MMSE receiver, the SC-CP system has the smallest BER and the OFDM system has the largest BER.

Proof of Theorem 7.2 As in the proof in Theorem 7.1, we can use the convexity of $h(y)$ to show that

$$\mathcal{P}_{mmse} \le \frac{1}{M} \sum_{i=0}^{M-1} \sum_{k=0}^{M-1} |p_{k,i}|^2 h\left(\frac{1}{1 + \gamma|C_k|^2}\right)$$

$$= \frac{1}{M} \sum_{k=0}^{M-1} h\left(\frac{1}{1 + \gamma|C_k|^2}\right) = \mathcal{P}_{ofdm}.$$

On the other hand, we can also apply the convexity of $h(y)$ to obtain,

$$\mathcal{P}_{mmse} \ge h\left(\frac{1}{M} \sum_{i=0}^{M-1} \sum_{k=0}^{M-1} \frac{|p_{k,i}|^2}{1 + \gamma|C_k|^2}\right)$$

$$= h\left(\frac{1}{M} \sum_{k=0}^{M-1} \frac{1}{1 + \gamma|C_k|^2}\right) = \mathcal{P}_{sc\text{-}cp,mmse}.$$

The above inequality becomes an equality if and only if the terms in the summation of (7.28) are equal, i.e. all the subchannels have the same SNR. This is true if and only if the subchannel NSRs $1/(1 + \beta(i))$ in (7.26) are equal, which means that the subchannel error variances $\sigma_{e_i}^2$ are the same.   ∎

Optimal channel independent precoders

Theorem 7.2 states that the minimum error rate $\mathcal{P}_{sc\text{-}cp,mmse}$ is achieved if and only if the subchannel error variances are equalized. Examining the expression of the subchannel error variances in (7.23), we see that such a property is achieved when the precoder has the equal magnitude property in (7.14). Again, the Hadamard matrix and the DFT matrix are examples of such precoders. The precoder that equalizes the subchannel error variances for the zero-forcing receiver also equalizes the subchannel error variances for the MMSE receiver. This means the transmitter does not need to know whether the receiver is zero-forcing or MMSE. The same transmitter can be optimal for different receivers. When the subchannel errors are equalized, all the subchannel error rates are the same, equal to that of the MMSE SC-CP system, and we have $\mathcal{P}_{mmse} = \mathcal{P}_{sc\text{-}cp,mmse}$.

7.4.3   Other modulation schemes

For modulations other than QPSK, we can use approximations of BER as in Section 7.3. The results will be stated without proof [78]. We will use $2b$-bit QAM as an example. We approximate the ith subchannel error rate as in (7.15), $\mathcal{P}_{mmse}(i) \approx aQ(\sqrt{\zeta\beta(i)})$, where $\beta(i)$ is the SNR given in (7.24), and the constants a and ζ are as given in (7.15). Let us define

$$g(y) = aQ(\sqrt{\zeta(y^{-1} - 1)}),\tag{7.31}$$

then $\mathcal{P}_{mmse}(i)$ can be conveniently approximated by

$$\mathcal{P}_{mmse}(i) \approx g\left(\frac{1}{1 + \beta(i)}\right),$$

where the argument of $g(\cdot)$ is the biased subchannel NSR. It can be shown (Problem 7.13) that the function $g(y)$ is convex over the intervals $\mathcal{S}_0$ and $\mathcal{S}_2$, and it is concave over $\mathcal{S}_1$. The intervals $\mathcal{S}_0$, $\mathcal{S}_1$, and $\mathcal{S}_2$ are, respectively,

$$\mathcal{S}_0 = (0, z_0], \ \mathcal{S}_1 = (z_0, z_1), \text{ and } \mathcal{S}_2 = [z_1, 1),$$

where $z_0 = \dfrac{3 + \zeta - \sqrt{\zeta^2 - 10\zeta + 9}}{8}$ and $z_1 = \dfrac{3 + \zeta + \sqrt{\zeta^2 - 10\zeta + 9}}{8}.$

$$\tag{7.32}$$

Note that for the OFDM system, when γ is large enough such that $1/(1 + \beta_{ofdm}(i)) \leq z_0$, the ith subchannel is operating in the convex region $\mathcal{S}_0$ of $g(\cdot)$. This requires

$$\gamma \geq \frac{z_0^{-1} - 1}{|C_i|^2}.$$

The condition for all the subchannels of the OFDM system to be operating in S_0 is

$$\gamma \geq \gamma_1', \quad \text{where} \quad \gamma_1' = \max_i \frac{z_0^{-1} - 1}{|C_i|^2}. \tag{7.33}$$

It can be shown (Problem 7.14) that when all the subchannels of the OFDM system are operating in S_0, so are the subchannels of the precoded system. Similarly, all the subchannels of the precoded system are operating in the convex region S_2 when

$$\gamma \leq \gamma_0', \quad \text{where} \quad \gamma_0' = \min_i \frac{z_1^{-1} - 1}{|C_i|^2}. \tag{7.34}$$

For convenience, we define

$$\mathcal{R}_{low}' = \{\gamma | \gamma \leq \gamma_0'\}, \ \mathcal{R}_{mid}' = \{\gamma | \gamma_0' < \gamma < \gamma_1'\}, \ \mathcal{R}_{high}' = \{\gamma | \gamma \geq \gamma_1'\}. \tag{7.35}$$

When the SNR is sufficiently high so that $\gamma \in \mathcal{R}_{high}'$, or when the SNR is low enough so that $\gamma \in \mathcal{R}_{low}'$, all the subchannels of the conventional OFDM system will operate in one of the two convex regions of $g(\cdot)$. In either case, we can invoke the convexity of $g(\cdot)$ to show that the MMSE SC-CP system is optimal. However, unlike the QPSK case, $\gamma \in \mathcal{R}_{mid}'$ does not mean that the OFDM system is optimal. In fact, there may not exist an SNR region in which all the subchannels are operating in the concave region of $g(\cdot)$ (Problem 7.16). Similar conclusions can be drawn for other modulation schemes whose subchannel error rates assume the form $aQ(\sqrt{\zeta\beta(i)})$.

The boundary points γ_0' and γ_1' of $\mathcal{R}_{low}'$ and $\mathcal{R}_{high}'$ are determined by the channel and the two scalars $(z_0^{-1} - 1)$ and $(z_1^{-1} - 1)$, which can be computed from the constellation size using (7.32) and (7.15). Table 7.1 gives the values of these two scalars for QAM constellations of sizes 4, 16, 64, and 256. As the size increases, the left boundary of $\mathcal{R}_{high}'$ is pushed further to the right while the right boundary of $\mathcal{R}_{low}'$ is pushed to the left. For the constellation of size 4, both scalars are equal to one, i.e. the function $g(y)$ is convex for all y. This is consistent with our earlier findings that for QPSK the MMSE SC-CP is optimal for all γ.

QAM constellation size	$z_1^{-1} - 1$	$z_0^{-1} - 1$
4	1.00	1.00
16	0.37	13.63
64	0.34	61.66
256	0.34	253.66

Table 7.1. The two scalars $(z_0^{-1} - 1)$ and $(z_1^{-1} - 1)$ for different QAM constellation sizes.

Compared to the BER behavior of the MMSE precoded system for the QPSK scheme, the non-QPSK cases are different. The MMSE SC-CP system is no longer optimal for all γ. It is optimal for low and high SNR ($\mathcal{R}'_{low}$ and $\mathcal{R}'_{high}$). In a simulation example shown later, we will see that $\mathcal{R}'_{low}$ corresponds to a BER range too large to be considered in practical applications. The BER discussion for the MMSE system in this section is, to some extent, similar to that for the zero-forcing case in Section 7.2. The solution of the optimal precoder depends on the given SNR and the SC-CP system is optimal for a higher SNR range.

7.5   Simulation examples

Example 7.1 Zero-forcing receivers. We will assume that the noise is AWGN with variance $\mathcal{N}_0$. The modulation symbols are QPSK with values equal to $\pm\sqrt{\mathcal{E}_s/2}\pm j\sqrt{\mathcal{E}_s/2}$ and $\gamma = \mathcal{E}_s/\mathcal{N}_0$. The number of subchannels M is 64. The length of the cyclic prefix is 3. The two four-tap channels in Example 6.1 will be used. The magnitude responses of the two channels $c_1(n)$ and $c_2(n)$ are as shown in Fig. 6.4.

For the channel $c_1(n)$, we compute the values of γ_0 and γ_1 defined in (7.11) and (7.10). They are, respectively, -0.51 dB and 14.74 dB. Figure 7.8(a) shows $\mathcal{P}_{ofdm}$ and $\mathcal{P}_{sc\text{-}cp}$ as functions of γ. For comparison, we also show an arbitrary example of BER when the transmitting matrix is unitary, but it is neither the identity matrix nor the IDFT matrix. The transmitting matrix $\mathbf{G}$ that is chosen here is a commonly known unitary matrix called a type II DCT matrix.[2] The BER is denoted as $\mathcal{P}_{dct}$ in the figure. In this case, the precoder $\mathbf{P}$ is $\mathbf{WG}$, whose elements do not have the unit magnitude property in (7.14). Whenever $\gamma = \mathcal{E}_s/\mathcal{N}_0$ is larger than $\gamma_1 = 14.74$ dB, the error rate of the SC-CP system becomes the minimum BER for any unitary precoder $\mathbf{P}$. For $\gamma \leq \gamma_0$, the conventional OFDM system is the optimal solution. When $\gamma = \gamma_0$, we observe from Fig. 7.8(a) that $\mathcal{P}_{ofdm} \approx 0.2$. That is, the OFDM system is optimal only for BER larger than 0.2. For either SNR range, $\gamma \leq \gamma_0$ or $\gamma \geq \gamma_1$, the performance of $\mathcal{P}_{dct}$ lies between $\mathcal{P}_{ofdm}$ and $\mathcal{P}_{sc\text{-}cp}$.

The channel $c_2(n)$ has a zero around 1.1π and its DFT coefficients around the spectral null are very small. The values of γ_0 and γ_1 are, respectively, 1.4 dB and 51.9 dB. Due to the zero of the channel that is close to the unit circle, γ_1 becomes very large. Figure 7.8(b) shows the BER performance of $\mathcal{P}_{ofdm}$, $\mathcal{P}_{dct}$, and $\mathcal{P}_{sc\text{-}cp}$. For all three systems, the BERs become small only when SNR is very large. Note that the crossing of $\mathcal{P}_{sc\text{-}cp}$ and $\mathcal{P}_{ofdm}$ occurs around $\gamma = 37$ dB and the BER corresponding to the crossing is approximately 6×10^{-3}. From this example we see that, for a reasonable BER, the SC-CP system outperforms the OFDM system. For example, when BER is 10^{-3}, the gains are 2 and 10 dB for $c_1(n)$ and $c_2(n)$, respectively.   ∎

---

[2]An $M \times M$ type II DCT matrix $\mathbf{C}_{II}$ is given by

$$[\mathbf{C}_{II}]_{mn} = \sqrt{\frac{2}{M}}\cos\left(\frac{\pi}{M}m(n+0.5)\right), 0 \leq m, n \leq M - 1.$$

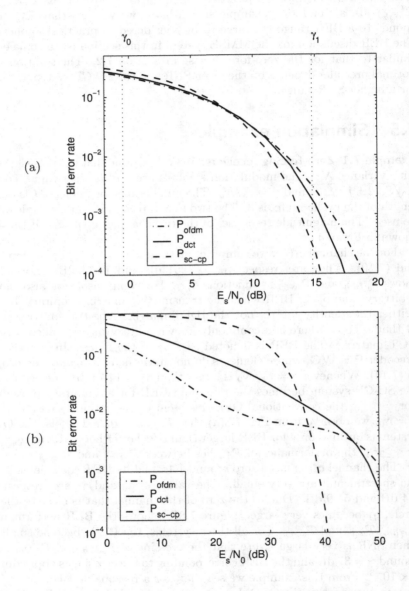

Figure 7.8. Example 7.1. Bit error rate performances of $\mathcal{P}_{ofdm}$, $\mathcal{P}_{dct}$, and $\mathcal{P}_{sc\text{-}cp}$ for (a) channel $c_1(n)$ and (b) channel $c_2(n)$.

Example 7.2 MMSE receivers. The simulation environment and parameters used in this example are the same as those in Example 7.1 except that the receiver is now MMSE rather than zero-forcing. For channel $c_1(n)$, Figure 7.9(a) shows the BER performances of two MMSE receivers, $\mathcal{P}_{dct,mmse}$ and $\mathcal{P}_{sc\text{-}cp,mmse}$, where $\mathcal{P}_{dct,mmse}$ corresponds to the case when the transmitting matrix $\mathbf{G}$ is a DCT matrix as in Example 7.1. We have also shown the error rates of the zero-forcing receivers, $\mathcal{P}_{dct}$ and $\mathcal{P}_{sc\text{-}cp}$. In either case, the BER of the MMSE receiver is lower than that of the zero-forcing receiver for all SNR. (For comparison, $\mathcal{P}_{ofdm}$ is also shown in the same plot. For the OFDM system, an MMSE receiver does not improve the BER, so the MMSE case is not shown.) We can see that the curve of the MMSE SC-CP system is below all the others. This corroborates the result that the MMSE SC-CP system has the smallest error rate for all γ.

For channel $c_2(n)$, Fig. 7.9(b) shows the five BER performance curves mentioned above. Due to the zero of the channel that is close to the unit circle, the error rates of the three zero-forcing systems are considerably affected. However, there is no serious performance degradation in the MMSE SC-CP system. Again we can observe that the curve of the MMSE SC-CP system stays below all the other curves for all γ. Comparing Fig. 7.9(a) and (b), we see that the gain of using an MMSE receiver is substantially more significant when the channel has a spectral null. ∎

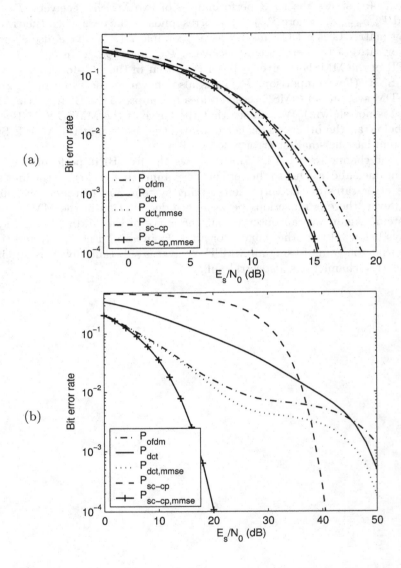

Figure 7.9. Example 7.2. Error rate performances of MMSE receivers for (a) channel $c_1(n)$ and (b) channel $c_2(n)$.

Example 7.3 Random channel. We use a random channel with four coefficients. The channel has an exponential power delay profile. The coefficients $c(n)$ are obtained from independent circular complex Gaussian random variables with zero-mean and variances given by 8/15, 4/15, 2/15, and 1/15 for $n = 0, 1, 2, 3$, respectively. The input symbols are QPSK. We compute the BER performances $\mathcal{P}_{ofdm}$, $\mathcal{P}_{dct,mmse}$, and $\mathcal{P}_{sc\text{-}cp,mmse}$ and average the results for 20000 random channels (Figure 7.10). For high SNR, the SC-CP system with the MMSE receiver requires a significantly smaller transmission power than the OFDM system for the same BER. The average performance of $\mathcal{P}_{dct,mmse}$ is between the other two systems for all SNR. ∎

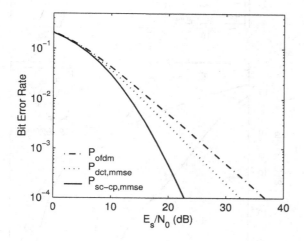

Figure 7.10. Example 7.3. Bit error rate performances $\mathcal{P}_{ofdm}$, $\mathcal{P}_{dct,mmse}$, and $\mathcal{P}_{sc\text{-}cp,mmse}$ over a four-tap random channel.

Example 7.4 QAM modulation. In this example the input modulation symbols are 4-bit QAM. The two channels in Example 7.1 will be used in our simulations. For channel $c_1(n)$, we compute the values of γ_0' and γ_1' defined, respectively, in (7.33) and (7.34). They are, respectively, -9.6 and 21.3 dB. For channel $c_2(n)$, the values of γ_0' and γ_1' are, respectively, -7.7 and 58.5 dB. The BER performances are shown in Fig. 7.11.

In Fig. 7.11(a), for both the SC-CP system and the system with the DCT matrix as the transmitter, the performance of the zero-forcing receiver is almost indistinguishable from that of the MMSE receiver. In both figures, we observe that the curve of the MMSE SC-CP system is no longer below all the other curves. For $c_1(n)$, it crosses the other curves at around $\gamma = 14$ dB, which corresponds to an error rate of around 0.07. From Section 7.4.3, we know the MMSE SC-CP has the lowest error rate when $\gamma \geq \gamma_1' = 21.3$. For $c_2(n)$, its curve crosses the others at around $\gamma = 23$ dB, a value much smaller than $\gamma_1' = 58.5$ dB. The BER corresponding to $\gamma = 23$ dB is around 0.02. The

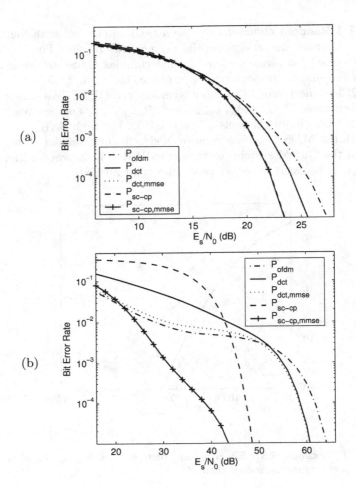

Figure 7.11. Example 7.4. Error rate performances of 4-bit QAM modulation symbols for (a) channel $c_1(n)$ and (b) channel $c_2(n)$.

simulations demonstrate that the BER of the MMSE SC-CP becomes lower than other systems at an SNR much smaller than γ_1'. Although it is not the optimal solution for all SNR, it outperforms the others for a more useful SNR range. ∎

Example 7.5 Coded error rates. In this example, the bits are coded using convolutional codes before they are mapped to QPSK symbols. We use the same four-tap random channel in Example 7.3. The BERs of the MMSE SC-CP and OFDM systems with code rates 1/2 and 1/3 are shown, respectively, in Fig. 7.12(a) and (b). In the uncoded case, Theorem 7.2 shows the MMSE SC-CP system achieves a lower error rate than the OFDM system. For the two coded cases in Fig. 7.12(a) and (b), we see that the two performance curves cross at a very high error rate. The OFDM system performs slightly better for very low SNR or high BER. The MMSE SC-CP system is better for a more useful range of error rate, and the gap between the two systems widens as SNR increases.

For the uncoded OFDM system, the subchannels near the null have a larger error rate while the bits on other subchannels are decoded correctly with high probability. Using an error correcting code, the bits on the bad subchannels are likely to be corrected with the aid of correct bits from the good subchannels. The error rate of the OFDM system can be considerably improved when an error correcting code is employed. For example, to achieve an error of 10^{-4}, the required SNR for the uncoded, 1/2 coded and 1/3 coded cases are respectively 37, 22.5 and 18 dB. We should also note that the reduction in SNR is achieved at the price of a decrease in transmission bit rates. The transmission rates of the two coded cases are respectively half and one-third that of the uncoded case. For the MMSE SC-CP system, the uncoded error rates are the same for all subchannels. For channels with a spectral null, the effect will be spread to all subchannels. There are no good subchannels to help with the correction of the bad subchannels. The improvement in error rates is not as much as the OFDM system. To achieve the same error rate 10^{-4}, the required SNR for the same three cases are respectively 23, 18 and 15 dB. For the uncoded case, the OFDM system requires 14 dB more SNR than the SC-CP system, while for 1/3 coded case the gap narrows to around 3 dB.

The block error rate (or packet error rate if each block is sent in one packet) is sometimes a more relevant measure because an uncorrected bit error usually means the whole block needs to be re-transmitted. The block error rates of the two systems for the code rate 1/3 are shown in Fig. 7.12(c). The MMSE SC-CP system has a lower block error rate than the OFDM system for all SNR. The difference between the two is more pronounced than in the BER plots. The reason is that errors mostly occur when the channel is not good, e.g. channels with spectral nulls. But this is even more so for the MMSE SC-CP system. For the same BER, the errors of the MMSE SC-CP system are even more concentrated in the cases of bad channels than the OFDM system. Therefore, for the same BER the MMSE SC-CP system is more likely to have error-less blocks than the OFDM system (i.e. a smaller block error rate). ■

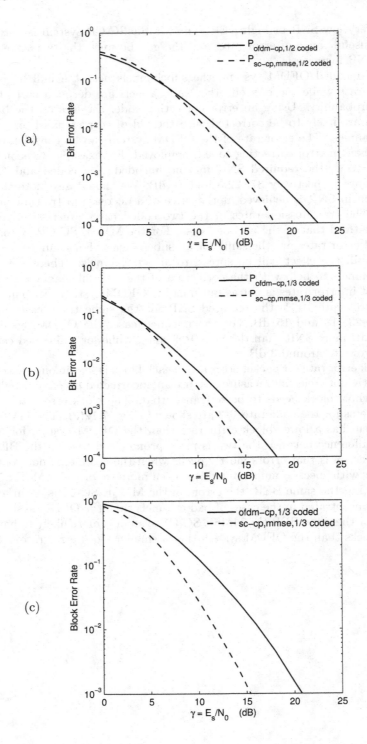

Figure 7.12. Example 7.5. Performance of the OFDM and MMSE SC-CP systems with channel coding. (a) BER for 1/2 coded symbols; (b) BER for 1/3 coded symbols; (c) block error rate for 1/3 coded symbols.

7.6 Further reading

Block transceivers with a general transmitting matrix, sometimes referred to
as precoding, have been studied extensively in the literature [5, 75, 126, 134,
135]. These transceivers are optimized for a given channel profile, including
the channel impulse response and the channel noise statistics. The solutions
are hence generally channel-dependent. Transceivers that are optimal in the
sense of minimum transmission power or minimum total noise power have been
developed [5, 75, 134, 135]. For the class of zero-padding transceivers, an optimal solution that minimizes the total output noise variance was found in [135].
The optimal receiver and zero-padded transmitter can be given in terms of an
appropriately defined channel matrix and the autocorrelation matrix of the
channel noise. Information rate optimized systems were considered in [5, 134].
In [75], optimal transmitters and receivers that minimize transmission power
for a given bit rate and error rate were derived under an optimal bit allocation. A fundamentally different useful unified approach to the joint design
of transmitters and receivers for MIMO channels using convex optimization
was developed in [106]. Several commonly used optimization criteria, such as
mean squared error and geometric mean of SINRs, were considered therein.
Optimal block transceivers can be obtained under this unified framework.

When the lengths of transmitting and receiving filters are not constrained
by the block size, the transmitter and receiver are no longer characterized
by constant matrices [47, 72, 74, 170]. Due to longer transmitting and receiving filters, there is more design freedom. A joint transmitter–receiver
optimization scheme using filter bank precoding was given in [72]. Optimal
ideal transmitting and receiving filters that minimize the mean squared error can be obtained in the frequency domain. In [74], the design freedom
was used to obtain transmitting and receiving filters with better frequency
selectivity. Frequency selectivity is important in applications where there is
certain constraint on the transmitter output and for suppressing narrowband
noise at the receiver.

Given the channel profile, transceivers have also been optimized for BER
minimization. In [33, 34], based on the zero-forcing solution given in [135],
a class of optimal precoders with power loading was considered. For a given
channel profile, the optimal precoder for BPSK modulation can be found in
a closed form when the SNR is sufficiently large. In [108], transceivers were
designed for subchannels with different constellations. For a given channel
profile and bit allocation, the transceiver was optimized for minimum BER
by using the primal decomposition. The problem of minimizing BER for
binary signaling over FIR MIMO channels was considered in [51]. An iter-
ative algorithm was given in [51] for jointly optimizing the FIR transmitter
and receiver. For the case when there is oversampling at the receiver, i.e.
fractionally spaced equalizer, transceiver optimization was developed in [176].
With oversampling, remarkable gain can be achieved for practical channels,
as demonstrated in [176].

7.7   Problems

7.1 Let $M = 4$ and let the length of cyclic prefix $\nu = 2$. Suppose the channel has two coefficients $c(0) = c(1) = 1/2$ and QPSK modulation is used. Compute γ_0 and γ_1. Find the SNR range for which the OFDM system has the smallest bit error rate among all zero-forcing precoded systems.

7.2 Suppose the setting is the same as in Problem 7.1 except that 4-bit QAM is used. Compute γ_0', and γ_1'. Find the SNR range for which the OFDM system has the smallest bit error rate among

   (a) all zero-forcing precoded systems;

   (b) repeat (a) for the SC-CP system.

7.3 Suppose the SNR quantity $\gamma = \mathcal{E}_s/\mathcal{N}_0$ is 10 dB in Problem 7.2. Find $\beta_{sc\text{-}cp}$ and $\beta_{sc\text{-}cp,mmse}$.

7.4 Let $M = 4$ and let the length of cyclic prefix $\nu = 2$. Suppose the channel is $C(z) = 1 + 0.5z^{-1}$. The transmitter input modulation symbols are QPSK. Find the SNR region where the SC-CP is optimal for the cases when

   (a) the receiver is zero-forcing;

   (b) the receiver is MMSE.

7.5 Repeat Problem 7.4 when the transmitter input modulation symbols are

   (a) 4-bit QAM;

   (b) 8-bit QAM.

7.6 Determine the subchannel SNR $\beta(i)$ given in (7.24) when $\mathbf{P} = \mathbf{I}_M$ for $M = 4$, $\nu = 2$, and the channel is as in Problem 7.1.

7.7 For the precoded OFDM system with a zero-forcing receiver, we showed that the subchannel SNRs are bounded as in (7.6). Show that for the MMSE case the subchannel SNR $\beta(i)$ continues to satisfy the bounds. That is,

$$\min_k \gamma |C_k|^2 \leq \beta(i) \leq \max_k \gamma |C_k|^2.$$

   (Hint. Use the relation in (7.26).)

7.8 Let $\mathbf{P}$ be a unitary matrix with the equal magnitude property in (7.14). Consider a matrix $\mathbf{P}'$ with

$$p'_{m,n} = e^{j(\theta_m + \alpha_n)} p_{m,n}, \quad 0 \leq m, n \leq M - 1,$$

   for arbitrary choices of real θ_m and α_n. Show that the new matrix $\mathbf{P}'$ is also unitary and that it also has the equal magnitude property.

7.9 In (7.3), we compute the SNR of the ith subchannel of the precoded OFDM system with a zero-forcing receiver. Equation (7.24) gives the ith subchannel SNR when the receiver is MMSE. Is the ith subchannel SNR given in (7.24) always larger than the ith subchannel SNR given in (7.3)?

7.10 Consider the MMSE precoded OFDM system in Fig. 7.5. Suppose the inputs of the transmitter are symmetric in the sense that the real and imaginary parts have equal variance $\mathcal{E}_s/2$. The channel noise $q(n)$ is circularly symmetric Gaussian. When the MMSE receiver is used, do the real and imaginary parts of the signal plus the noise term τ_i in (7.21) also have equal variance?

7.11 Derive the first derivative $h'(y)$ and the second derivative $h''(y)$ of the function $h(y)$ defined in (7.27). Show that $h(y)$ is convex with $h'(y) > 0$ and $h''(y) \geq 0$ for $0 < y < 1$.

7.12 In Section 7.3, we consider optimal precoders for modulation symbols other than QPSK. The derivation is valid for any modulation schemes in which the subchannel error probability can be either approximated or expressed as $aQ\left(\sqrt{\zeta\beta(i)}\right)$ for some constants a and ζ that are independent of subchannels. PSK modulation is such an example. Determine the SNR region in which the SC-CP system is optimal for the channel $C(z) = 1 + 0.5z^{-1}$.

Note: A b-bit PSK symbol with symbol power $\mathcal{E}_s$ is of the form

$$s = \sqrt{\mathcal{E}_s}\cos\left(\frac{2\pi}{2^b}m\right) + j\sqrt{\mathcal{E}_s}\sin\left(\frac{2\pi}{2^b}m\right), 0 \leq m < 2^b.$$

When it is transmitted over an AWGN channel with noise variance $\mathcal{N}_0$, the symbol error rate can be approximated by

$$\mathcal{P}_{SER} = 2Q\left(\sqrt{\frac{2\mathcal{E}_s}{\mathcal{N}_0}}\sin\left(\frac{\pi}{2^b}\right)\right).$$

When a Gray code is used in the mapping, the BER is well approximated by $\mathcal{P}_{SER}/b$.

7.13 Show that the function $g(y)$ defined in (7.31) is convex in intervals $\mathcal{S}_0$ and $\mathcal{S}_2$ and concave in $\mathcal{S}_1$.

7.14 Show that when $\gamma \in \mathcal{R}'_{high}$ in (7.35), all the subchannels of the OFDM system are operating in the convex region $\mathcal{S}_0$ of $g(\cdot)$ and all the subchannels of the MMSE precoded OFDM system are also operating in the convex region $\mathcal{S}_0$ of $g(\cdot)$.

7.15 Show that when the precoder satisfies the equal magnitude property in (7.14), it is optimal for $\gamma \in \mathcal{R}'_{high}$ in (7.35).

7.16 In the zero-forcing precoded OFDM system with QPSK modulation, we saw that the SC-CP and OFDM systems are optimal for different SNR regions. The SC-CP system is optimal for the SNR region $\mathcal{R}_{high}$ while the OFDM system is optimal for $\mathcal{R}_{low}$. For the MMSE case with general QAM modulation, there are also SNR regions ($\mathcal{R}'_{low}$ and $\mathcal{R}'_{high}$ given in (7.35)) for which the SC-CP system is optimal. Show that $\gamma \in \mathcal{R}'_{mid}$ does not mean that all the subchannels are operating in the concave region of $g(\cdot)$; thus the optimality of the OFDM system is not guaranteed in this SNR region. Give a counterexample to show that there may not exist an SNR region in which all subchannels are operating in the concave region of $g(\cdot)$.

8

Transceiver design with channel information at the transmitter

From Chapter 6, we know that an OFDM system converts an LTI channel into a set of parallel subchannels. When the channel is frequency-selective, these subchannels can have very different subchannel gains. The system performance can be severely limited by a few bad subchannels. One solution to this problem is to use a precoder as demonstrated in Chapter 7. In many applications, the transmission environment does not change frequently. These applications include most wired transmission schemes, such as ADSL and VDSL systems, and many wireless transmission schemes, such as fixed wireless access systems and wireless LAN system where the end users are not mobile. Under these environments, if the channel state information is known at the transmitter, one can exploit this information to carry out bit allocation. By optimally assigning the bits to the subchannels, the system performance can be substantially improved, especially when the channel is highly frequency-selective. In this chapter, we will derive the optimal zero-forcing transceivers when there is bit allocation at the transmitter.

8.1   Zero-forcing block transceivers

The DFT-based transceivers studied in Chapter 6 employ either DFT or IDFT operations at the transmitter and receiver. The filter bank formulation shows that their polyphase matrices are constant matrices related to the DFT/IDFT operations. In this chapter, we study general block transceivers where the polyphase matrices can be arbitrary constant matrices. Figure 8.1 shows a block transceiver system. The $M \times 1$ input vector $\mathbf{s}$ is processed by the $N \times M$ transmitting matrix $\mathbf{G}_0$ to produce an $N \times 1$ output vector

$$\mathbf{x} = \mathbf{G}_0 \mathbf{s},$$

which is converted to a sequence $x(n)$ and transmitted over the channel. At the receiver, the received sequence is blocked into vectors of size N. The $N \times 1$ received vector $\mathbf{r}$ is then processed by the $M \times N$ receiving matrix $\mathbf{S}_0$ to obtain the $M \times 1$ output vector

$$\widehat{\mathbf{s}} = \mathbf{S}_0 \mathbf{r}.$$

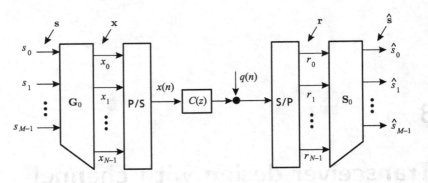

Figure 8.1. Block-based transceiver system.

In this chapter, we consider only zero-forcing block transceiver; that is, the transfer matrix from the transmitter input $\mathbf{s}$ to the receiver output $\hat{\mathbf{s}}$ is an identity matrix. Because there is no interblock interference (IBI) in a zero-forcing transceiver and the processing in a block transceiver is done in a block by block manner, only one-shot transmission is considered. Therefore in Fig. 8.1, we omit the time index n of the input and output vectors.

Assume that the integers N and M satisfy $N \geq M$ so that zero-forcing equalization is possible. The number

$$\nu = N - M$$

represents the number of redundant samples inserted per input block. In this chapter, it is assumed that the channel is an LTI FIR filter with order $L \leq \nu$:

$$C(z) = c(0) + c(1)z^{-1} + \cdots + c(\nu)z^{-\nu}.$$

The last few coefficients are zero when $L < \nu$. The channel noise $q(n)$ is a zero-mean WSS Gaussian process with known autocorrelation coefficients. Unlike in Chapter 7, $q(n)$ is not restricted to be white and it can be colored as in most wired environments. The channel impulse response and the noise statistics are known to both the transmitter and receiver.

In this section, we will derive the necessary and sufficient conditions on the matrices $\mathbf{G}_0$ and $\mathbf{S}_0$ such that the block transceiver system achieves zero-forcing. Recall from Chapter 5 that the N-input N-output system from $\mathbf{x}$ to $\mathbf{r}$ is an LTI system and the transfer matrix is the pseudocirculant $\mathbf{C}_{ps}(z)$. The overall transfer matrix from $\mathbf{s}$ to $\hat{\mathbf{s}}$ is an $M \times M$ matrix given by

$$\mathbf{T}(z) = \mathbf{S}_0 \mathbf{C}_{ps}(z) \mathbf{G}_0. \tag{8.1}$$

The block transceiver is free from IBI if $\mathbf{T}(z)$ is a constant matrix independent of z. As the channel order L satisfies $L \leq \nu < N$, the $N \times N$ matrix $\mathbf{C}_{ps}(z)$ has the special form given in (5.5); only the $\nu \times \nu$ submatrix at the top right corner of $\mathbf{C}_{ps}(z)$ is z-dependent. From Chapter 5, we know that we can avoid or remove IBI by setting either the last ν rows of $\mathbf{G}_0$ to zero or the first

ν columns of $\mathbf{S}_0$ to zero. The former is equivalent to padding zeros at the transmitter and the latter is equivalent to discarding or zero jamming the IBI corrupted samples at the receiver. Thus the former is referred to as a **zero-padded (ZP) system** and the latter is called a **zero-jamming (ZJ) system** [162] (also known as a leading zero system [135]). In the following, we shall derive the zero-forcing solutions for the ZP and ZJ transceivers.

8.1.1 Zero-forcing ZP systems

For a ZP system, we pad ν zeros for every block of M data samples. The transmitting and receiving matrices are, respectively, of the forms

$$\mathbf{G}_0 = \begin{bmatrix} \mathbf{G}_{zp} \\ \mathbf{0} \end{bmatrix} \quad \text{and} \quad \mathbf{S}_0 = \mathbf{S}_{zp},$$

where $\mathbf{G}_{zp}$ is of dimension $M \times M$, the bottom matrix $\mathbf{0}$ is of dimension $\nu \times M$, and $\mathbf{S}_{zp}$ is an $M \times N$ constant matrix. Substituting the above relation into (8.1), the transfer matrix $\mathbf{T}(z)$ becomes a constant matrix:

$$\mathbf{T} = \mathbf{S}_{zp}\mathbf{C}_{low}\mathbf{G}_{zp},$$

where $\mathbf{C}_{low}$ is the $N \times M$ lower triangular Toeplitz matrix given in (5.13), which we reproduce below:

$$\mathbf{C}_{low} = \begin{bmatrix} c(0) & 0 & \cdots & 0 & \cdots & 0 \\ c(1) & c(0) & & & & 0 \\ \vdots & & \ddots & & & \vdots \\ c(\nu) & & & c(0) & & \\ 0 & \ddots & & & \ddots & \\ \vdots & & c(\nu) & & \cdots & c(0) \\ 0 & \cdots & 0 & c(\nu) & & c(1) \\ \vdots & & \vdots & & \ddots & \vdots \\ 0 & \cdots & 0 & & & c(\nu) \end{bmatrix}. \tag{8.2}$$

Using this matrix representation, we can redraw the transceiver system in Fig. 8.1 as Fig. 8.2, where the noise vector $\mathbf{q}_{zp}$ is an $N \times 1$ vector obtained by blocking $q(n)$. Due to the padded zeros, there is no overlap between adjacent blocks after channel filtering. Processing can be done on a block by block basis. The zero-forcing condition becomes $\mathbf{S}_{zp}\mathbf{C}_{low}\mathbf{G}_{zp} = \mathbf{I}_M$, which implies that the matrices $\mathbf{S}_{zp}$, $\mathbf{C}_{low}$, and $\mathbf{G}_{zp}$ have full rank. Since the matrix $\mathbf{G}_{zp}$ is $M \times M$, it is invertible. Thus the zero-forcing condition can be rewritten as

$$\mathbf{G}_{zp}\mathbf{S}_{zp}\mathbf{C}_{low} = \mathbf{I}_M.$$

This means that $\mathbf{G}_{zp}\mathbf{S}_{zp}$ is a left inverse of $\mathbf{C}_{low}$. Since $\mathbf{C}_{low}$ is $N \times M$, its left inverse is not unique when $N > M$. To characterize all the left inverses of $\mathbf{C}_{low}$, let us apply the singular value decomposition (SVD) to $\mathbf{C}_{low}$:

$$\mathbf{C}_{low} = \mathbf{U}_{zp} \begin{bmatrix} \boldsymbol{\Lambda} \\ \mathbf{0} \end{bmatrix} \mathbf{V}_{zp}^\dagger, \tag{8.3}$$

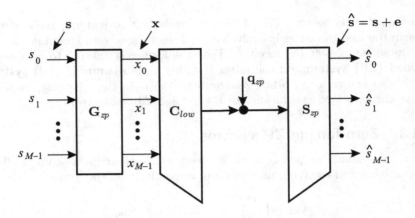

Figure 8.2. Simplified representation of the zero-padded system.

where $\mathbf{U}_{zp}$ and $\mathbf{V}_{zp}$ are, respectively, $N \times N$ and $M \times M$ unitary matrices and $\mathbf{\Lambda}$ is an $M \times M$ diagonal matrix whose diagonal entries are the singular values of $\mathbf{C}_{low}$. As $\mathbf{C}_{low}$ has full rank, all M singular values are positive and hence $\mathbf{\Lambda}$ is invertible. Using the SVD of $\mathbf{C}_{low}$, its left inverse $\mathbf{B}_{zp}$ can be expressed as

$$\mathbf{B}_{zp} = \mathbf{V}_{zp}\mathbf{\Lambda}^{-1}[\mathbf{I}_M \ \ \mathbf{A}_{zp}]\mathbf{U}_{zp}^{\dagger}, \tag{8.4}$$

where $\mathbf{A}_{zp}$ is an arbitrary $M \times \nu$ matrix. In summary, we have shown that the ZP system is zero-forcing if and only if

(a) the $M \times M$ matrix $\mathbf{G}_{zp}$ is invertible;

(b) for a given transmitting matrix $\mathbf{G}_{zp}$, the receiving matrix is $\mathbf{S}_{zp} = \mathbf{G}_{zp}^{-1}\mathbf{B}_{zp}$.

Note that the matrices $\mathbf{U}_{zp}$, $\mathbf{V}_{zp}$, and $\mathbf{\Lambda}$ are determined by the channel matrix $\mathbf{C}_{low}$. After we impose the zero-forcing condition, the free parameters available in the transceiver design are the invertible matrix $\mathbf{G}_{zp}$ and the $M \times \nu$ matrix $\mathbf{A}_{zp}$, which is completely arbitrary.

8.1.2   Zero-forcing ZJ systems

In a ZJ system, the IBI is removed by dropping the ν corrupted received samples at the receiver. The transmitting and receiving matrices are, respectively,

$$\mathbf{G}_0 = \mathbf{G}_{zj}, \ \ \mathbf{S}_0 = [\mathbf{0} \ \ \mathbf{S}_{zj}],$$

where $\mathbf{G}_{zj}$, $\mathbf{0}$, and $\mathbf{S}_{zj}$ are of dimensions $N \times M$, $M \times \nu$, and $M \times M$, respectively. The transmitted sequence is obtained by unblocking the vector $\mathbf{x} = \mathbf{G}_{zj}\mathbf{s}$. As $\mathbf{G}_{zj}$ is arbitrary, the redundant samples of a ZJ system are not necessarily zeros or a cyclic prefix. They are embedded in the transmitted sequence as a linear combination of the input symbols s_i. Using the above expression, we can show that the transfer matrix $\mathbf{T}(z)$ in (8.1) becomes a constant matrix,

$$\mathbf{T} = \mathbf{S}_{zj}\mathbf{C}_{up}\mathbf{G}_{zj},$$

where $\mathbf{C}_{up}$ is the bottom $M \times N$ submatrix of the pseudocirculant channel matrix $\mathbf{C}_{ps}(z)$. It is an upper triangular Toeplitz matrix given by

$$\mathbf{C}_{up} = \begin{bmatrix} c(\nu) & c(\nu-1) & \cdots & c(0) & 0 & \cdots & 0 \\ 0 & c(\nu) & c(\nu-1) & \cdots & c(0) & \cdots & 0 \\ \vdots & \vdots & \ddots & \ddots & & \ddots & \vdots \\ 0 & \cdots & 0 & c(\nu) & c(\nu-1) & \cdots & c(0) \end{bmatrix}. \quad (8.5)$$

The subscript "up" indicates that it is upper triangular.

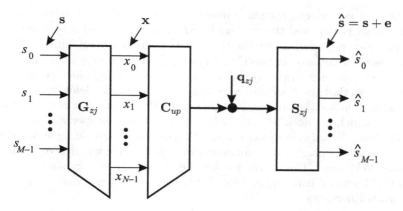

Figure 8.3. Simplified representation of the zero-jamming system.

The block diagram of a ZJ system can be drawn as Fig. 8.3. Note that the noise vector $\mathbf{q}_{zj}$ is an $M \times 1$ vector obtained by blocking $q(n)$. The zero-forcing condition $\mathbf{T} = \mathbf{I}_M$ implies that the square matrix $\mathbf{S}_{zj}$ is invertible. The zero-forcing condition can be rewritten as

$$\mathbf{C}_{up}\mathbf{G}_{zj}\mathbf{S}_{zj} = \mathbf{I}_M.$$

Thus the ZJ system is zero-forcing if and only if the product $\mathbf{G}_{zj}\mathbf{S}_{zj}$ is a right inverse of $\mathbf{C}_{up}$. Let the SVD of $\mathbf{C}_{up}$ be

$$\mathbf{C}_{up} = \mathbf{U}_{zj}[\mathbf{\Lambda} \ \ 0]\mathbf{V}_{zj}^{\dagger}, \quad (8.6)$$

where $\mathbf{U}_{zj}$ and $\mathbf{V}_{zj}$ are, respectively, unitary matrices of dimensions $M \times M$ and $N \times N$. The $M \times M$ matrix $\mathbf{\Lambda}$ is diagonal with diagonal entries equal to the singular values of $\mathbf{C}_{up}$. It can be shown (Problem 8.3) that the singular values of $\mathbf{C}_{up}$ are identical to those of $\mathbf{C}_{low}$. The right inverses of $\mathbf{C}_{up}$ have the form

$$\mathbf{B}_{zj} = \mathbf{V}_{zj}\begin{bmatrix} \mathbf{I}_M \\ \mathbf{A}_{zj} \end{bmatrix}\mathbf{\Lambda}^{-1}\mathbf{U}_{zj}^{\dagger}, \quad (8.7)$$

where $\mathbf{A}_{zj}$ is an arbitrary $\nu \times M$ matrix. Similar to the ZP case, one can solve for the zero-forcing solutions for the ZJ system in Fig. 8.3. The ZJ system is zero-forcing if and only if

(a) the $M \times M$ matrix $\mathbf{S}_{zj}$ is invertible;

(b) for a given receiving matrix $\mathbf{S}_{zj}$, the transmitting matrix is $\mathbf{G}_{zj} = \mathbf{B}_{zj}\mathbf{S}_{zj}^{-1}$.

Note that the free parameters in a zero-forcing ZJ system are the invertible matrix $\mathbf{S}_{zj}$ and the arbitrary $\nu \times M$ matrix $\mathbf{A}_{zj}$.

8.2   Problem formulation

In a transceiver system, the performance is characterized by the error rate, the transmitted power, and the average bit rate. In transceiver design we can fix two parameters and optimize the third one. There are many different criteria for transceiver designs. In this chapter, we consider the case when the average bit rate and the error rate are fixed, and the zero-forcing block transceiver is optimized so that the transmitted power $\mathcal{E}$ is minimized. Below we will derive the expression for the transmitted power for the block transceiver shown in Fig. 8.1, which includes both the ZP and ZJ systems as special cases.

The modulation symbols s_k can be PAM or QAM, depending on whether it is a baseband or passband communication scheme. We will derive the results for the PAM case. The derivation for the QAM case is very similar. Let s_k be a PAM symbol carrying b_k bits of information. Then the average bit rate per symbol becomes[1]

$$b = \frac{1}{M} \sum_{k=0}^{M-1} b_k. \tag{8.8}$$

Each input block $\mathbf{s}$ carries Mb bits. In the following derivation, we assume that the input symbols have zero-mean and they are uncorrelated. That is, the autocorrelation matrix of $\mathbf{s}$ is a diagonal matrix:

$$\mathbf{R}_s = \begin{bmatrix} \mathcal{E}_{s,0} & 0 & \cdots & 0 \\ 0 & \mathcal{E}_{s,1} & \cdots & 0 \\ \vdots & & \ddots & \vdots \\ 0 & \cdots & 0 & \mathcal{E}_{s,M-1} \end{bmatrix}. \tag{8.9}$$

The assumption is usually reasonable with proper bit interleaving. Note that the kth signal power $\mathcal{E}_{s,k}$ can be different for different k.

Consider Fig. 8.1. The transmitted power $\mathcal{E} = E[\|x(n)\|^2]$ can be computed by averaging the power of x_i, and it is given by

$$\mathcal{E} = \frac{1}{N} \sum_{n=0}^{N-1} \sigma_{x_n}^2 = \frac{1}{N} \, trace(\mathbf{R}_x).$$

---

[1]For the QAM case, if the kth symbol carries $2b_k$ bits, the average bit rate per symbol becomes

$$b = \frac{1}{M} \sum_{k=0}^{M-1} 2b_k.$$

Using the fact that $\mathbf{R}_x = \mathbf{G}_0 \mathbf{R}_s \mathbf{G}_0^\dagger$, we can write

$$\mathcal{E} = \frac{1}{N} \, trace\left(\mathbf{G}_0 \mathbf{R}_s \mathbf{G}_0^\dagger\right) = \frac{1}{N} \, trace\left(\mathbf{R}_s \mathbf{G}_0^\dagger \mathbf{G}_0\right),$$

where we have used $trace(\mathbf{AB}) = trace(\mathbf{BA})$. Using (8.9), we can write the transmitted power as

$$\mathcal{E} = \frac{1}{N} \sum_{k=0}^{M-1} \mathcal{E}_{s,k} \left[\mathbf{G}_0^\dagger \mathbf{G}_0\right]_{kk}, \qquad (8.10)$$

where the notation $[\mathbf{A}]_{kk}$ denotes the kth diagonal entry of the matrix $\mathbf{A}$. Note that the kth column vector of $\mathbf{G}_0$ corresponds to the kth transmitting filter. Thus the quantity $[\mathbf{G}_0^\dagger \mathbf{G}_0]_{kk}$ is simply the energy of the kth transmitting filter.

In the following sections, we will study the design of zero-forcing ZP and ZJ transceivers. For a fixed bit rate b and a fixed error rate, the transceiver is optimized so that the transmitted power $\mathcal{E}$ is minimized. The optimization process can be decomposed into two steps. Firstly, the bits b_k are optimally allocated to minimize the transmitted power for any given pair of transmitting and receiving matrices $\mathbf{G}_0$ and $\mathbf{S}_0$. Then, under the optimal bit allocation, we will find the transmitting and receiving matrices that minimize the transmitted power. In the second step, the free parameters are either $(\mathbf{G}_{zp}, \mathbf{A}_{zp})$ or $(\mathbf{S}_{zj}, \mathbf{A}_{zj})$ depending on whether it is a ZP or a ZJ system.

8.3   Optimal bit allocation

In a ZP or ZJ system, the scalar LTI channel $C(z)$ is converted to a set of M parallel subchannels. In many applications, these subchannels can have very different SNRs. In Section 2.4, we have learnt how to exploit this to do bit loading when the peak power allowed on each subchannel is constrained. Here we shall consider the bit allocation problem for a different setting. Under the assumption that all the subchannels have the same fixed error rate, we will show how we can employ bit allocation to minimize the transmitted power needed for a fixed bit rate. Let $\widehat{s}_k$ be the kth output of the receiver. Then the kth output error is $e_k = \widehat{s}_k - s_k$. As the transceiver is zero-forcing, the error comes entirely from the channel noise $q(n)$. Define the $M \times 1$ output noise vector as $\mathbf{e} = \begin{bmatrix} e_0 & e_1 & \cdots & e_{M-1} \end{bmatrix}^T$; then

$$\mathbf{e} = \mathbf{Sq},$$

where $\mathbf{S} = \mathbf{S}_{zp}$ and $\mathbf{q} = \mathbf{q}_{zp}$ for the ZP case and $\mathbf{S} = \mathbf{S}_{zj}$ and $\mathbf{q} = \mathbf{q}_{zj}$ for the ZJ case.

Recall from (2.14) that for a b_k-bit PAM symbol, the bit number b_k, the symbol power $\mathcal{E}_{s,k}$ and the noise power $\sigma_{e_k}^2$ are related by

$$b_k = \frac{1}{2} \log_2 \left(1 + \frac{\mathcal{E}_{s,k}/\sigma_{e_k}^2}{\Gamma_k}\right),$$

where Γ_k is the SNR gap for PAM symbols given in (2.15). It is a quantity that depends only on the error rate, and the values of Γ_k are listed for some typical error rates in Table 2.1. As all the subchannels have the same error rate, the SNR gap is the same for all subchannels, i.e.

$$\Gamma_k = \Gamma, \quad \text{for all } k.$$

By rearranging the terms in the expression for b_k, we get

$$\mathcal{E}_{s,k} = \Gamma(2^{2b_k} - 1)\sigma_{e_k}^2.$$

Assume that b_k is large enough so that $2^{2b_k} - 1 \approx 2^{2b_k}$, then we have[2]

$$\mathcal{E}_{s,k} \approx \Gamma 2^{2b_k} \sigma_{e_k}^2. \tag{8.11}$$

The above equation gives an approximation of the required signal power for transmitting b_k bits on the kth subchannel when the noise variance is $\sigma_{e_k}^2$. For a given b_k and $\sigma_{e_k}^2$, one can allocate symbol power $\mathcal{E}_{s,k}$ according to the expression given in (8.11). Substituting (8.11) into (8.10), the transmitted power becomes

$$\mathcal{E} = \frac{\Gamma}{N} \sum_{k=0}^{M-1} 2^{2b_k} \sigma_{e_k}^2 \left[\mathbf{G}_0^\dagger \mathbf{G}_0\right]_{kk}. \tag{8.12}$$

Applying the arithmetic mean (AM) and geometric mean (GM) inequality (Appendix A), we obtain

$$\mathcal{E} \geq \frac{\Gamma M}{N} 2^{2b} \prod_{k=0}^{M-1} \left(\sigma_{e_k}^2 \left[\mathbf{G}_0^\dagger \mathbf{G}_0\right]_{kk}\right)^{1/M} \triangleq \mathcal{E}_{bit}. \tag{8.13}$$

Equality holds if and only if the quantity $2^{2b_k}\sigma_{e_k}^2[\mathbf{G}_0^\dagger \mathbf{G}_0]_{kk}$ is the same for $k = 0, 1, \ldots, M-1$. From (8.11), we can conclude that, when the bits are allocated optimally, the quantities

$$\mathcal{E}_{s,k}[\mathbf{G}_0^\dagger \mathbf{G}_0]_{kk} = \text{a constant} \quad \text{for } k = 0, 1, \ldots, M-1. \tag{8.14}$$

are the same for all k. Note that $\mathcal{E}_{bit}$ depends only on b, $\sigma_{e_k}^2$, and $[\mathbf{G}_0^\dagger \mathbf{G}_0]_{kk}$, where $\sigma_{e_k}^2$ is determined once the receiver is known and $[\mathbf{G}_0^\dagger \mathbf{G}_0]_{kk}$ is determined once the transmitter is given. Therefore when the transceiver is given and the bit rate per symbol b is fixed, $\mathcal{E}_{bit}$ is a fixed quantity and it is the lower bound on the transmitted power, independent of the bit allocation $\{b_k\}_{k=0}^{M-1}$. The lower bound $\mathcal{E}_{bit}$ is achieved if and only if bits are allocated such that $2^{2b_k}\sigma_{e_k}^2[\mathbf{G}_0^\dagger \mathbf{G}_0]_{kk}$ are equalized. This condition implies that

$$2^{2b_k}\sigma_{e_k}^2 \left[\mathbf{G}_0^\dagger \mathbf{G}_0\right]_{kk} = 2^{2b} \prod_{i=0}^{M-1} \left(\sigma_{e_i}^2 \left[\mathbf{G}_0^\dagger \mathbf{G}_0\right]_{ii}\right)^{1/M}, \quad \text{for all } k.$$

That is, the quantity on the left is independent of k. Solving the above equation for the optimal b_k, we have

$$b_k = b - \frac{1}{2} \log_2\left(\sigma_{e_k}^2 \left[\mathbf{G}_0^\dagger \mathbf{G}_0\right]_{kk}\right) + \frac{1}{2M} \log_2\left(\Pi_{\ell=0}^{M-1} \sigma_{e_\ell}^2 \left[\mathbf{G}_0^\dagger \mathbf{G}_0\right]_{\ell\ell}\right). \tag{8.15}$$

---

[2]For the $2b_k$-bit QAM case, the signal power $\mathcal{E}_{s,k}$ and the noise variance $\sigma_{e_k}^2$ also satisfy the same relation as (8.11), but the SNR gap for QAM is given by the formula in (2.21).

Note that in general the optimal b_k is not an integer. Moreover the right-hand side of (8.15) could be negative. We will address this issue later in this section.

We can quantify the performance improvement of bit allocation by comparing $\mathcal{E}_{bit}$ with the transmitted power when there is no bit allocation. If we do not apply bit allocation, every subchannel carries b bits. The required transmitted power can be obtained by setting all $b_k = b$ in (8.12), and it is given by

$$\mathcal{E}_{no\text{-}bit} = \frac{\Gamma 2^{2b}}{N} \sum_{k=0}^{M-1} \sigma_{e_k}^2 \left[\mathbf{G}_0^\dagger \mathbf{G}_0 \right]_{kk}.$$

The bit allocation gain is thus given by

$$\mathcal{G}_{bit} = \frac{\frac{1}{M} \sum_{k=0}^{M-1} \sigma_{e_k}^2 \left[\mathbf{G}_0^\dagger \mathbf{G}_0 \right]_{kk}}{\prod_{k=0}^{M-1} \left(\sigma_{e_k}^2 \left[\mathbf{G}_0^\dagger \mathbf{G}_0 \right]_{kk} \right)^{1/M}}. \tag{8.16}$$

In many cases, the transmitter has the same $[\mathbf{G}_0^\dagger \mathbf{G}_0]_{kk}$ for all k. For example, the CP-OFDM, ZP-OFDM, and SC-ZP systems[3] studied in Chapter 6 satisfy such a condition. In this case the optimal bit allocation reduces to

$$b_k = b - \frac{1}{2} \log_2 \left(\sigma_{e_k}^2 \right) + \frac{1}{2M} \log_2 \left(\Pi_{\ell=0}^{M-1} \sigma_{e_\ell}^2 \right). \tag{8.17}$$

More bits are assigned to subchannels with a smaller $\sigma_{e_k}^2$ (*good subchannels*) and fewer bits are assigned to subchannels with a larger $\sigma_{e_k}^2$ (*bad subchannels*). Note that in this case the optimal bit allocation is such that $\mathcal{E}_{s,k} = \Gamma 2^{2b_k} \sigma_{e_k}^2$ are equalized for all k; all the subchannels have the same signal power $\mathcal{E}_{s,k}$. In other words, when the bits are allocated optimally, *no power allocation is needed*. The bit allocation gain is given by

$$\mathcal{G}_{bit} = \frac{\frac{1}{M} \sum_{k=0}^{M-1} \sigma_{e_k}^2}{\prod_{k=0}^{M-1} \left(\sigma_{e_k}^2 \right)^{1/M}}. \tag{8.18}$$

The gain $\mathcal{G}_{bit}$ is equal to the ratio of the AM over the GM of the output noise variances $\sigma_{e_k}^2$. As the AM is always larger than equal to the GM, the gain can never be smaller than one and it is equal to one if and only if all the subchannels have the same noise variance. The gain comes from the disparity among $\sigma_{e_k}^2$. For transmission scenarios that have a large difference in the output noise variances, this gain can be substantial, as we shall demonstrate later. Also observe that the bit allocation gain is independent of the bit rate per symbol b and the SNR gap Γ, which is determined by the error rate. Moreover if all the variances $\sigma_{e_k}^2$ are scaled by a common factor, the gain remains the same.

---

[3]Note that in a SC-CP system, the last ν symbols, $s_{M-\nu}, \ldots, s_{M-1}$, are duplicated as the cyclic prefix. For this reason, $[\mathbf{G}_0^\dagger \mathbf{G}_0]_{kk}$ of the first $(M - \nu)$ subchannels are different from those of the last ν subchannels. This will be elaborated later.

OFDM and single-carrier systems   We now analyze the system performance when the optimal bit allocation is applied to the previously studied OFDM and single-carrier systems (SC-CP and SC-ZP). Let us assume that the channel noise $q(n)$ is white with variance $\mathcal{N}_0$. Firstly let us look at the OFDM system. From Chapter 6, we know that the kth output error variance of both the CP-OFDM system and ZP-OFDM system (with the efficient receiver in (6.15)) have the form

$$\sigma_{e_k}^2 = \alpha \frac{\mathcal{N}_0}{|C_k|^2},$$

where $C_k = C(e^{j2\pi k/M})$ is the kth DFT coefficient of the channel $c(n)$. The parameter α is a constant equal to one for the CP-OFDM system and equal to N/M for the ZP-OFDM system. From (8.17), we get the optimal bit allocation as

$$b_k = b + \log_2(|C_k|) - \frac{1}{M} \log_2 \left(\Pi_{\ell=0}^{M-1} |C_\ell| \right).$$

The last term is independent of k. So more bits are allocated to subchannels with a larger gain. The corresponding bit allocation gain is

$$\mathcal{G}_{ofdm} = \frac{\frac{1}{M} \sum_{k=0}^{M-1} \frac{1}{|C_k|^2}}{\Pi_{k=0}^{M-1} \left(\frac{1}{|C_k|^2} \right)^{1/M}}.$$

It is equal to the ratio of AM over GM of $1/|C_k|^2$. The gain satisfies $\mathcal{G}_{ofdm} \geq 1$ with equality if and only if $|C_k|$ are the same for all k. When the channel is highly frequency-selective, the quantity $1/|C_k|^2$ can vary greatly with respect to k and the bit allocation can improve the performance significantly. Also note that $\mathcal{G}_{ofdm}$ is not necessarily an increasing function of M. It is not difficult to construct examples where the bit allocation gain decreases when the number of subchannels, M, increases. As M approaches infinity, one can derive the asymptotic bit allocation gain. For a very large M, the AM can be approximated as

$$\frac{1}{M} \sum_{k=0}^{M-1} \frac{1}{|C_k|^2} \approx \int_0^{2\pi} \frac{1}{|C(e^{j\omega})|^2} \frac{d\omega}{2\pi}.$$

Similarly, we can write the GM as

$$\prod_{k=0}^{M-1} (1/|C_k|^2)^{\frac{1}{M}} = \exp \left(\frac{1}{M} \sum_{k=0}^{M-1} \ln 1/|C_k|^2 \right).$$

When M is large, we have the following approximation:

$$\prod_{k=0}^{M-1} \left(\frac{1}{|C_k|^2} \right)^{\frac{1}{M}} \approx \exp \left(\int_0^{2\pi} \ln \frac{1}{|C(e^{j\omega})|^2} \frac{d\omega}{2\pi} \right).$$

If we define the ratio

$$\xi = \frac{\exp \left(\int_0^{2\pi} \ln \frac{1}{|C(e^{j\omega})|^2} \frac{d\omega}{2\pi} \right)}{\int_0^{2\pi} \frac{1}{|C(e^{j\omega})|^2} \frac{d\omega}{2\pi}},$$

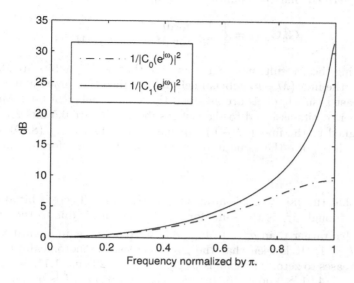

Figure 8.4. Plots of $1/|C_0(e^{j\omega})|^2$ and $1/|C_1(e^{j\omega})|^2$, where $C_0(z) = \frac{1}{3}(1+2z^{-1})$ and $C_1(z) = \frac{1}{1.95}(1+0.95z^{-1})$.

then the bit allocation gain $\mathcal{G}_{ofdm}$ asymptotically approaches $1/\xi$. The quantity ξ is a measure of the **flatness** of the curve $1/|C(e^{j\omega})|^2$ [159, 164]. We have $\xi = 1$ when $1/|C(e^{j\omega})|^2$ is increasingly flat and it becomes smaller when $1/|C(e^{j\omega})|^2$ becomes nonflat. In other words, a smaller ξ implies that the channel $c(n)$ is more frequency-selective. For highly frequency-selective channels, ξ can be very small and the asymptotic gain can be very large. For example, consider the two channels

$$C_0(z) = \frac{1}{3}(1+2z^{-1}) \quad \text{and} \quad C_1(z) = \frac{1}{1.95}(1+0.95z^{-1}).$$

The flatness measure is given by $\xi = 0.75$ for $1/|C_0(e^{j\omega})|^2$ and $\xi = 0.0975$ for $1/|C_1(e^{j\omega})|^2$. We can also see from Fig. 8.4 that $1/|C_0(e^{j\omega})|^2$ is much flatter than $1/|C_1(e^{j\omega})|^2$. As a result, the asymptotic gain for $C_1(z)$, which is equal to 10.26 (about 10 dB), is much larger than that of $C_0(z)$, which is only 1.33 (about 1.2 dB).

Secondly, let us consider the single-carrier systems. For the SC-CP system, the kth output error variance $\sigma_{e_k}^2$ is the same for all k. The transmitting matrix $\mathbf{G}_0$ of the SC-CP system is

$$\mathbf{G}_0 = \begin{bmatrix} \mathbf{0} & \mathbf{I}_\nu \\ \mathbf{I}_{M-\nu} & \mathbf{0} \\ \mathbf{0} & \mathbf{I}_\nu \end{bmatrix}.$$

It can be verified that the quantities $[\mathbf{G}_0^\dagger \mathbf{G}_0]_{kk}$ satisfy

$$[\mathbf{G}_0^\dagger \mathbf{G}_0]_{kk} = \begin{cases} 1, & \text{for } 0 \le k < M - \nu; \\ 2, & \text{for } M - \nu \le k < M. \end{cases} \tag{8.19}$$

Substituting these results into (8.15), we find that when the bits are allocated optimally, the first $(M - \nu)$ subchannels are assigned the same number of bits and the last ν subchannels are assigned the same number of bits. Moreover the number of bits assigned to the last ν subchannels are 0.5 bits fewer than that assigned to the first $(M - \nu)$ subchannels. Substituting (8.19) and the fact that all $\sigma_{e_k}^2$ are the same into the bit allocation gain formula in (8.16), we have

$$\mathcal{G}_{sc\text{-}cp} = 2^{\nu/M}(1 + \nu/M).$$

Observe that the gain $\mathcal{G}_{sc\text{-}cp}$ is always larger than one. There is bit allocation gain even though $\sigma_{e_k}^2$ is the same for all k. The gain is due to the disparity of $[\mathbf{G}_0^\dagger \mathbf{G}_0]_{kk}$ rather than $\sigma_{e_k}^2$. In Fig. 8.5, we plot the gain against ν/M for $0 \le \nu/M \le 1$. It is seen that the gain decreases monotonically to one as ν/M decreases to zero. The gain is $\mathcal{G}_{sc\text{-}cp} = 1.49$, 1.23, and 1.11, respectively, for $\nu/M = 1/4$, 1/8, and 1/16. In practice, the ratio ν/M is usually a small number for the reason of spectral efficiency. In this case, the gain is negligible. Thus bit allocation is not used in the SC-CP system. The same is also true for the SC-ZP system.

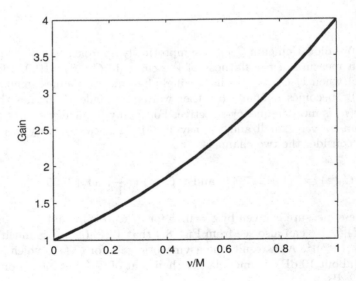

Figure 8.5. Bit allocation gain of the SC-CP system.

Optimal bit allocation subject to positivity   In the optimal bit allocation formulas, the right-hand side of (8.15) and (8.17) could be negative. When some b_j is negative, one can set it to zero. This is consistent with the Karush–Kuhn–Tucker (KKT) optimality conditions and b_k remains optimal subject to positivity [16]. Below we will show how to do this for the case of equal $[G_0^\dagger G_0]_{kk}$. Recall that the optimal bit allocation formula is given by (8.17). We can rewrite the optimal bit allocation as $b_k = D - \log_2 \sigma_k$ for some constant D. By setting the negative bits to zero, we have

$$b_k^+ = \begin{cases} D - \log_2 \sigma_{e_k}, & \text{if this is positive;} \\ 0, & \text{otherwise.} \end{cases} \tag{8.20}$$

The constant D is chosen such that the following bit constraint is satisfied:

$$\frac{1}{M} \sum_{k=0}^{M-1} b_k^+ = b.$$

The above formulas have a beautiful water-filling interpretation. It is like pouring water into a tank with an uneven floor $\log_2 \sigma_{e_k}$ (see Fig. 8.6 for an example). The number of bits b_k allocated to each subchannel is such that the water height

$$D = b_k^+ + \log_2 \sigma_{e_k}$$

is a constant. Equation (8.20) says that when $\log_2(\sigma_{e_k})$ is large, we "pour" fewer bits into that subchannel and vice versa. If for some subchannel the bottom of the tank $\log_2 \sigma_{e_k}$ is too high ($\log_2 \sigma_{e_3}$ in Fig. 8.6), then no bit is allocated to that subchannel ($b_3 = 0$ in the figure).

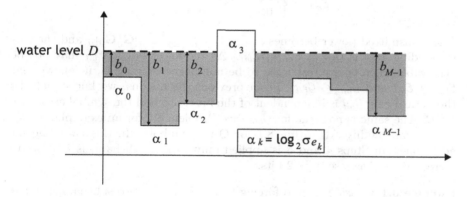

Figure 8.6. Water-filling interpretation of optimal bit allocation.

Algorithm 8.1.   Non-negative integer bit allocation   The optimal bit allocation formula in (8.17) may not yield a non-negative integer b_k. To obtain a non-negative integer bit allocation for PAM symbols, one can follow the following procedure.

(1) Set negative b_k to zero and round each b_k to the nearest integer. Let the resulting non-negative integer be $\bar{b}_k$.

(2) Compute the bit rate $\bar{b} = 1/M \sum_{i=0}^{M-1} \bar{b}_k$. Let $\rho = M(\bar{b} - b)$.

(3) If $\rho = 0$, then $\bar{b}_k$ is the solution.

(4) If $\rho < 0$, then find the ρ subchannels corresponding to the ρ smallest

$$\Gamma 2^{2\bar{b}_k} \sigma_{e_k}^2 \left[\mathbf{G}_0^\dagger \mathbf{G}_0 \right]_{kk}.$$

For each of these ρ subchannels, increase the corresponding $\bar{b}_k$ by one. The new $\bar{b}_k$ is the solution.

(5) If $\rho > 0$, then find the ρ subchannels corresponding to the ρ largest

$$\Gamma 2^{2\bar{b}_k} \sigma_{e_k}^2 \left[\mathbf{G}_0^\dagger \mathbf{G}_0 \right]_{kk},$$

where $\bar{b}_k \geq 1$. For each of these ρ subchannels, decrease the corresponding $\bar{b}_k$ by one. The new $\bar{b}_k$ is the solution.

In [167], the authors propose an efficient algorithm for optimal bit allocation of a non-negative integer. Moreover, the complexity of the proposed algorithm is independent of the total number of bits.

Note that when $\bar{b}_i = 0$ for some i, no information bit is transmitted in the ith subchannel. It is intuitive to set $\mathcal{E}_{s,i} = 0$. Therefore the signal power of the kth subchannel is

$$\bar{\mathcal{E}}_{s,k} = \begin{cases} \Gamma 2^{2\bar{b}_k} \sigma_{e_k}^2, & \text{if } \bar{b}_k \neq 0; \\ 0, & \text{otherwise.} \end{cases}$$

The transmitted power becomes $\bar{\mathcal{E}}_{bit} = 1/N \sum_{i=0}^{M-1} \bar{\mathcal{E}}_{s,i} [\mathbf{G}_0^\dagger \mathbf{G}_0]_{ii}$ and the corresponding gain is $\bar{\mathcal{G}}_{bit} = \mathcal{E}_{no\text{-}bit}/\bar{\mathcal{E}}_{bit}$. As the lower bound $\mathcal{E}_{bit}$ may not be achievable when b_k are restricted to be non-negative integers, it follows that $\bar{\mathcal{E}}_{bit} \geq \mathcal{E}_{bit}$ and $\bar{\mathcal{G}}_{bit} \leq \mathcal{G}_{bit}$. From previous discussions, we know that the theoretical gain $\mathcal{G}_{bit}$ is independent of the bit rate b and the symbol error rate SER. The same is not true for $\bar{\mathcal{G}}_{bit}$. See Problem 8.1 for an example.

We can modify Algorithm 8.1 for QAM symbols. In this case, the bit adjustment in Steps 4 and 5 is applied only to $\rho/2$ subchannels but $\bar{b}_k$ are increased (or decreased) by 2 bits.

Example 8.1 Consider a zero-forcing transceiver with two subchannels (that is, $M = 2$). Suppose that $\left[\mathbf{G}_0^\dagger \mathbf{G}_0 \right]_{00} = \left[\mathbf{G}_0^\dagger \mathbf{G}_0 \right]_{11} = 1$. Let the noise variances of the two subchannels be

$$\sigma_{e_0}^2 = \frac{1}{1 + \alpha^2}, \quad \sigma_{e_1}^2 = \frac{\alpha^2}{1 + \alpha^2},$$

for some $0 < \alpha \leq 1$. So the ratio and the sum of the two noise variances are $1/\alpha^2$ and 1, respectively. The theoretical bit allocation gain is therefore given by

$$\mathcal{G}_{bit} = \frac{1 + \alpha^2}{2\alpha} = 0.5(\alpha + 1/\alpha).$$

Observe that the gain $\mathcal{G}_{bit}$ has a minimum of one when $\alpha = 1$, in which case the two subchannels have the same noise variance of $1/2$. As α decreases, the disparity of the two noise variances increases and the gain increases. To see how the bits are allocated, let us consider $b = 3$. In Table 8.1, we list the optimal bit allocation and the corresponding gain for both the theoretical optimal case and the suboptimal case of non-negative integer bit allocation as obtained by Algorithm 8.1. As we can see from the table, when α decreases, more bits are assigned to Subchannel #1. This is consistent with the fact that Subchannel #1 is increasingly better than Subchannel #0 as α approaches zero. Note that, although the gain for the suboptimal case is smaller, the gain can still be quite substantial even when the high bit rate assumption is not valid (the average bit rate b is only 3 in this example). ∎

α^2	b_0, b_1	$\mathcal{G}_{bit}$	$\bar{b}_0, \bar{b}_1$	$\bar{\mathcal{G}}_{bit}$
1	3, 3	1	3, 3	1
10^{-1}	2.17, 3.83	1.74	2, 4	1.69
10^{-2}	1.34, 4.66	5.05	1, 5	4.54
10^{-3}	0.51, 5.49	15.83	1, 5	12.75

Table 8.1. **Example 8.1.** Comparison of theoretical and non-negative integer bit allocations

Example 8.2 In this example, we demonstrate the effect of bit allocation on the BER performance of CP-OFDM system. The following two channels are considered:

$$C_0(z) = 1 + 2z^{-1},$$
$$C_1(z) = 1 + 0.95z^{-1}.$$

The number of subchannels is $M = 64$ and the CP length is $\nu = 1$. The average bit rate per symbol is $b = 4$. The modulation symbols s_k are QAM symbols with Gray code mapping. The symbol s_k satisfies the conjugate symmetric property in (6.33) so that the transmitted signal has real value. Non-negative integer bit allocation is applied (Algorithm 8.1). The transmitted power is $\mathcal{E}$ and the channel noise is assumed to be white and Gaussian with variance $\mathcal{N}_0$. So the SNR is $\mathcal{E}/\mathcal{N}_0$. Figure 8.7(a) shows the number of bits allocated to the subchannels. Due to the conjugate symmetric property, the kth and $(M - k)$th subchannels are allocated the same number of bits. From the plot, one realizes that for $C_0(z)$ all the subchannels have the same $\bar{b}_k$ because this channel is fairly flat. On the other hand, $C_1(z)$ is a lowpass filter; its gain in the high-frequency region is much smaller than that in the low-frequency region. As a result, no bit is transmitted over the 31st, 32nd, and 33rd subchannels, which correspond to the frequency bins centered at $\pi - \pi/32$, π, and $\pi + \pi/32$, respectively. Figures 8.7(b) and (c) show the BER results for $C_0(z)$ and $C_1(z)$, respectively. For the purpose of comparison, we have also included the BER curve for the zero-forcing SC-CP system with no bit allocation in these figures (as $M = 64$ is much larger than $\nu = 1$, the bit allocation gain is negligible for the SC-CP system). From these figures, we observe that, for $C_0(z)$, the SC-CP system has a better performance for a moderate BER, as explained in Chapter 7. For $C_1(z)$, the bit allocation significantly improves the BER performance of the CP-OFDM system. For a BER of 10^{-6}, the gain is more than 10 dB.

It is interesting to note that for highly frequency-selective channels, *the CP-OFDM system outperforms the SC-CP system when bit allocation is applied*. Moreover, the bit allocation gain can be quite substantial. ∎

Under optimal bit allocation, the minimized transmitted power $\mathcal{E}_{bit}$ depends on $[\mathbf{G}_0^\dagger \mathbf{G}_0]_{kk}$ and $\sigma_{e_k}^2$, where the former is determined by the transmitter and the latter is related to the receiver. In the following, we will show how to choose the transmitter and receiver so that $\mathcal{E}_{bit}$ is minimized. Because the ZP and ZJ systems (Figs. 8.2 and 8.3) have different structures, we will discuss the optimization of these systems separately in the following two sections.

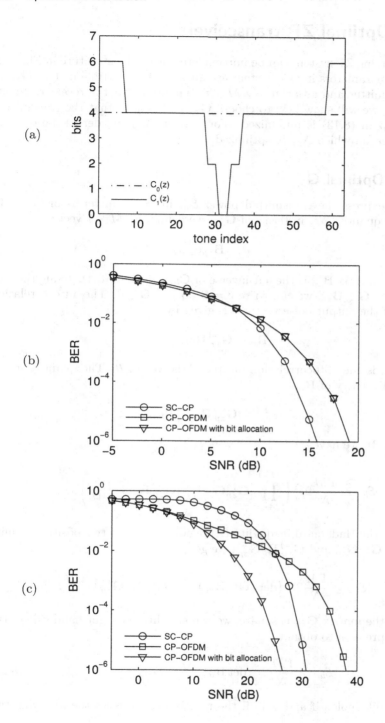

Figure 8.7. Example 8.2. (a) Bit allocation; (b) BER for $C_0(z) = 1 + 2z^{-1}$; (c) BER for $C_1(z) = 1 + 0.95z^{-1}$.

8.4 Optimal ZP transceivers

A zero-forcing ZP system can be implemented using the structure in Fig. 8.2. The free parameters in this system are an $M \times M$ invertible matrix $\mathbf{G}_{zp}$ at the transmitter and an arbitrary $M \times \nu$ matrix $\mathbf{A}_{zp}$ at the receiver. In the following, we will show how to choose $\mathbf{G}_{zp}$ and $\mathbf{A}_{zp}$ so that the transmitted power $\mathcal{E}_{bit}$ in (8.13) is minimized. For a given $\mathbf{A}_{zp}$, we derive the optimal $\mathbf{G}_{zp}$, based on which $\mathbf{A}_{zp}$ is optimized.

8.4.1 Optimal $\mathbf{G}_{zp}$

We first express the transmitted power $\mathcal{E}_{bit}$ in (8.13) in terms of $\mathbf{G}_{zp}$. To write the quantity $\sigma_{e_k}^2$ in terms of $\mathbf{G}_{zp}$, we define the $M \times 1$ vector

$$\boldsymbol{\theta} = \mathbf{B}_{zp}\mathbf{q}_{zp},$$

where the matrix $\mathbf{B}_{zp}$ is the left inverse of $\mathbf{C}_{low}$ given in (8.4). Using the fact that $\mathbf{S}_{zp} = \mathbf{G}_{zp}^{-1}\mathbf{B}_{zp}$, we can write $\mathbf{e} = \mathbf{S}_{zp}\mathbf{q}_{zp} = \mathbf{G}_{zp}^{-1}\boldsymbol{\theta}$. The autocorrelation matrix of the output noise vector $\mathbf{e}$ is given by

$$\mathbf{R}_e = \mathbf{G}_{zp}^{-1}\mathbf{R}_\theta\mathbf{G}_{zp}^{-\dagger},$$

where $\mathbf{R}_\theta$ is the autocorrelation matrix of the vector $\boldsymbol{\theta}$. The quantity $\sigma_{e_k}^2$ is the (k,k)th entry of $\mathbf{R}_e$:

$$\sigma_{e_k}^2 = [\mathbf{G}_{zp}^{-1}\mathbf{R}_\theta\mathbf{G}_{zp}^{-\dagger}]_{kk}.$$

From (8.13), we have

$$\mathcal{E}_{bit} = \frac{\Gamma M}{N}2^{2b}\left(\prod_{k=0}^{M-1}[\mathbf{G}_{zp}^\dagger\mathbf{G}_{zp}]_{kk}\,[\mathbf{G}_{zp}^{-1}\mathbf{R}_\theta\mathbf{G}_{zp}^{-\dagger}]_{kk}\right)^{1/M}.$$

Applying the Hadamard inequality (Appendix A) to the two positive definite matrices $\mathbf{G}_{zp}^\dagger\mathbf{G}_{zp}$ and $\mathbf{G}_{zp}^{-1}\mathbf{R}_\theta\mathbf{G}_{zp}^{-\dagger}$, we get

$$\mathcal{E}_{bit} \geq \frac{\Gamma M}{N}2^{2b}\left(\det\left(\mathbf{G}_{zp}^\dagger\mathbf{G}_{zp}\right)\det\left(\mathbf{G}_{zp}^{-1}\mathbf{R}_\theta\mathbf{G}_{zp}^{-\dagger}\right)\right)^{1/M}.$$

Because the matrix $\mathbf{G}_{zp}$ is square, we can simplify the right-hand side of the above expression to obtain

$$\mathcal{E}_{bit} \geq \frac{\Gamma M}{N}2^{2b}(\det \mathbf{R}_\theta)^{1/M} \triangleq \mathcal{E}_{bit,G_{zp}}. \tag{8.21}$$

The equality holds if and only if the matrix $\mathbf{G}_{zp}$ satisfies the following two conditions:

(i) $\mathbf{G}_{zp}^\dagger\mathbf{G}_{zp}$ is diagonal;

(ii) $\mathbf{G}_{zp}^{-1}\mathbf{R}_\theta\mathbf{G}_{zp}^{-\dagger}$ is diagonal.

Note that the lower bound $\mathcal{E}_{bit,G_{zp}}$ does not depend on the matrix $\mathbf{G}_{zp}$ because $\boldsymbol{\theta}$ is independent of $\mathbf{G}_{zp}$. This lower bound is achieved if and only if $\mathbf{G}_{zp}$ satisfies both conditions (i) and (ii). The first condition implies that the columns of $\mathbf{G}_{zp}$ are orthogonal, whereas the second condition means that $\mathbf{G}_{zp}^{-1}$ decorrelates the noise vector $\boldsymbol{\theta}$. As $\mathbf{R}_\theta$ is positive semidefinite, it is always diagonalizable by some unitary matrix. Let us decompose the matrix $\mathbf{R}_\theta$ as

$$\mathbf{R}_\theta = \mathbf{Q}_{zp}\boldsymbol{\Sigma}\mathbf{Q}_{zp}^\dagger,$$

where $\mathbf{Q}_{zp}$ is an $M \times M$ unitary matrix whose columns are the eigenvectors of $\mathbf{R}_\theta$ and $\boldsymbol{\Sigma}$ is a diagonal matrix consisting of the corresponding eigenvalues. Combining conditions (i) and (ii), we conclude that $\mathbf{G}_{zp}$ is a matrix of the form $\mathbf{G}_{zp} = \mathbf{Q}_{zp}\mathbf{D}$, where $\mathbf{D}$ is an arbitrary invertible diagonal matrix. Since $\mathbf{G}_{zp} = \mathbf{Q}_{zp}\mathbf{D}$ achieves the lower bound for any invertible $\mathbf{D}$, without loss of generality we can choose $\mathbf{D} = \mathbf{I}$ and we can choose the matrix $\mathbf{G}_{zp}$ as

$$\mathbf{G}_{zp} = \mathbf{Q}_{zp}.$$

With this choice of $\mathbf{G}_{zp}$ and using the fact that $\mathbf{S}_{zp} = \mathbf{G}_{zp}^{-1}\mathbf{B}_{zp}$, we can express the receiving matrix $\mathbf{S}_{zp}$ as

$$\mathbf{S}_{zp} = \mathbf{Q}_{zp}^\dagger\mathbf{V}_{zp}\boldsymbol{\Lambda}^{-1}[\mathbf{I}_M \ \ \mathbf{A}_{zp}]\mathbf{U}_{zp}^\dagger. \tag{8.22}$$

Observe that when $\mathbf{G}_{zp}$ is chosen optimally, the autocorrelation matrix of the output noise vector $\mathbf{e}$ becomes

$$\mathbf{R}_e = \mathbf{G}_{zp}^{-1}\mathbf{R}_\theta\mathbf{G}_{zp}^{-\dagger} = \boldsymbol{\Sigma}.$$

That is, $E[e_i e_j^\dagger] = 0$ for $i \neq j$; the output noise components are uncorrelated. In summary, the optimal transmitting matrix $\mathbf{G}_{zp}$ is a unitary matrix whose inverse decorrelates the output noise vector.

Note that the transmitted power is equal to $\mathcal{E}_{bit,G_{zp}}$ when the matrix $\mathbf{G}_{zp}$ is chosen as $\mathbf{Q}_{zp}$ and the bits are optimally allocated according to the formula in (8.17). The quantity $\mathcal{E}_{bit,G_{zp}}$ depends on $\mathbf{R}_\theta$. Recall that $\boldsymbol{\theta} = \mathbf{B}_{zp}\mathbf{q}_{zp}$, where $\mathbf{B}_{zp}$ depends on $\mathbf{A}_{zp}$ as given in (8.4). Next we are going to show how to choose $\mathbf{A}_{zp}$ so that $\mathcal{E}_{bit,G_{zp}}$ is minimized.

8.4.2   Optimal $\mathbf{A}_{zp}$

It follows from (8.21) that minimizing $\mathcal{E}_{bit,G_{zp}}$ is equivalent to minimizing $\det(\mathbf{R}_\theta)$. To derive the optimal $\mathbf{A}_{zp}$ that minimizes $\det(\mathbf{R}_\theta)$, let us decompose the $N \times N$ unitary matrix $\mathbf{U}_{zp}$ in (8.3) as

$$\mathbf{U}_{zp} = [\mathbf{U}_0 \ \ \mathbf{U}_1], \tag{8.23}$$

where $\mathbf{U}_0$ and $\mathbf{U}_1$ consist of the first M and the last ν columns of $\mathbf{U}_{zp}$, respectively. Using this decomposition, we can redraw the receiver as Fig. 8.8, in which we show only the noise component. Let the noise vectors $\boldsymbol{\mu}_0$, $\boldsymbol{\mu}_1$, $\boldsymbol{\rho}_0$, and $\boldsymbol{\tau}$ be, respectively, defined as in Fig. 8.8. Then the autocorrelation matrices of $\boldsymbol{\theta}$ and $\boldsymbol{\tau}$ are related by

$$\mathbf{R}_\theta = \mathbf{V}_{zp}\mathbf{R}_\tau\mathbf{V}_{zp}^\dagger.$$

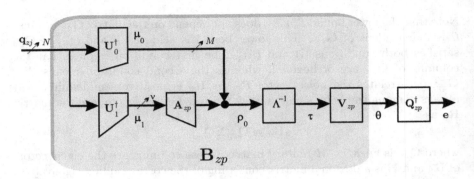

Figure 8.8. Noise path at the receiver of a ZP system.

As $\mathbf{V}_{zp}$ is unitary, we have $\det(\mathbf{R}_\theta) = \det(\mathbf{R}_\tau)$. It follows from Fig. 8.8 that

$$\det(\mathbf{R}_\tau) = \det(\mathbf{\Lambda}^{-2}) \det(\mathbf{R}_{\rho_0}),$$

where $\mathbf{R}_{\rho_0}$ is the autocorrelation matrix of ρ_0. As the matrix $\mathbf{\Lambda}^{-2}$ is independent of $\mathbf{A}_{zp}$, minimizing $\det(\mathbf{R}_\tau)$ (and hence $\det(\mathbf{R}_\theta)$) is equivalent to minimizing $\det(\mathbf{R}_{\rho_0})$. To solve such a problem, let us define the $\nu \times 1$ vector ρ_1, the $(M + \nu) \times 1$ vectors ρ, and μ, respectively, as

$$\rho_1 = \mu_1, \quad \rho = \begin{bmatrix} \rho_0 \\ \rho_1 \end{bmatrix}, \quad \text{and} \quad \mu = \begin{bmatrix} \mu_0 \\ \mu_1 \end{bmatrix}.$$

Then, using the fact that $\rho_0 = \mu_0 + \mathbf{A}_{zp}\mu_1$, we have

$$\rho = \begin{bmatrix} \mathbf{I}_M & \mathbf{A}_{zp} \\ \mathbf{0} & \mathbf{I}_\nu \end{bmatrix} \mu,$$

which implies that the autocorrelation matrices of ρ and μ are related by

$$\mathbf{R}_\rho = \begin{bmatrix} \mathbf{I}_M & \mathbf{A}_{zp} \\ \mathbf{0} & \mathbf{I}_\nu \end{bmatrix} \mathbf{R}_\mu \begin{bmatrix} \mathbf{I}_M & \mathbf{A}_{zp} \\ \mathbf{0} & \mathbf{I}_\nu \end{bmatrix}^\dagger. \tag{8.24}$$

Moreover, from the definitions of the vectors ρ, ρ_0, and ρ_1, we know that the autocorrelation matrix of ρ can also be written as

$$\mathbf{R}_\rho = \begin{bmatrix} \mathbf{R}_{\rho_0} & \mathbf{R}_{\rho_0\rho_1} \\ \mathbf{R}_{\rho_0\rho_1}^\dagger & \mathbf{R}_{\rho_1} \end{bmatrix},$$

where $\mathbf{R}_{\rho_0\rho_1} = E[\rho_0\rho_1^\dagger]$ is the cross-correlation matrix of ρ_0 and ρ_1. Using the Fischer inequality (Appendix A) for positive definite matrices, we have

$$\det(\mathbf{R}_{\rho_0}) \geq \frac{\det(\mathbf{R}_\rho)}{\det(\mathbf{R}_{\rho_1})},$$

with equality if and only if $\mathbf{R}_{\rho_0\rho_1} = \mathbf{0}$. Using (8.24) and the fact that $\mu = \mathbf{U}_{zp}^\dagger \mathbf{q}_{zp}$, we have $\det(\mathbf{R}_\rho) = \det(\mathbf{R}_\mu) = \det(\mathbf{R}_{q_{zp}})$ since the matrix

$\mathbf{U}_{zp}$ is unitary. Because $\rho_1 = \mu_1$, we have $\det(\mathbf{R}_{\rho_1}) = \det(\mathbf{R}_{\mu_1})$. Using these relations, the above equation can be expressed as

$$\det(\mathbf{R}_{\rho_0}) \geq \frac{\det(\mathbf{R}_{q_{zp}})}{\det(\mathbf{R}_{\mu_1})}.$$

Because both $\mathbf{q}_{zp}$ and μ_1 are independent of $\mathbf{A}_{zp}$, the above lower bound is also independent of $\mathbf{A}_{zp}$. Moreover, this lower bound can be obtained if and only if $\mathbf{A}_{zp}$ is chosen such that

$$\mathbf{R}_{\rho_0 \rho_1} = E\big[(\mu_0 + \mathbf{A}_{zp}\mu_1)\mu_1^\dagger\big] = \mathbf{0}. \tag{8.25}$$

Solving the above equation, the optimal $\mathbf{A}_{zp}$ is uniquely given by

$$\mathbf{A}_{zp} = -\mathbf{U}_0^\dagger \mathbf{R}_{q_{zp}} \mathbf{U}_1 \left(\mathbf{U}_1^\dagger \mathbf{R}_{q_{zp}} \mathbf{U}_1\right)^{-1}. \tag{8.26}$$

It can be shown (Problem 8.4) that when $\mathbf{A}_{zp}$ is chosen as above, the matrix $-\mathbf{A}_{zp}$ is the optimal estimator of μ_0 given the observation of μ_1. In fact, the solution of $\mathbf{A}_{zp}$ in (8.26) minimizes not only $\det(\mathbf{R}_e)$ (which is also equal to $\det(\mathbf{R}_\theta)$ because $\mathbf{Q}_{zp}$ is unitary), but also the total output noise power given by $\sum_{k=0}^{M-1} \sigma_{e_k}^2$ or $trace(\mathbf{R}_e)$. To see this, we consider Fig. 8.8. Since the vectors $\mathbf{e}$ and $\boldsymbol{\tau}$ are related through the unitary matrix $\mathbf{Q}_{zp}^\dagger \mathbf{V}_{zp}$, which preserves traces, we have

$$trace(\mathbf{R}_\tau) = trace(\mathbf{R}_e).$$

On the other hand, because $\mathbf{\Lambda}$ is diagonal, the diagonal entries of $\mathbf{R}_\tau$ and $\mathbf{R}_{\rho_0}$ are related as

$$[\mathbf{R}_\tau]_{kk} = [\mathbf{R}_{\rho_0}]_{kk}/[\mathbf{\Lambda}]_{kk}^2.$$

Thus the minimization of $trace(\mathbf{R}_\tau)$ can be achieved if each term $[\mathbf{R}_{\rho_0}]_{kk}$ can be individually minimized. As $-\mathbf{A}_{zp}$ is the best estimator of μ_0, the kth row of $-\mathbf{A}_{zp}$ minimizes $[\mathbf{R}_{\rho_0}]_{kk}$ for $0 \leq k < M$. Therefore, the solution of $\mathbf{A}_{zp}$ given in (8.26) is also optimal for minimizing the total output noise $trace(\mathbf{R}_e)$.

8.4.3 Summary and discussions

In the following, we summarize the design procedure for the optimal ZP system.

(1) Form the $N \times M$ matrix $\mathbf{C}_{low}$ in (8.2) and compute its SVD $\mathbf{C}_{low} = \mathbf{U}_{zp}[\mathbf{\Lambda} \ \mathbf{0}]^T \mathbf{V}_{zp}^\dagger$. Partition the unitary matrix $\mathbf{U}_{zp}$ as (8.23).

(2) Given the noise autocorrelation matrix $\mathbf{R}_{q_{zp}}$, calculate the optimal $M \times \nu$ matrix $\mathbf{A}_{zp}$ in (8.26).

(3) Compute the matrix $\mathbf{B}_{zp}$ in (8.4). Find a unitary matrix $\mathbf{Q}_{zp}$ such that the matrix $\mathbf{Q}_{zp}^\dagger \mathbf{B}_{zp} \mathbf{R}_{q_{zp}} \mathbf{B}_{zp}^\dagger \mathbf{Q}_{zp}$ is diagonal. The optimal transmitting matrix is $\mathbf{G}_{zp} = \mathbf{Q}_{zp}$.

(4) The optimal receiving matrix is $\mathbf{S}_{zp} = \mathbf{Q}_{zp}^\dagger \mathbf{B}_{zp}$.

(5) The optimal bit allocation is given by (8.17) or by Algorithm 8.1.

The minimum transmitted power of the optimal ZP transceiver is given by

$$\mathcal{E}_{zp,opt} = \frac{\Gamma M}{N} 2^{2b} \left[\det(\mathbf{\Lambda}^{-2}) \frac{\det(\mathbf{R}_{q_{zp}})}{\det(\mathbf{U}_1^\dagger \mathbf{R}_{q_{zp}} \mathbf{U}_1)} \right]^{1/M}. \tag{8.27}$$

Some remarks on the optimal ZP systems are listed below.

(1) As the optimal $\mathbf{G}_{zp} = \mathbf{Q}_{zp}$ is unitary, the transmitting matrix $\mathbf{G}_0 = [\mathbf{G}_{zp}^T \ \mathbf{0}]^T$ satisfies $\mathbf{G}_0^\dagger \mathbf{G}_0 = \mathbf{I}_M$, i.e. the columns of $\mathbf{G}_0$ are orthonormal. This means that the modulation symbols s_i are transmitted using orthonormal vectors. The optimal ZP transceiver has an **orthonormal transmitter**. There is no loss of generality in using orthonormal transmitters.

(2) As $[\mathbf{G}_0^\dagger \mathbf{G}_0]_{kk} = 1$ for all k, from (8.14) we can conclude that all modulation symbols s_k have the same signal power $\mathcal{E}_{s,k}$; no power allocation is needed for the optimal ZP system.

(3) **White noise case.** When the channel noise $q(n)$ is white, the autocorrelation matrix of the noise vector $\mathbf{q}_{zp}$ becomes $\mathbf{R}_{q_{zp}} = \mathcal{N}_0\mathbf{I}$. In this case, the optimal $\mathbf{A}_{zp}$ given in (8.26) becomes

$$\mathbf{A}_{zp} = -\mathbf{U}_0^\dagger \mathbf{U}_1 \left(\mathbf{U}_1^\dagger \mathbf{U}_1 \right)^{-1} = \mathbf{0},$$

where we have used $\mathbf{U}_0^\dagger \mathbf{U}_1 = \mathbf{0}$. Substituting $\mathbf{A}_{zp} = \mathbf{0}$ into (8.4), we have

$$\mathbf{B}_{zp} = \mathbf{V}_{zp} \mathbf{\Lambda}^{-1} \left[\ \mathbf{I}_M \quad \mathbf{0} \ \right] \mathbf{U}_{zp}^\dagger.$$

One can verify that $\mathbf{B}_{zp}$ is in fact equal to $(\mathbf{C}_{low}^\dagger \mathbf{C}_{low})^{-1}\mathbf{C}_{low}^\dagger$, the pseudo-inverse of $\mathbf{C}_{low}$. The noise vector $\boldsymbol{\theta}$ has the autocorrelation matrix

$$\mathbf{R}_\theta = \mathcal{N}_0(\mathbf{C}_{low}^\dagger \mathbf{C}_{low})^{-1} = \mathbf{V}_{zp}(\mathbf{\Lambda}^\dagger \mathbf{\Lambda})^{-1}\mathbf{V}_{zp}^\dagger.$$

As the optimal $\mathbf{G}_{zp}$ is such that $\mathbf{G}_{zp}^{-1}$ decorrelates $\mathbf{R}_\theta$, we can choose $\mathbf{G}_{zp} = \mathbf{V}_{zp}$. Therefore, when $q(n)$ is white, the optimal transmitting and receiving matrices simplify to

$$\mathbf{G}_{zp} = \mathbf{V}_{zp}, \quad \mathbf{S}_{zp} = \mathbf{\Lambda}^{-1}[\mathbf{I}_M \ \mathbf{0}]\mathbf{U}_{zp}^\dagger. \tag{8.28}$$

From the SVD expression $\mathbf{C}_{low} = \mathbf{U}_{zp} \begin{bmatrix} \mathbf{\Lambda} \\ \mathbf{0} \end{bmatrix} \mathbf{V}_{zp}^\dagger$, we can see how the optimal transceiver works. The transmitting matrix $\mathbf{G}_{zp}$ cancels the effect of $\mathbf{V}_{zp}^\dagger$ in $\mathbf{C}_{low}$, whereas the receiving matrix $\mathbf{S}_{zp}$ nullifies the effects of $\mathbf{U}_{zp}$ and $\mathbf{\Lambda}$ in $\mathbf{C}_{low}$. The minimized transmitted power is given by

$$\mathcal{E}_{zp,opt} = \frac{\Gamma M}{N} 2^{2b} \mathcal{N}_0 \left[\det(\mathbf{\Lambda}^{-2}) \right]^{1/M}.$$

It can be shown that the diagonal matrix $\mathbf{\Lambda}$ is independent of the phase of $C(e^{j\omega})$ [163]. This implies that the minimized transmitted power is independent of the phase of $C(e^{j\omega})$; the channel phase does not affect the performance of the optimal ZP system when the noise is white. For example, if $C(z)$ has the factor $(1 - 0.5z^{-1})$ and we replace it by $(0.5 - z^{-1})$, the performance of the optimal transceiver remains the same.

(4) When the channel noise $q(n)$ is white and the channel $c(n)$ is frequency-nonselective, the optimal transceiver is not unique (see Problem 8.8). One of the solutions is given by $\mathbf{U}_{zp} = \mathbf{I}_N$ and $\mathbf{V}_{zp} = \mathbf{I}_M$. In this case, the optimal ZP transceiver reduces to the SC-ZP system with $\mathbf{G}_{zp} = \mathbf{I}_M$ and $\mathbf{S}_{zp} = [\mathbf{I}_M \ \ \mathbf{0}]$.

Example 8.3 Optimal ZP systems. In this example, we evaluate the performance of the optimal ZP transceiver in an ADSL environment. The channel $C(z)$ is the equivalent channel after a time-domain equalizer (TEQ) is used to shorten DSL Loop 6 [7]. The TEQ is a fourth order MSSNR TEQ and the shortened channel has order $\nu = 4$. The noise is colored and it includes AWGN, NEXT and FEXT [7]. The magnitude response of the channel $C(z)$ and the power spectrum of the colored noise $q(n)$ used are shown, respectively, in Fig. 8.9(a) and (b). The symbol error rate is $SER = 10^{-6}$ and the average bit rate per sample is $R_b = bM/N = 2$. We compare the transmitted powers of ZP systems optimized under three different scenarios: (i) the optimal ZP system; (ii) the ZP system optimized under the constraint $\mathbf{A}_{zp} = \mathbf{0}$; (iii) the ZP system optimized under the assumption that the noise is white although the actual noise is colored. The three transmitted powers are respectively denoted by $\mathcal{E}_{zp,opt}$, $\mathcal{E}_{zp,\mathbf{A}=\mathbf{0}}$, and $\mathcal{E}_{zp,wh}$. Figure 8.10 shows the results. As expected, the transmitted power of the optimal ZP system $\mathcal{E}_{zp,opt}$ is the smallest. By comparing $\mathcal{E}_{zp,opt}$ and $\mathcal{E}_{zp,\mathbf{A}=\mathbf{0}}$, we see that, by optimizing $\mathbf{A}_{zp}$, we get an improvement of around 2 dB for $M = 10$ and 0.5 dB for $M = 50$. This gain reduces as the number of subchannels M increases because, for large M, the dimension M of the signal subspace is almost as large as the dimension $(M + \nu)$ of the received signal. Also observe from the figure that $\mathcal{E}_{zp,opt}$ is much smaller than $\mathcal{E}_{zp,wh}$. The gain can still be as large as 5 dB for $M = 64$. Thus, by exploiting the noise correlation, we can significantly improve the system performance. ∎

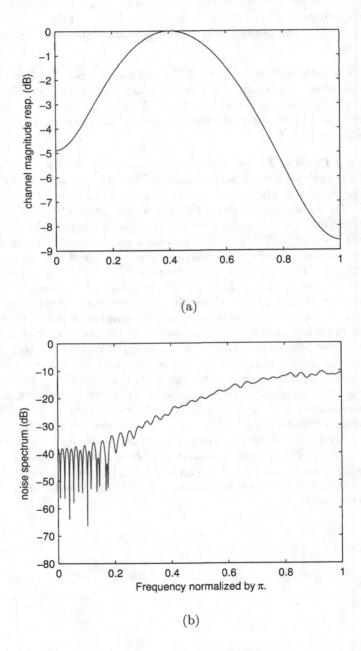

(a)

(b)

Figure 8.9.   Example 8.3. (a) Magnitude response $|C(e^{j\omega})|$ of a TEQ shortened channel from an ADSL environment; (b) the noise spectrum.

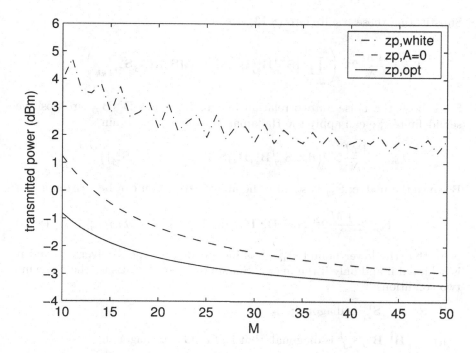

Figure 8.10. Example 8.3. Comparison of transmitted power for three different ZP systems.

8.5 Optimal zero-jamming (ZJ) transceivers

In this section, we derive the optimal zero-forcing ZJ system. The block diagram of a ZJ system was shown in Fig. 8.3. Recall from Section 8.1.2 that the free parameters that can be designed in a zero-forcing ZJ system are an $M \times M$ invertible receiving matrix $\mathbf{S}_{zj}$ and a $\nu \times M$ matrix $\mathbf{A}_{zj}$ at the transmitter. For a given $\mathbf{A}_{zj}$, we will derive the optimal $\mathbf{S}_{zj}$, based on which $\mathbf{A}_{zj}$ is optimized.

8.5.1 Optimal $\mathbf{S}_{zj}$

We will first express the transmitted power $\mathcal{E}_{bit}$ in (8.13) in terms of $\mathbf{S}_{zj}$. As $\mathbf{e} = \mathbf{S}_{zj}\mathbf{q}_{zj}$, the noise variance $\sigma_{e_k}^2$ of the kth subchannel is given by

$$\sigma_{e_k}^2 = \left[\mathbf{S}_{zj}\mathbf{R}_{q_{zj}}\mathbf{S}_{zj}^{\dagger}\right]_{kk}.$$

And since $\mathbf{G}_0 = \mathbf{G}_{zj} = \mathbf{B}_{zj}\mathbf{S}_{zj}^{-1}$, we have

$$\left[\mathbf{G}_0^{\dagger}\mathbf{G}_0\right]_{kk} = \left[\mathbf{S}_{zj}^{-\dagger}\mathbf{B}_{zj}^{\dagger}\mathbf{B}_{zj}\mathbf{S}_{zj}^{-1}\right]_{kk}.$$

Substituting these results into (8.13), we get

$$\mathcal{E}_{bit} = \frac{\Gamma M}{N} 2^{2b} \left(\prod_{k=0}^{M-1} \left[\mathbf{S}_{zj}^{-\dagger} \mathbf{B}_{zj}^{\dagger} \mathbf{B}_{zj} \mathbf{S}_{zj}^{-1} \right]_{kk} \left[\mathbf{S}_{zj} \mathbf{R}_{q_{zj}} \mathbf{S}_{zj}^{\dagger} \right]_{kk} \right)^{1/M}.$$

Since both the noise autocorrelation matrix $\mathbf{R}_{q_{zj}}$ and $\mathbf{B}_{zj}^{\dagger} \mathbf{B}_{zj}$ are positive semidefinite, we can apply the Hadamard inequality to obtain

$$\mathcal{E}_{bit} \geq \frac{\Gamma M}{N} 2^{2b} \left(\det[\mathbf{S}_{zj}^{-\dagger} \mathbf{B}_{zj}^{\dagger} \mathbf{B}_{zj} \mathbf{S}_{zj}^{-1}] \det[\mathbf{S}_{zj} \mathbf{R}_{q_{zj}} \mathbf{S}_{zj}^{\dagger}] \right)^{1/M}.$$

Because the matrix $\mathbf{S}_{zj}$ is square, the above expression can be written as

$$\mathcal{E}_{bit} \geq \frac{\Gamma M}{N} 2^{2b} \left(\det[\mathbf{B}_{zj}^{\dagger} \mathbf{B}_{zj}] \det[\mathbf{R}_{q_{zj}}] \right)^{1/M} \triangleq \mathcal{E}_{bit,S_{zj}}. \qquad (8.29)$$

Note that the lower bound $\mathcal{E}_{bit,S_{zj}}$ is independent of the receiver $\mathbf{S}_{zj}$ and it is achieved if and only if the matrix $\mathbf{S}_{zj}$ simultaneously satisfies the following two conditions:

(i) $\mathbf{S}_{zj} \mathbf{R}_{q_{zj}} \mathbf{S}_{zj}^{\dagger}$ is diagonal;

(ii) $\mathbf{S}_{zj}^{-\dagger} \mathbf{B}_{zj}^{\dagger} \mathbf{B}_{zj} \mathbf{S}_{zj}^{-1}$ is diagonal; that is, $\mathbf{G}_{zj}^{\dagger} \mathbf{G}_{zj}$ is diagonal.

The first condition means that the receiver $\mathbf{S}_{zj}$ decorrelates the noise vector $\mathbf{q}_{zj}$, whereas the second condition implies that the transmitter $\mathbf{G}_{zj}$ is orthogonal. To derive the optimal $\mathbf{S}_{zj}$ that satisfies both conditions, we first decompose the positive definite matrix[4] $\mathbf{R}_{q_{zj}}$ as

$$\mathbf{R}_{q_{zj}} = \mathbf{R}_{q_{zj}}^{1/2} \mathbf{R}_{q_{zj}}^{1/2},$$

for some positive definite matrix $\mathbf{R}_{q_{zj}}^{1/2}$. One way to identify $\mathbf{R}_{q_{zj}}^{1/2}$ is the Cholesky decomposition. The matrix $\mathbf{R}_{q_{zj}}^{1/2}$ is also known as the Hermitian square root of $\mathbf{R}_{q_{zj}}$. Using the above decomposition, it can be verified by direct substitution that Condition (i) is satisfied if $\mathbf{S}_{zj}$ has the following form:

$$\mathbf{S}_{zj} = \mathbf{D} \mathbf{Q}_{zj} \mathbf{R}_{q_{zj}}^{-1/2}, \qquad (8.30)$$

where $\mathbf{Q}_{zj}$ is any $M \times M$ unitary matrix and $\mathbf{D}$ is any invertible diagonal matrix. Using the above expression, we see that Condition (ii) is equivalent to saying that the matrix

$$\mathbf{D}^{-\dagger} \mathbf{Q}_{zj} \mathbf{R}_{q_{zj}}^{1/2} \mathbf{B}_{zj}^{\dagger} \mathbf{B}_{zj} \mathbf{R}_{q_{zj}}^{1/2} \mathbf{Q}_{zj}^{-1} \mathbf{D}^{-1}$$

is diagonal. This is true if the unitary matrix $\mathbf{Q}_{zj}$ diagonalizes the positive semidefinite matrix $\mathbf{R}_{q_{zj}}^{1/2} \mathbf{B}_{zj}^{\dagger} \mathbf{B}_{zj} \mathbf{R}_{q_{zj}}^{1/2}$. Therefore Condition (ii) holds if the unitary matrix $\mathbf{Q}_{zj}$ satisfies

$$\mathbf{R}_{q_{zj}}^{1/2} \mathbf{B}_{zj}^{\dagger} \mathbf{B}_{zj} \mathbf{R}_{q_{zj}}^{1/2} = \mathbf{Q}_{zj}^{-1} \Sigma \mathbf{Q}_{zj}$$

---

[4] To avoid degenerated cases, we assume $\mathbf{R}_{q_{zj}}$ is positive definite. In practice, this is almost always true as $\mathbf{R}_{q_{zj}}$ is the autocorrelation matrix of a noise vector.

for some diagonal matrix $\mathbf{\Sigma}$. Therefore we conclude that the lower bound $\mathcal{E}_{bit,S_{zj}}$ given in (8.29) is achievable, and it is achieved if the receiving matrix $\mathbf{S}_{zj}$ is chosen as in (8.30) and the transmitting matrix is selected as

$$\mathbf{G}_{zj} = \mathbf{B}_{zj}\mathbf{R}_{q_{zj}}^{1/2}\mathbf{Q}_{zj}^{\dagger}\mathbf{D}^{-1}. \qquad (8.31)$$

Because this lower bound is achieved for any diagonal $\mathbf{D}$, one can choose $\mathbf{D}$ such that $\mathbf{G}_{zj}^{\dagger}\mathbf{G}_{zj} = \mathbf{I}_M$. That is, *there is no loss of generality in choosing an orthonormal transmitter*, as in the ZP case. It is shown in Problem 8.11 that this normalization condition is satisfied when $\mathbf{D} = \mathbf{\Sigma}^{1/2}$.

Note that once the channel and noise are given, the lower bound $\mathcal{E}_{bit,S_{zj}}$ in (8.29) depends only on $\mathbf{B}_{zj}$, whose only free parameters are the elements of the matrix $\mathbf{A}_{zj}$. In what follows, we will optimize $\mathbf{A}_{zj}$ so that this lower bound is minimized.

8.5.2   Optimal $\mathbf{A}_{zj}$

From the expression for $\mathcal{E}_{bit,S_{zj}}$ in (8.29), we see that the matrix $\mathbf{A}_{zj}$ should be chosen such that $\det[\mathbf{B}_{zj}^{\dagger}\mathbf{B}_{zj}]$ is minimized. Recall from (8.7) that

$$\mathbf{B}_{zj} = \mathbf{V}_{zj}\begin{bmatrix}\mathbf{I}\\\mathbf{A}_{zj}\end{bmatrix}\mathbf{\Lambda}^{-1}\mathbf{U}_{zj}^{\dagger},$$

where $\mathbf{U}_{zj}$ and $\mathbf{V}_{zj}$ are square unitary matrices defined in (8.6). Substituting the above expression into $\det[\mathbf{B}_{zj}^{\dagger}\mathbf{B}_{zj}]$, we have

$$\det[\mathbf{B}_{zj}^{\dagger}\mathbf{B}_{zj}] = \det[\mathbf{\Lambda}^{-2}]\det[\mathbf{I} + \mathbf{A}_{zj}^{\dagger}\mathbf{A}_{zj}].$$

As $\mathbf{A}_{zj}^{\dagger}\mathbf{A}_{zj}$ is positive semidefinite, it can be shown (Problem 8.17) that

$$\det[\mathbf{I} + \mathbf{A}_{zj}^{\dagger}\mathbf{A}_{zj}] \geq 1$$

with equality if and only if $\mathbf{A}_{zj} = \mathbf{0}$. Hence the transmitted power satisfies

$$\mathcal{E}_{bit,S_{zj}} \geq \frac{\Gamma M}{N}2^{2b}\left[\det(\mathbf{\Lambda}^{-2})\det(\mathbf{R}_{q_{zj}})\right]^{1/M} \stackrel{\triangle}{=} \mathcal{E}_{zj,opt}. \qquad (8.32)$$

Equality holds if and only if $\mathbf{A}_{zj} = \mathbf{0}$. Therefore, the optimal transmitter is given by

$$\mathbf{G}_{zj} = \mathbf{V}_0\mathbf{\Lambda}^{-1}\mathbf{U}_{zj}^{\dagger}\mathbf{S}_{zj}^{-1},$$

where $\mathbf{V}_0$ is the $N \times M$ matrix consisting of the first M columns of $\mathbf{V}_{zj}$.

8.5.3   Summary and discussions

The design procedure for the optimal zero-forcing ZJ system is summarized as follows.

(1) Form the matrix $\mathbf{C}_{up}$ in (8.2) and compute its SVD $\mathbf{C}_{up} = \mathbf{U}_{zj}[\mathbf{\Lambda}\ \mathbf{0}]\mathbf{V}_{zj}^{\dagger}$. Find its right inverse $\mathbf{B}_{zj} = \mathbf{V}_0\mathbf{\Lambda}^{-1}\mathbf{U}_{zj}^{\dagger}$, where $\mathbf{V}_0$ is the $N \times M$ matrix consisting of the first M columns of $\mathbf{V}_{zj}$.

(2) Form the matrix $\mathbf{\Psi} = \mathbf{R}_{q_{zj}}^{1/2}\mathbf{B}_{zj}^{\dagger}\mathbf{B}_{zj}\mathbf{R}_{q_{zj}}^{1/2}$, where $\mathbf{R}_{q_{zj}}$ is the $M \times M$ autocorrelation matrix of $\mathbf{q}_{zj}$.

(3) Decompose the matrix $\mathbf{\Psi}$ as $\mathbf{\Psi} = \mathbf{Q}_{zj}^{\dagger}\mathbf{\Sigma}\mathbf{Q}_{zj}$, where $\mathbf{Q}_{zj}^{\dagger}$ is a unitary matrix consisting of the eigenvectors of $\mathbf{\Psi}$ and $\mathbf{\Sigma}$ is a diagonal matrix consisting of the corresponding eigenvalues.

(4) Calculate the optimal transmitting matrix $\mathbf{G}_{zj} = \mathbf{B}_{zj}\mathbf{R}_{q_{zj}}^{1/2}\mathbf{Q}_{zj}^{\dagger}\mathbf{\Sigma}^{-1/2}$.

(5) Calculate the optimal receiving matrix $\mathbf{S}_{zj} = \mathbf{\Sigma}^{1/2}\mathbf{Q}_{zj}\mathbf{R}_{q_{zj}}^{-1/2}$.

(6) Allocate the bits b_k as in (8.17) or by Algorithm 8.1.

Then the minimum transmitted power $\mathcal{E}_{zj,opt}$ is as given in (8.32). In the following, some remarks on the optimal ZJ system are in order.

(1) In the optimal ZJ system, the transmitting matrix $\mathbf{G}_{zj}$ can be chosen to be orthonormal without loss of generality. As a result, we can conclude from (8.14) that no power allocation is needed for the optimal ZJ system and all s_k have the same signal power.

(2) Substituting the expression $\mathbf{S}_{zj} = \mathbf{\Sigma}^{1/2}\mathbf{Q}_{zj}\mathbf{R}_{q_{zj}}^{-1/2}$ into $\mathbf{e} = \mathbf{S}_{zj}\mathbf{q}_{zj}$, one immediately finds that the autocorrelation matrix $\mathbf{R}_e$ of the output noise vector $\mathbf{e}$ is a diagonal matrix. In the optimal ZJ system, the output noise is also decorrelated.

(3) **White noise case.** When the noise is white, $\mathbf{R}_{q_{zj}} = \mathcal{N}_0\mathbf{I}$. It can be shown (Problem 8.12) that the optimal transmitting and receiving matrices reduce to

$$\mathbf{G}_{zj} = \mathbf{V}_0, \quad \mathbf{S}_{zj} = \mathbf{\Lambda}^{-1}\mathbf{U}_{zj}^{\dagger}, \tag{8.33}$$

respectively. From the expression $\mathbf{C}_{up} = \mathbf{U}_{zj}[\mathbf{\Lambda} \ \mathbf{0}]\mathbf{V}_{zj}^{\dagger}$, we see that the receiving matrix cancels the effect of $\mathbf{U}_{zj}$ and $\mathbf{\Lambda}$, whereas the transmitting matrix equalizes the effect of $\mathbf{V}_{zj}^{\dagger}$.

(4) Similar to the ZP case, when the noise is white and the channel is frequency-nonselective, the optimal ZJ system is not unique. One of the solutions is given by

$$\mathbf{G}_{zj} = \begin{bmatrix} \mathbf{I}_M \\ \mathbf{0} \end{bmatrix} \quad \text{and} \quad \mathbf{S}_{zj} = \mathbf{I}_M.$$

In this case, the optimal ZJ system becomes a ZP system and it reduces to the SC-ZP system.

(5) **Comparison with the ZP systems.** In ZJ systems, nonzero prefix samples are padded. There is IBI in the received signal due to the channel. To remove IBI, the receiver retains only M samples of each block. Although the transmitter sends out N samples for every M input symbols, the receiver uses only M samples for decoding. On the other hand, in ZP systems, zeros are padded at the end of every M samples. After passing through the channel, samples are spread to nonoverlapping blocks of length N. As there is no IBI, all the N samples can be used for

decoding. There are more observations than unknowns; the dimension of the signal subspace is M, whereas the received signal has dimension $N = M + \nu$. The eigenstructure of the signal subspace can be exploited to our advantage in ZP systems. Therefore, the performance of ZP systems is generally better than that of ZJ system, as demonstrated in the example below.

(6) **On the optimal solution of the matrices $\mathbf{A}_{zp}$ and $\mathbf{A}_{zj}$.** For ZP systems, the free parameter $\mathbf{A}_{zp}$ can be exploited to improve further the performance. On the other hand, it is found that, for ZJ systems, the optimal $\mathbf{A}_{zj} = \mathbf{0}$. To explain this difference, let us first consider the ZP system. Note that the matrix $\mathbf{A}_{zp}$ is at the receiver. When the channel noise is colored, the matrix $\mathbf{A}_{zp}$ at the receiver can be used to reduce the noise variance (and hence achieve a smaller transmission power) when the noise components are correlated. When the noise is white, the optimal $\mathbf{A}_{zp}$ reduces to a zero matrix. For ZJ systems, the matrix $\mathbf{A}_{zj}$ is at the transmitter. If the input symbols are correlated, then $\mathbf{A}_{zj}$ can be chosen to decorrelate the input. However, in our derivation the input symbols s_k are assumed to be uncorrelated, thus $\mathbf{A}_{zj}$ cannot be exploited to reduce the transmission power further.

Optimal cyclic-prefixed (CP) transceivers. In ZJ systems, the redundant samples are implicitly embedded in the transmission block. One important special case of ZJ systems is the CP system studied in Section 5.4.2. For a CP system, the transmitting and receiving matrices become

$$\mathbf{G}_0 = \mathbf{F}_{cp}\mathbf{G}_{cp} \quad \text{and} \quad \mathbf{S}_0 = \mathbf{S}_{cp}[\mathbf{0} \ \mathbf{I}_M],$$

where both $\mathbf{G}_{cp}$ and $\mathbf{S}_{cp}$ are $M \times M$ matrices and $\mathbf{F}_{cp}$ is the $N \times M$ matrix describing the action of adding CP (see (5.14)), which we reproduce below:

$$\mathbf{F}_{cp} = \begin{bmatrix} \mathbf{0} & \mathbf{I}_\nu \\ \mathbf{I}_{M-\nu} & \mathbf{0} \\ \mathbf{0} & \mathbf{I}_\nu \end{bmatrix}.$$

From Section 5.4.2, we know that the operations of adding CP at the transmitter and removing CP at the receiver convert the LTI channel $C(z)$ to an $M \times M$ circulant matrix $\mathbf{C}_{circ}$ of the form in (5.19). Therefore the CP transceiver is zero-forcing if and only if

$$\mathbf{S}_{cp}\mathbf{C}_{circ}\mathbf{G}_{cp} = \mathbf{I}_M.$$

The above condition implies that all three matrices $\mathbf{S}_{cp}$, $\mathbf{C}_{circ}$, and $\mathbf{G}_{cp}$ are invertible. We know that $\mathbf{C}_{circ}$ is invertible if and only if all the DFT coefficients of the channel are nonzero. Let us assume that $\mathbf{C}_{circ}$ is invertible. For a given invertible receiving matrix $\mathbf{S}_{cp}$, the zero-forcing condition infers that the transmitting matrix is

$$\mathbf{G}_{cp} = (\mathbf{S}_{cp}\mathbf{C}_{circ})^{-1}.$$

Therefore, in a CP system, the only free parameter that can be designed is the invertible matrix $\mathbf{S}_{cp}$. By repeating the earlier optimization process, we can obtain the optimal zero-forcing CP system. This is left as an exercise (Problem 8.18). The design procedure for optimal CP systems is as follows.

(1) Form the matrix $\boldsymbol{\Psi} = (\mathbf{R}_{q_{cp}}^{1/2}\mathbf{C}_{circ}^{-\dagger}\mathbf{F}_{cp}^{\dagger}\mathbf{F}_{cp}\mathbf{C}_{circ}^{-1}\mathbf{R}_{q_{zj}}^{1/2})$, where $\mathbf{R}_{q_{zj}}$ is the autocorrelation matrix of the noise vector $\mathbf{q}_{zj}$.

(2) Decompose the matrix $\boldsymbol{\Psi}$ as $\boldsymbol{\Psi} = \mathbf{Q}_{cp}^{\dagger}\boldsymbol{\Sigma}\mathbf{Q}_{cp}$, where $\mathbf{Q}_{cp}^{\dagger}$ is a unitary matrix consisting of the eigenvectors of $\boldsymbol{\Psi}$, and $\boldsymbol{\Sigma}$ is a diagonal matrix consisting of the corresponding eigenvalues.

(3) Obtain the optimal receiving matrix $\mathbf{S}_0 = [\mathbf{0}\ \mathbf{I}_M]\mathbf{S}_{cp} = \boldsymbol{\Sigma}^{1/2}\mathbf{Q}_{cp}[\mathbf{0}\ \mathbf{I}_M]$.

(4) Obtain the optimal transmitting matrix $\mathbf{G}_0 = \mathbf{F}_{cp}\mathbf{C}_{circ}^{-1}\mathbf{S}_{cp}^{-1}$.

(5) Allocate the bits b_k as in (8.17) or by Algorithm 8.1.

The minimum transmitted power is given by

$$\mathcal{E}_{cp,opt} = \frac{\Gamma M}{N}2^{2b}\left(2^{\nu}\det[\boldsymbol{\Gamma}^{-\dagger}\boldsymbol{\Gamma}^{-1}]\det[\mathbf{R}_{q_{cp}}]\right)^{1/M},$$

where $\boldsymbol{\Gamma}$ is an $M \times M$ diagonal matrix consisting of the DFT coefficients $C_k = C(e^{j2\pi k/M})$. Comparing the above expression with (8.32), one immediately realizes that the ratio of $\mathcal{E}_{cp,opt}$ over $\mathcal{E}_{zj,opt}$ is given by

$$\frac{\mathcal{E}_{cp,opt}}{\mathcal{E}_{zj,opt}} = 2^{\nu/M}\left[\frac{\det[\boldsymbol{\Gamma}^{-\dagger}\boldsymbol{\Gamma}^{-1}]}{\det[\boldsymbol{\Lambda}^{-2}]}\right]^{1/M}.$$

The factor $2^{\nu/M}$ is always larger than unity and it approaches one when M increases. It is found numerically that the ratio

$$\left(\det[\boldsymbol{\Gamma}^{-\dagger}\boldsymbol{\Gamma}^{-1}]/\det[\boldsymbol{\Lambda}^{-2}]\right)^{1/M}$$

is also always larger than unity and it decreases as M increases. *The optimal CP system always needs more transmitted power than the optimal ZJ system.* Moreover, the difference becomes smaller when M increases. This is consistent with the fact that the CP system is a special case of the ZJ system. It can be shown (Problem 8.18) that the optimal transmitting matrix $\mathbf{G}_0$ also satisfies $\mathbf{G}_0^{\dagger}\mathbf{G}_0 = \mathbf{I}_M$; the transmitter is *orthonormal*. Furthermore, the output noise vector of the optimal CP system is also decorrelated.

The CP-OFDM system belongs to the class of CP systems. As the optimal CP system achieves the minimum transmitted power among all zero-forcing systems with a CP, we have $\mathcal{E}_{zj,opt} \leq \mathcal{E}_{cp,opt} \leq \mathcal{E}_{cp\text{-}ofdm}$. In fact, $\mathcal{E}_{cp,opt}$ can be much smaller than $\mathcal{E}_{cp\text{-}ofdm}$, as we shall see in Example 8.4. It is interesting to note that, when the noise is white, the optimal CP transceiver becomes the CP-OFDM system (see Problem 8.13).

Example 8.4 Optimal ZJ systems. The transmission channel and the transmission environment settings are the same as those in Example 8.3. We compare the transmitted power of three ZJ systems: (i) $\mathcal{E}_{zj,opt}$, (ii) $\mathcal{E}_{cp,opt}$, and (iii) $\mathcal{E}_{cp\text{-}ofdm}$. Figure 8.11 shows the results for $M = 10$ to 50. As a

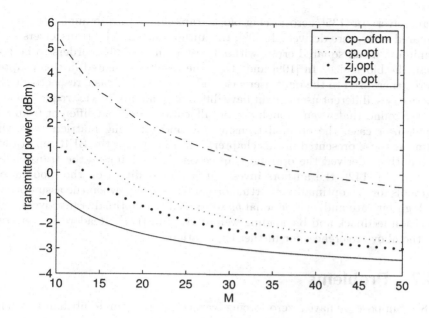

Figure 8.11. **Example 8.4.** Comparison of transmitted power.

comparison, the transmitted power for the optimal ZP system $\mathcal{E}_{zp,opt}$ is also shown in the plot. The optimal ZJ system has approximately 2.5 dB gain ($M = 50$) to 4 dB gain ($M = 10$) over the CP-OFDM systems; the gain of $\mathcal{E}_{zj,opt}$ over $\mathcal{E}_{cp,opt}$ is approximately 1.2 dB for $M = 10$ and 0.3 dB for $M = 50$.

We also see in this example that the optimal ZP system performs better than the optimal ZJ system. The difference can be as large as 2 dB when $M = 10$. The reason is that in ZP systems all the samples in the received block of size N can be employed at the receiver, whereas in ZJ systems only M samples per received block are available. In numerical experiments, it is found that it is true that the optimal ZP system outperforms the optimal ZJ system for most cases. However, as ZP and ZJ systems are two different classes of systems, it is possible to construct toy examples (see Problem 8.19) such that a ZJ system is better. ∎

8.6 Further reading

There have been many reports in the literature on the design of transceivers that are optimized for a variety of different criteria. The design methods described in this chapter were adopted from [73, 75]. One of the early designs of optimal multicarrier transceiver systems was given by [63]. The zero-forcing ZP block transceiver was optimized under the assumption that the channel noise is white and the resulting optimal transceiver is called a vector coding transceiver. In [135], zero-forcing block transceivers that minimize the mean squared error were derived. In [92], the author derived FIR transceivers with minimum mean squared error, without restricting the filter orders to be less than the block size. In [109] and [110], the results presented in this chapter were generalized to the multiuser case and the multiflow case, respectively. In these cases, different users might have different quality of service requirements. It was found that, even though the bit allocation might be different from the single-user case, the optimal transmitting and receiving matrices were the same as those presented in this chapter. In [186], under the BER constraints the authors derived the optimal transceiver that minimizes the transmitted power. In [142], the authors investigated the condition on the modulation symbols for the optimality of orthonormal transceivers when the assumptions of high bit rate and/or fractional b_k were relaxed. Optimal transceivers with decision feedback and bit loading were derived in [177]. Readers are referred to the above references for further exploration.

8.7 Problems

8.1 Suppose we have a zero-forcing transceiver with three subchannels. The output noise variances are, respectively

$$\sigma_{e_0}^2 = 0.1, \quad \sigma_{e_1}^2 = 0.01, \quad \sigma_{e_2}^2 = 0.001.$$

Suppose that the input symbols are PAM and $[\mathbf{G}_0^\dagger \mathbf{G}_0]_{kk} = 1$ for all k. Find the optimal b_k, the non-negative integer $\bar{b}_k$, and the corresponding bit allocation gains for the following cases:

(a) the bit rate per symbol is $b = 3$ and the symbol error rate is $SER = 10^{-6}$;

(b) the bit rate per symbol is $b = 2$ and the symbol error rate is $SER = 10^{-6}$;

(c) the bit rate per symbol is $b = 2$ and the symbol error rate is $SER = 10^{-4}$.

From this example, we see that for the case of non-negative integer bit allocation, the gain depends on both the bit rate and the SER.

8.2 Let $\mathbf{J}_k$ be the $k \times k$ reversal matrix. Prove that the matrices $\mathbf{C}_{low}$ and $\mathbf{C}_{up}$ are related by $\mathbf{C}_{up} = \mathbf{J}_M \mathbf{C}_{low}^T \mathbf{J}_N$.

8.3 Let the SVDs of the $N \times M$ matrix $\mathbf{C}_{low}$ and the $M \times N$ matrix $\mathbf{C}_{up}$ be denoted, respectively, by

$$\mathbf{C}_{low} = \mathbf{U}_{low} \begin{bmatrix} \mathbf{\Lambda}_{low} \\ \mathbf{0} \end{bmatrix} \mathbf{V}_{low}^\dagger, \quad \mathbf{C}_{up} = \mathbf{U}_{up} [\mathbf{\Lambda}_{up} \;\; \mathbf{0}] \mathbf{V}_{up}^\dagger.$$

Using the result in Problem 8.2, show that

(a) the unitary matrices $\mathbf{U}_{low}$, $\mathbf{V}_{low}$, $\mathbf{U}_{up}$, and $\mathbf{V}_{up}$ can be chosen such that

$$\mathbf{U}_{up} = \mathbf{J}_M \mathbf{V}_{low}^*, \quad \mathbf{V}_{up}^\dagger = \mathbf{U}_{low}^T \mathbf{J}_N;$$

(b) when the unitary matrices are chosen so that the above relations are satisfied, the two diagonal matrices Λ_{low} and Λ_{up} in the two SVD expressions are the same. In other words, show that $\mathbf{C}_{low}$ and $\mathbf{C}_{up}$ have identical singular values.

8.4 Consider Fig. 8.8. Let $\omega = -\mathbf{A}_{zp}\mu_1$. Use the orthogonality principle to show that, if $\mathbf{A}_{zp}$ is chosen optimally as (8.26), then ω is the best estimate of μ_0 given the observation of μ_1.

8.5 Give one example to show that the bit allocation gain $\mathcal{G}_{bit}$ of the CP-OFDM system can decrease when the number of subchannels M increases.

8.6 Consider the ZP-OFDM system. Suppose that the computationally efficient zero-forcing receiver in (6.15) is employed for symbol recovery. Assume that the noise $q(n)$ is white with variance $\mathcal{N}_0$. Compute the bit allocation gain and express your answer in terms of the DFT coefficients C_k of the channel $c(n)$. Under what condition is the gain equal to unity?

8.7 Let the channel $c(n)$ and the noise $q(n)$ both be real. Show that the output noise variances of the CP-OFDM system satisfy

$$\sigma_{e_k}^2 = \sigma_{e_{M-k}}^2,$$

for $k = 1, 2, \ldots, M/2$ (M even). Prove that the optimal bit allocation also satisfies $b_k = b_{M-k}$, for $k = 1, 2, \ldots, M/2$. This implies that there is no loss of optimality in restricting the symbols s_k to satisfy the conjugate symmetric property in (6.33) when the system is a baseband communication system.

8.8 Suppose that, in a ZP system, the noise $q(n)$ is white and the channel has only one nonzero tap. Let the SVD of the channel matrix $\mathbf{C}_{low}$ be $\mathbf{U}_{zp}\Lambda\mathbf{V}_{zp}^\dagger$. Show that the optimal $\mathbf{V}_{zp}$ can be an arbitrary unitary matrix. Express the corresponding $\mathbf{U}_{zp}$ in terms of $\mathbf{V}_{zp}$.

8.9 Consider a zero-forcing ZP system with $M = 2$ and $\nu = 1$. Let the channel be $C(z) = 1 + z^{-1}$. The noise is colored with the autocorrelation coefficients given by $r_q(k) = \mathcal{N}_0 0.5^{|k|}$. Find the optimal transmitting and receiving matrices. Suppose $b = 4$ and $\mathcal{N}_0 = 0.1$. What is the minimum transmission power needed for an SER of 10^{-6}?

8.10 Repeat the above problem for a zero-forcing ZJ system.

8.11 Show that, when $\mathbf{D} = \Sigma^{1/2}$, the transmitting matrix $\mathbf{G}_{zj}$ in (8.31) of the optimal ZJ system satisfies $\mathbf{G}_{zj}^\dagger \mathbf{G}_{zj} = \mathbf{I}_M$.

8.12 Show that, when the noise is white, the transmitting and receiving matrices of the optimal ZJ system reduce to those given in (8.33). Find the expression for the minimum transmitted power in this case.

8.13 Show that, when the noise is white, the optimal CP transceiver is the CP-OFDM system.

8.14 *Pre-whitening approach.* Consider the ZP system in Fig. 8.2. Let $\mathbf{R}_{q_{zp}}$ be the $N \times N$ autocorrelation matrix of the noise vector $\mathbf{q}_{zp}$. Suppose that we apply the pre-whitening matrix $\mathbf{R}_{q_{zp}}^{-1/2}$ to whiten the noise as in Fig. P8.14. Thus we have a new channel matrix $\mathbf{C}'_{low} = \mathbf{R}^{-1/2}\mathbf{C}_{low}$, and the new noise vector $\mathbf{q}'_{zp} = \mathbf{R}_{q_{zp}}^{-1/2}\mathbf{q}_{zp}$ is white.

  (a) Find the new optimal matrices $\mathbf{G}'_{zp}$ and $\mathbf{S}'_{zp}$. How are the new optimal transmitting and receiving matrices related to the optimal $\mathbf{G}_{zp}$ and $\mathbf{S}_{zp}$ given in Section 8.4.3?

  (b) Show that the minimum transmitted powers of the two optimal transceivers (with and without the pre-whitening matrix) are the same. In other words, one can employ this pre-whitening approach for the design of optimal ZP transceiver for the colored noise case.

The same approach can be adopted for the design of ZJ systems.

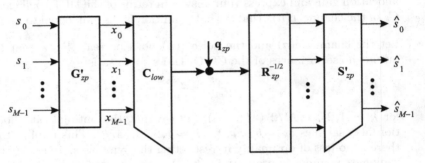

Figure P8.14. ZP system with a pre-whitening matrix.

8.15 Suppose that the settings are identical to those in Problem 8.9. Use the pre-whitening approach to design the optimal ZP transceiver. Compute the minimum transmitted power and show that it is the same as that obtained in Problem 8.9.

8.16 Repeat Problem 8.15 for the ZJ system.

8.17 Let λ_k, for $0 \leq k \leq M-1$, be the eigenvalues of the matrix $(\mathbf{I}+\mathbf{A}_{zj}^{\dagger}\mathbf{A}_{zj})$.

  (a) Prove that $\lambda_k \geq 1$; then show that $\lambda_k = 1$ for all k if and only if $\mathbf{A}_{zj} = \mathbf{0}$.

  (b) Show that $\det[\mathbf{I}+\mathbf{A}_{zj}^{\dagger}\mathbf{A}_{zj}] \geq 1$, with equality if and only if $\mathbf{A}_{zj} = \mathbf{0}$.

8.18 Derive the expressions for $\mathbf{S}_{cp}$, $\mathbf{G}_0$, and $\mathcal{E}_{cp,opt}$ of the optimal CP system. Also show the following:

  (a) the transmitting matrix $\mathbf{G}_0$ satisfies $\mathbf{G}_0^{\dagger}\mathbf{G}_0 = \mathbf{I}_M$;

  (b) the autocorrelation matrix $\mathbf{R}_e$ of the output noise $\mathbf{e} = \hat{\mathbf{s}} - \mathbf{s}$ is diagonal; that is, the receiving matrix decorrelates the noise vector.

8.19 Let the channel and noise power spectrum be given, respectively, by

$$C(z) = 1 + 2z^{-1} + z^{-2} \quad \text{and} \quad S_q(e^{j\omega}) = |C(e^{j\omega})|^2.$$

Suppose that $M = 2$ and $\nu = 2$. Compute the ratio $\mathcal{E}_{zp,opt}/\mathcal{E}_{zj,opt}$. Verify that this ratio is larger than one, proving that the optimal ZJ system outperforms the optimal ZP system in this case.

9

DMT systems with improved frequency characteristics

The topic of frequency responses of the transmitting and receiving filters has not come into our earlier discussions of transceiver design. The frequency characteristics of the filters are also an important aspect of transceiver designs. The stopband attenuation of the transmitting (receiving) filters determines how well separated the subchannels are in the frequency domain at the transmitter (receiver). Frequency separation at the transmitter side is important for the control of spectral leakage, i.e. undesired out-of-band spectral components. Poor separation will lead to significant spectral leakage. This could pose a problem in applications where the power spectrum of the transmitted signal is required to have a large rolloff in certain frequency bands. Wired applications with frequency division multiplexing, e.g. ADSL and VDSL, are such examples [7, 8]. The power spectrum of the transmitted signal should be properly attenuated in the transmission bands of the opposite direction to avoid interference. The power spectrum should also be attenuated in amateur radio bands to reduce interference to radio transmission, called *egress emission* [8]. On the other hand, poor frequency separation at the receiver side results in poor out-of-band rejection. In ADSL and VDSL applications, some of the frequency bands are also used by radio transmission systems such as amplitude-modulation stations and amateur radio. The radio frequency signals can be coupled into the wires and this introduces *radio frequency interference* (RFI) or *ingress* [29]. Poor frequency selectivity of the receiving filters means many neighboring tones can be affected. The signal to interference noise ratios of these tones are reduced and the total transmission rate decreased.

We found in Chapter 6 that in the DMT transceiver the transmitting and receiving filters come from rectangular windows. The spectral sidelobes of these filters are often inadequate to provide sufficient subchannel separation. In this chapter, we will use a filter bank approach to improving frequency separation among subchannels. Based on the filter bank representation of the DMT transceiver, we will introduce what we call *subfilters* in the subchannels. We can include the subfilters without changing the ISI-free property of the DMT system by using a cyclic prefix slightly larger than the channel order. For the transmitter side, the subfilters can improve the spectral rolloff of the transmitted spectrum while having little effect on the error rate performance. For the receiver side RFI can be suppressed and the transmission rate can be improved considerably. Moreover, when the subfilters form a DFT bank, they can be tied nicely to *windowing*, a very useful method that improves the frequency characteristics of the DMT system. This way the windows used in windowing can also be optimized through the design of subfilters. In the literature many windowing and non-windowing methods have been proposed to achieve better subchannel frequency separation. Interested readers are referred to Section 9.8 for more references on this topic.

9.1 Sidelobes matter!

The sidelobes of the transmitting and receiving filters are both important in transceiver designs. We will first examine the transmitted power spectrum and see how it is affected by the frequency characteristics of the transmitting filters. Figure 9.1 shows the transmitter with a D/C converter. Recall that the power spectrum of the transmitted signal can be obtained by summing the power spectra at the outputs of the transmitting filters (Section 6.5), i.e.

$$S_x(e^{j\omega}) = \frac{1}{N} \sum_{k \in \mathcal{A}} \mathcal{E}_{s,k} |F_k(e^{j\omega})|^2,$$

where $\mathcal{A}$ is the set of subchannels that are actually used for transmission and $\mathcal{E}_{s,k}$ is the power of the kth subchannel signal. Each transmitting filter $F_k(z)$ is a bandpass filter centered at $2k\pi/M$. After the D/C converter (Figure 9.1),

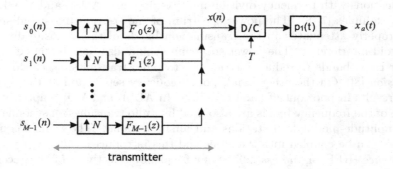

Figure 9.1. Transmitter shown with a continuous-time output $x_a(t)$.

the spectrum of the continuous-time transmitted signal $x_a(t)$ is

$$S_{x_a}(j\Omega) = \frac{1}{T} S_x(e^{j\Omega T}) |P_1(j\Omega)|^2 = \frac{1}{NT} \sum_{k \in \mathcal{A}} \mathcal{E}_{s,k} |F_k(e^{j\Omega T})|^2 |P_1(j\Omega)|^2,$$

where T is the underlying sample spacing and $P_1(j\Omega)$ is the transmitting pulse. As $F_k(e^{j\omega})$ and $P_1(j\Omega)$ are not ideal filters, the spectrum is nonzero not only in the frequency bins of the subchannels that are used, but also in other frequency bands; this is referred to as **spectral leakage**. An example of $S_x(e^{j\omega})$ with $M = 512$ is plotted for a full period $[0, 2\pi]$ in Fig. 9.2(a). The tones that are used for transmission are 38–89, and 111–255. The corresponding $S_{x_a}(j\Omega)$ is shown in Fig. 9.2(b) for $1/T = 2.208$ MHz. The transmitting pulse is an elliptic filter of order 4 and the stopband attenuation is 45 dB. We can see from Fig. 9.2(a) that, although tones 90–110 (corresponding to ω around 0.4π in the plot) are not used, $S_x(e^{j\omega})$ is not zero due to the finite stopband attenuation of $F_k(e^{j\omega})$. As a result, the spectrum $S_{x_a}(j\Omega)$ is not zero in the frequency range corresponding to these tones and there is spectral leakage. There is also spectral leakage for $\Omega >$ Nyquist frequency (which is equal to 1.104 MHz in this example) due to the finite stopband attenuation of $P_1(j\Omega)$. The transmitting pulse $P_1(j\Omega)$ will help to attenuate the transmitted power spectrum beyond the Nyquist frequency. For the band within the Nyquist frequency, which falls into the passband of $P_1(j\Omega)$, spectrum shaping relies on the transmitting filters $F_k(z)$. The sidelobes of the transmitting filters will directly affect the amount of spectral leakage within the Nyquist frequency.

At the receiver, the sidelobe of the receiving filters is also important, but for a different reason. In Section 6.6, we discussed different types of impairments in a DMT system. One impairment is RFI (radio frequency interference). These radio interfering signals have a much narrower bandwidth. To understand how a narrowband interference can affect the output of the receiver, let us consider a simple continuous-time exponential interference $\mu \exp(j(\Omega_0 t + \theta))$ with frequency Ω_0 and amplitude μ. Upon sampling with period T at the receiver, the discrete-time exponential interference signal is

$$v(n) = \mu \exp(j(\omega_0 n + \theta)), \quad \text{where } \omega_0 = \Omega_0 T.$$

The interference, being a part of the received signal, is passed to the receiving filters. At the output of the kth receiving filter (Fig. 9.3), the interference term $u_k(n)$ is also an exponential,

$$u_k(n) = \mu_k \exp(j(\omega_0 n + \theta)), \quad \text{where} \quad \mu_k = \mu H_k(e^{j\omega_0}).$$

We see that the amplitude is scaled by the receiving filters. The interference cannot be completely eliminated due to the finite stopband attenuation of the receiving filters. A higher stopband attenuation of the receiving filters means better suppression of RFI. One simple approach to improving the frequency selectivity of the transmitting and receiving filters is by including short subfilters. When an additional guard interval is available, these subfilters will enhance the stopband of the transmitting/receiving filters without incurring additional ISI. The following section discusses the overall transfer matrix of the transceiver and gives a simple observation that will be very useful in incorporating subfilters later.

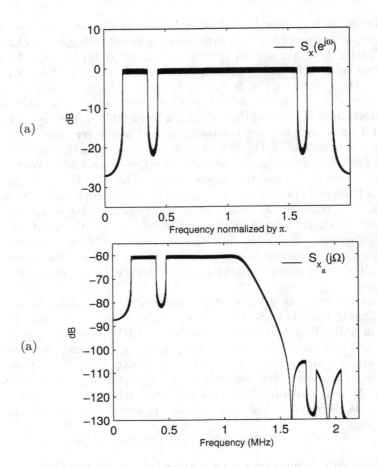

Figure 9.2. Example of spectral leakage: (a) the spectrum $S_x(e^{j\omega})$ of the discrete-time transmitted signal $x(n)$; (b) the spectrum $S_{x_a}(j\Omega)$ of the continuous-time transmitted signal $x_a(t)$.

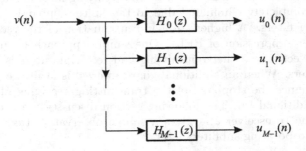

Figure 9.3. Receiving bank with an interference-only input.

9.2   Overall transfer matrix

Figure 9.4 shows the filter bank representation of the DMT transceiver with M subchannels. Recall from Section 6.5, the first transmitting filter $F_0(z)$ is a rectangular window of length $N = M + \nu$, where ν is the cyclic prefix length. All the other transmitting filters are scaled and frequency-shifted versions of the prototype $F_0(z)$,

$$F_k(z) = W^{\nu k} F_0(zW^k), \quad k = 0, 1, \ldots, M - 1. \tag{9.1}$$

For the receiving side, the prototype filter $H_0(z)$ is a rectangular window of length M and all the other receiving filters are scaled and frequency-shifted versions of the first filter,

$$H_k(z) = W^{-\nu k} H_0(zW^k), \quad k = 0, 1, \ldots, M - 1. \tag{9.2}$$

For convenience of notation, we have used $c(n)$ in Fig. 9.4 to represent the equivalent channel, possibly shortened with time domain equalization. Assume the channel $c(n)$ is an FIR filter of order $L \le \nu$. The frequency domain equalizers $1/\lambda_k$ are chosen as $\lambda_k = C_k$, where C_k are the M-point DFT of the channel impulse response $c(n)$. The receiver is a zero-forcing receiver. In the absence of channel noise, the kth receiver output is

$$\hat{s}_k(n) = s_k(n), \quad k = 0, 1, \ldots, M - 1. \tag{9.3}$$

Using Theorem 4.1, we know that the system from the ith transmitter input $s_i(n)$ to the kth signal $y_k(n)$ at the receiver is an LTI system and that the transfer function is

$$T_{ki}(z) = \Big[H_k(z) C(z) F_i(z) \Big]_{\downarrow N}, \quad 0 \le k, i \le M - 1, \tag{9.4}$$

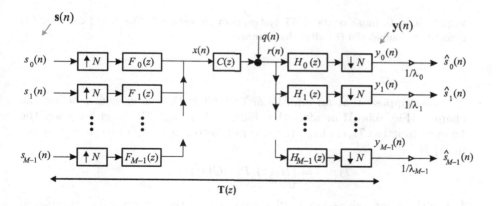

Figure 9.4. Filter bank representation of the DMT system.

where the notation $[A(z)]_{\downarrow N}$ denotes the N-fold decimated version of $A(z)$ as defined in (4.8). In view of (9.3), we can see that the system from $s_i(n)$ to $\hat{s}_k(n)$ is LTI with transfer function $\delta(k-i)$. As $\hat{s}_k(n)$ differs from $y_k(n)$ only in the scalar $1/\lambda_k$, we can conclude that $T_{ki}(z) = \lambda_k \delta(k-i)$. Summarizing, we can obtain the following lemma.

Lemma 9.1 Consider the system in Fig. 9.4 with filters as defined as (9.1) and (9.2). The transfer function $T_{ki}(z)$ from the ith transmitter input $s_i(n)$ to the kth signal $y_k(n)$ at the receiver is given by

$$T_{ki}(z) = \lambda_k \delta(k-i), \quad 0 \le k, i \le M-1. \tag{9.5}$$

The result holds for any FIR filter $C(z)$ of order $L \le \nu$, where $\nu = N - M$ is the cyclic prefix length. The constant λ_k are the M-point DFT of $c(n)$, i.e. $\lambda_k = C(z)|_{z=e^{j2\pi k/M}}$. ■

So long as the order of $C(z)$ is not larger than ν, the system is free from interblock interference and inter subchannel interference.

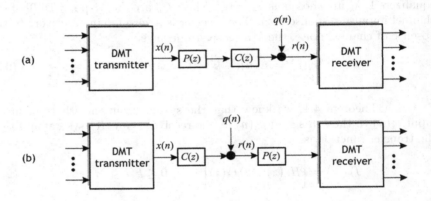

Figure 9.5. Example of the DMT system with an additional filter $P(z)$ cascaded (a) before the channel and (b) after the channel.

Now suppose that an additional FIR filter $P(z)$ is cascaded before the channel (Fig. 9.5(a)) or after the channel (Fig. 9.5(b)). In either case, the transfer function $T_{ki}(z)$ from the ith transmitter input $s_i(n)$ to the kth signal $y_k(n)$ is

$$T_{ki}(z) = \Big[H_k(z) \big(P(z)C(z) \big) F_i(z) \Big]_{\downarrow N}.$$

It has the same expression as (9.4) except that the channel $C(z)$ is replaced by the product $P(z)C(z)$. The lemma implies that, as long as the product $P(z)C(z)$ has order no larger than ν, the overall system remains ISI-free. The idea of introducing an extra filter in Fig. 9.5 will be extended in later sections to include a set of filters, called subfilters, one for each subchannel. One can design these subfilters for various purposes. Our goal here is to shape the frequency responses of the transmitting/receiving filters so that better frequency separation among the subchannels can be achieved.

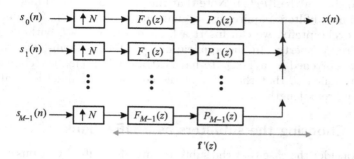

Figure 9.6. Transmitting bank with subfilters.

9.3   Transmitters with subfilters

Figure 9.6 shows the transmitter with additional FIR filters $P_k(z)$ in each subchannel. Suppose the orders of $P_k(z)$ are α,

$$P_k(z) = \sum_{n=0}^{\alpha} p_k(n)z^{-n}.$$

These additional filters will be called *subfilters* as they generally have small orders. With the subfilters, the kth effective transmitting filter is

$$F_k'(z) = F_k(z)P_k(z).$$

Now the transfer function from the ith transmitter input $s_i(n)$ to the kth signal $y_k(n)$ at the receiver (Fig. 9.4) becomes

$$T_{ki}(z) = \Big[H_k(z)(P_i(z)C(z))F_i(z) \Big]_{\downarrow N}.$$

It is the same as (9.4) except that $C(z)$ is replaced by $P_i(z)C(z)$. Using the result in Lemma 9.1, we know the overall system remains ISI-free as long as the order of the product $P_i(z)C(z)$ is not larger than ν. The condition for this is

$$\alpha + L \leq \nu, \tag{9.6}$$

where α is the order of $P_k(z)$. Also, the transfer function is $T_{ki}(z) = \lambda_k \delta(k-i)$, the same form as (9.5). Now λ_k are the kth DFT coefficient of $p_k(n) * c(n)$,

$$\lambda_k = P_k(z)C(z)\Big|_{z=e^{j2k\pi/M}}. \tag{9.7}$$

The new transmitting filters $F_k'(z)$ are of length $N + \alpha$, as $F_k(z)$ are of length N. We know that for the reason of bandwidth efficiency M is usually much larger than ν and hence also much larger than α. So the subfilters $P_k(z)$ are much shorter than $F_k(z)$. Note that the condition in (9.6) means that the insertion of subfilters requires extra guard interval $(\nu > L)$. Whenever there is extra cyclic prefix, we can insert a filter of order $\nu - L$ without affecting the ISI-free property. In the applications of DMT systems, the channel is typically shortened by a TEQ. To have extra cyclic prefix, the channel needs to be shortened so that the order of the equivalent channel is smaller than the cyclic prefix length.

9.3.1   Choosing the subfilters as a DFT bank

Let us consider the case that the subfilters are also shifted versions of the first subfilter,

$$P_k(z) = P_0(zW^k). \tag{9.8}$$

The kth new transmitting filter becomes $F_k'(z) = W^{\nu k} F_0(zW^k) P_0(zW^k)$. We can also write it as

$$F_k'(z) = W^{\nu k} F_0'(zW^k), \quad \text{where} \quad F_0'(z) = F_0(z) P_0(z). \tag{9.9}$$

They are also shifted versions of the new prototype filter $F_0'(z)$ except for some scalars. The new prototype is of length $N + \alpha$ and is no longer a rectangular window. Let us denote its coefficients by $a_i/\sqrt{M}$, then

$$F_0'(z) = \frac{1}{\sqrt{M}} \sum_{i=0}^{N+\alpha-1} a_i z^{-i}.$$

We call a_i the **transmitter window coefficients** for reasons that will become clear later. These coefficients come from the convolution of a rectangular window and $p_0(n)$. As the order of $P_0(z)$ is much smaller than N, most of the window coefficients are of the same value. A typical plot of the window is shown in Fig. 9.7. When we choose the subfilters as a DFT bank, these new transmitting filters again form a DFT bank and thus can be implemented efficiently, as we will see below.

9.3.2   DFT bank implementation

Using (9.9), we have

$$F_k'(z) = \frac{1}{\sqrt{M}} W^{\nu k} \sum_{i=0}^{N+\alpha-1} a_i W^{-ki} z^{-i}.$$

We can write it in a matrix form as follows:

$$F_k'(z) = \frac{1}{\sqrt{M}} \begin{bmatrix} 1 & z^{-1} & \cdots & z^{-N+1} \end{bmatrix} \mathbf{A}(z^N) \begin{bmatrix} W^{-k(-\nu)} \\ W^{-k(1-\nu)} \\ \vdots \\ W^{-k(N+\alpha-1-\nu)} \end{bmatrix},$$

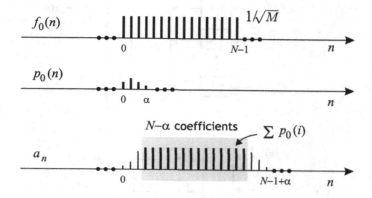

Figure 9.7. Example of the transmitter window.

where $\quad \mathbf{A}(z) = \begin{bmatrix} \mathbf{A}_0 & 0 & \mathbf{A}_2 z^{-1} \\ 0 & \mathbf{A}_1 & 0 \end{bmatrix}.$

The matrices $\mathbf{A}_0$, $\mathbf{A}_1$, and $\mathbf{A}_2$ are diagonal,

$$\mathbf{A}_0 = \mathrm{diag}\begin{bmatrix} a_0 & a_1 & \cdots & a_{\alpha-1} \end{bmatrix}, \mathbf{A}_1 = \mathrm{diag}\begin{bmatrix} a_\alpha & a_{\alpha+1} & \cdots & a_{N-1} \end{bmatrix},$$

$$\mathbf{A}_2 = \mathrm{diag}\begin{bmatrix} a_N & a_{N+1} & \cdots & a_{N+\alpha-1} \end{bmatrix}.$$

Using $W^M = 1$, the last column vector in the above expression can be written as

$$\begin{bmatrix} W^{-k(-\nu)} \\ W^{-k(1-\nu)} \\ \vdots \\ W^{-k(N+\alpha-1-\nu)} \end{bmatrix} = \mathbf{F}_1 \begin{bmatrix} 1 \\ W^{-k} \\ \vdots \\ W^{-k(M-1)} \end{bmatrix}, \quad \text{where } \mathbf{F}_1 = \begin{bmatrix} 0 & 0 & \mathbf{I}_\nu \\ \mathbf{I}_\alpha & 0 & 0 \\ 0 & \mathbf{I}_{M-\nu-\alpha} & 0 \\ 0 & 0 & \mathbf{I}_\nu \\ \mathbf{I}_\alpha & 0 & 0 \end{bmatrix}.$$

Therefore, we have

$$F_k'(z) = \frac{1}{\sqrt{M}} \begin{bmatrix} 1 & z^{-1} & \cdots & z^{-N+1} \end{bmatrix} \mathbf{A}(z^N) \mathbf{F}_1 \begin{bmatrix} 1 \\ W^{-k} \\ \vdots \\ W^{-k(M-1)} \end{bmatrix}. \qquad (9.10)$$

The $1 \times M$ new transmitting bank $\mathbf{f}'(z)$, as indicated in Fig. 9.6, can be obtained by putting $F_k'(z)$ together in a row vector. In (9.10), only the last column vector depends on k; putting these column vectors together gives rise to the IDFT matrix $\sqrt{M}\mathbf{W}^\dagger$. The transmitting bank is thus given by

$$\mathbf{f}'(z) = \begin{bmatrix} F_0'(z) & F_1'(z) & \cdots & F_{M-1}'(z) \end{bmatrix} = \begin{bmatrix} 1 & z^{-1} & \cdots & z^{-N+1} \end{bmatrix} \mathbf{G}(z^N),$$

$$(9.11)$$

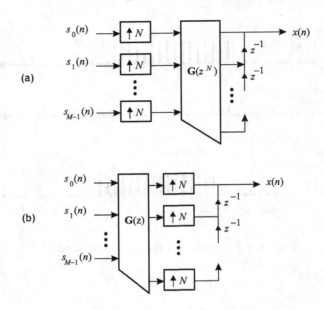

Figure 9.8. Polyphase implementation of the transmitting bank.

where $\mathbf{G}(z) = \mathbf{A}(z)\mathbf{F}_1\mathbf{W}^\dagger$.

Note that $\mathbf{G}(z)$ is the polyphase matrix of the transmitting filter bank. It is a polynomial matrix in z^{-1}, not a constant matrix like those in Chapters 6–8. With the polyphase representation, the transmitting bank has the implementation in Fig. 9.8(a). Using the noble identity for exchanging LTI filters and expanders in Section 4.1.2, we can move $\mathbf{G}(z^N)$ to the left of the expanders, as shown in the figure. We can further redraw it as in Fig. 9.9 using the expression of $\mathbf{G}(z)$ in (9.11).

Let us examine the implementation in Fig. 9.9 step by step. For each input block, M-point IDFT is applied, followed by prefixing and suffixing. The matrix $\mathbf{F}_1$ inserts a cyclic prefix of length ν and also a suffix of length α to the vector $\mathbf{u}(n)$. An illustration of prefixing and suffixing is given in Fig. 9.10. The prefixed and suffixed vector $\mathbf{t}(n)$, as indicated in Fig. 9.9, is of size $N+\alpha$. The samples of the vector $\mathbf{t}(n)$ are multiplied by the coefficients $a_0, a_1, \ldots, a_{N+\alpha-1}$ as demonstrated in Fig. 9.11(a). (This is why these coefficients are called window coefficients.) The resulting vector $\mathbf{v}(n)$ will be the output due to the nth input block $\mathbf{s}(n)$. Then the last α samples of $\mathbf{v}(n-1)$ are added to the first α samples of $\mathbf{v}(n)$ (overlap-and-add operation), as shown in Fig. 9.11(b). The output samples due to the nth and $(n+1)$th input blocks overlap by α samples. When we inspect the transmitter output, there will be $(N-\alpha)$ samples due solely to the nth input block $\mathbf{s}(n)$. The implementation in Fig. 9.9 is the same as the transmitter windowing given in [8].

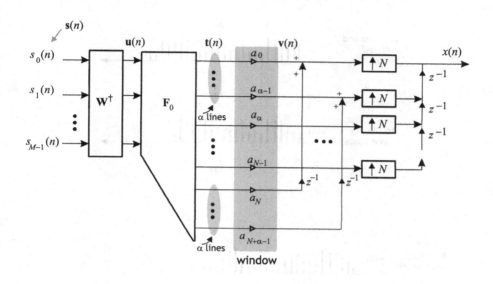

Figure 9.9. Efficient DFT implementation of the transmitting bank.

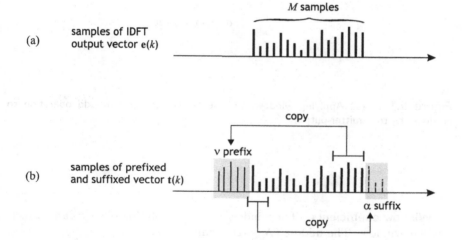

Figure 9.10. Prefixing and suffixing: (a) the samples of the IDFT output vector $\mathbf{u}(n)$; (b) the samples of the prefixed and suffixed vector $\mathbf{t}(n)$.

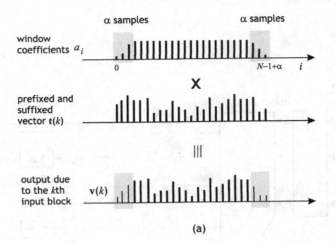

(a)

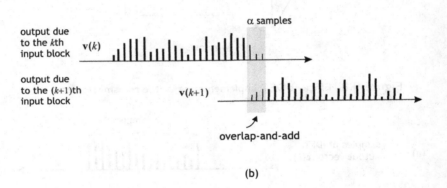

overlap-and-add

(b)

Figure 9.11. (a) Applying window coefficients; (b) overlap-and-add operation to produce the transmitter output.

Window coefficients The window is the convolution of $p_0(n)$ and a rectangular window. The middle $(N - \alpha)$ coefficients are

$$a_i = \sum_{i=0}^{\alpha} p_0(i), \quad i = \alpha, \alpha + 1, \ldots, N - 1.$$

They are equal to the DC value of $P_0(z)$, as shown Fig. 9.7. We can write the α coefficients on the two ends of the window as follows:

$$
\begin{aligned}
a_i &= p_0(0) + p_0(1) + \cdots + p_0(i), \\
a_{N+i} &= p_0(i + 1) + p_0(i + 2) + \cdots + p_0(\alpha),
\end{aligned}
\tag{9.12}
$$

for $i = 0, 1, \ldots, \alpha - 1$. Note that $a_i + a_{N+i}$ is also equal to the DC value of $P_0(z)$. As a result, the shifts of the window add up to a constant,

$$\sum_{\ell=-\infty}^{\infty} a_{i+N\ell} = \text{a constant.} \tag{9.13}$$

This has been referred to as the time-domain **Nyquist(N)** property in [97, 98]. A sequence that satisfies the time domain Nyquist(N) property has regular zero crossings in the frequency domain (Problem 9.4). When the DC value of $P_0(z)$ is normalized to one,

$$\sum_{i=0}^{\alpha} p_0(i) = 1, \tag{9.14}$$

the window coefficients in the middle are equal to one and the shifts of the window add up to one, i.e. $\sum_{\ell=-\infty}^{\infty} a_{i+N\ell} = 1$.

Implementation complexity From Fig. 9.9 we can observe the complexity of the transmitting bank. In this implementation, only the IDFT operation and the part of applying windowing require computations. The complexity for windowing one block is that of the $N + \alpha$ window coefficients plus α additions due to overlap-and-add. When the DC value of $P_0(z)$ is normalized to one as in (9.14), the middle $(N - \alpha)$ coefficients are equal to one and no multiplications are needed. Only the 2α coefficients at the two ends of the window require multiplications. The complexity of each block is equivalent to one IDFT plus 2α multiplications and α additions. Usually the window coefficients a_i are real in practice. Compared with the conventional case, only 2α more multiplications and α more additions per block are needed, which is a small overhead.

FEQ coefficients We can choose the subfilter coefficients so that the FEQ coefficients $1/\lambda_k$ have the same values after the subfilters are included, i.e. $\lambda_k = C_k$. In view of (9.7), the subfilters $P_k(z)$ need to satisfy

$$P_k(e^{j2\pi k/M}) = 1, \quad k = 0, 1, \ldots, M - 1. \tag{9.15}$$

This requires that the coefficients of each $P_k(z)$ be normalized. In the special case when the subfilters are shifted versions of the first subfilter, as in (9.8), this condition reduces to

$$P_k(e^{j2\pi k/M}) = P_0(e^{j2\pi k/M}W^k) = P_0(e^{j0}) = 1, \tag{9.16}$$

which is the same as the normalization condition in (9.14). In this case there is no need to modify the FEQ coefficients in the original DMT receiver. The FEQ coefficients remain the same after the subfilters are included. The unity DC value condition can be easily satisfied by a simple normalization. We can first design $P_0(z)$ without constraints and then normalize the coefficients.

Transmitted power spectrum Using the filter bank representation in Fig. 9.6, we can express the transmitted power spectrum in terms of the transmitting filters and thus in terms of the subfilters to be optimized. We know that the DFT filters $F_k(z)$ have the conjugate-pair property $(f_{M-k}(n) = f_k^*(n))$. Suppose the subfilters $P_k(z)$ also have the conjugate-pair property (not necessarily the frequency-shifting property in (9.8)). Then the equivalent transmitting filters continue to have such a property, i.e.

$$f'_{M-k}(n) = f_k'^*(n).$$

In the derivation of the transmitted power spectrum in (6.35), the only assumption on the transmitting filters is that the filters are in conjugate pairs. Therefore, the transmitted power spectrum with subfilters can be obtained directly from (6.35),

$$S_x(e^{j\omega}) = \frac{1}{N} \sum_{k \in \mathcal{A}} \mathcal{E}_{s,k} |F_k'(e^{j\omega})|^2, \qquad (9.17)$$

where $\mathcal{A}$ is the collection of the subchannels that are used for transmission. The spectral leakage is directly related to the spectral rolloff of the new transmitting filters. If the subfilters form a DFT bank, so do the new transmitting filters, and

$$S_x(e^{j\omega}) = \frac{1}{N} \sum_{k \in \mathcal{A}} \mathcal{E}_{s,k} |F_0'(e^{j(\omega - 2\pi k/M)})|^2, \qquad (9.18)$$

which consists of the shifts of $|F_0'(e^{j\omega})|^2$. In this case the spectral rolloff depends only on the prototype filter $F_0'(z)$.

9.4   Design of transmit subfilters

The addition of subfilters in Section 9.3 allows us to modify the frequency characteristics of the transmitting filters. We now design the subfilters $P_k(z)$ to shape the transmitted power spectrum and minimize the spectral leakage. We will first consider the general case of unconstrained subfilters and then the special case of $P_k(z)$ with the frequency-shifting property in (9.8).

Unconstrained case When the subfilters are not constrained to form a DFT bank, the transmitted power spectrum is as given in (9.17). The total spectral leakage is

$$S = \int_{\omega \in \mathcal{O}} S_x(e^{j\omega}) d\omega = \frac{1}{N} \sum_{k \in \mathcal{A}} \mathcal{E}_{s,k} \int_{\omega \in \mathcal{O}} |F_k'(e^{j\omega})|^2 \, d\omega, \qquad (9.19)$$

where $\mathcal{O}$ denotes the band in which leakage is undesired. The total leakage S can be minimized if we can minimize the individual contribution S_k from each subchannel,

$$S_k = \int_{\omega \in \mathcal{O}} |F_k'(e^{j\omega})|^2 \, d\omega.$$

The filter $F_k'(z)$ is the product of $F_k(z)$ and $P_k(z)$; we can write its Fourier transform as

$$F_k'(e^{j\omega}) = \boldsymbol{\tau}_k(\omega) \mathbf{p}_k,$$

where $\mathbf{p}_k$ is an $(\alpha + 1) \times 1$ vector consisting of the coefficients of $p_k(n)$ and

$$\boldsymbol{\tau}_k(\omega) = F_k(e^{j\omega}) \begin{bmatrix} 1 & e^{-j\omega} & \cdots & e^{-ja\omega} \end{bmatrix}. \tag{9.20}$$

We can then write $\mathcal{S}_k$ in the following quadratic form:

$$\mathcal{S}_k = \mathbf{p}_k^\dagger \boldsymbol{\Phi}_k \mathbf{p}_k, \quad \text{where} \quad \boldsymbol{\Phi}_k = \int_{\omega \in \mathcal{O}} \boldsymbol{\tau}_k^\dagger(\omega) \boldsymbol{\tau}_k(\omega) d\omega. \tag{9.21}$$

The spectral leakage due to the kth subchannel can be minimized if we can choose the kth subfilter $\mathbf{p}_k$ to minimize $\mathcal{S}_k$. The optimization problem can be cast as

$$\text{minimize} \quad \mathcal{S}_k, \quad \text{subject to} \quad \mathbf{p}_k^\dagger \mathbf{p}_k = 1.$$

The above optimization is the well-known eigenfilter problem [114, 159]. As $\mathcal{S}_k$ represents the energy of the kth transmitting filter in the frequency band $\mathcal{O}$, and the FIR transmitting filters cannot have zero energy in $\mathcal{O}$, $\mathcal{S}_k$ is always positive. Therefore the matrix $\boldsymbol{\Phi}_k$ is positive definite. To minimize $\mathcal{S}_k$, we can choose $\mathbf{p}_k$ as the eigenvector associated with the smallest eigenvalue of $\boldsymbol{\Phi}_k$. If the subfilters do not form a DFT bank, neither do the new transmitting filters (Problem 9.3).

Constrained case When $P_k(z)$ satisfy the frequency-shifting property in (9.8), the transmitting filters form a DFT bank. Now we do not have the flexibility of optimizing each transmitting filter separately. Nonetheless, from the expression in (9.18), we see that there is no spectral leakage if the prototype filter $F_0'(z)$ is an ideal lowpass filter. The undesired leakage comes from the finite stopband attenuation of the prototype. The total leakage can be reduced by minimizing the stopband energy of the prototype filter, which is given by

$$\phi = \int_{\omega \in \mathcal{O}_0} |F_0'(e^{j\omega})|^2 \, d\omega, \tag{9.22}$$

where $\mathcal{O}_0$ denotes the stopband of the prototype filter. For real a_i, a typical choice of $\mathcal{O}_0$ is $[\pi/2M + \epsilon, \pi]$, where ϵ is a small number. Following a procedure similar to that of deriving $\mathcal{S}_k$, we can write the stopband energy ϕ as

$$\phi = \mathbf{p}_0^\dagger \boldsymbol{\Phi} \mathbf{p}_0, \quad \text{where} \quad \boldsymbol{\Phi} = \int_{\omega \in \mathcal{O}_0} \boldsymbol{\tau}_0^\dagger(\omega) \boldsymbol{\tau}_0(\omega) d\omega,$$

where $\boldsymbol{\tau}_0(\omega)$ is as given in (9.20). Thus the optimization problem of the subfilter coefficients $\mathbf{p}_0$ can be formulated as

$$\text{minimize} \quad \phi = \mathbf{p}_0^\dagger \boldsymbol{\Phi} \mathbf{p}_0, \quad \text{such that} \quad \mathbf{p}_0^\dagger \mathbf{p}_0 = 1.$$

The optimal $\mathbf{p}_0$ is the eigenvector associated with the smallest eigenvalue of $\boldsymbol{\Phi}$.

Example 9.1 Transmitter subfilters for spectral leakage suppression. The block size is $M = 512$ and the prefix length is $\nu = 40$. The channel used in this example is VDSL loop#1 (4500 ft) [8] and it is shortened by a TEQ. The equivalent channel has order 26, smaller than ν so that subfilters can be included.   The tones used are 38–90 and 111–255, and the sampling frequency $1/T = 2.208$ MHz. We assume that the subfilters are shifted versions of the first subfilter $P_0(z)$, and thus the transmitting filters $F'_k(z)$ form a DFT bank. The order α of the subfilters is 14. We compute the positive definite matrix Φ and the eigenvector corresponding to the smallest eigenvalue to obtain $\mathbf{p}_0$. The magnitude response of the subfilter $P_0(e^{j\omega})$ is shown in Fig. 9.12(a). The coefficients of $P_0(z)$ are normalized so that the DC value is equal to one. Figure 9.12(b) shows the magnitude response of $F'_0(z)$ normalized with respect to its maximum. For comparison, we have also shown the magnitude response of the original prototype $F_0(z)$. The new prototype $F'_0(z)$ has a better attenuation in the stopband. Figure 9.12(c) shows the spectrum of the transmitter output with and without subfilters. We see that the spectrum of the windowed output has a much smaller spectral leakage in the unused bands. ∎

A note on BER performance   When we include subfilters in the transmitter, we are effectively changing the transmitting filters. This in general will affect the transmission power, the subchannel SNRs, and hence the BER performance of the overall system. It turns out that the subfilters have little effect on any of these as we now explain. When the subfilters are normalized as in (9.15), the subchannel gains will stay the same. As the receiver is not changed, the subchannel SNRs will not be changed. Also note that the orders of the subfilters are very small compared to the number of subchannels M. The main lobe of $P_k(z)$ is much wider than that of $F_k(e^{j\omega})$, as we can see from Figure 9.12(a). (The first zero of $P_0(z)$ is around 0.1π, while the first zero of the rectangular window $F_0(z)$ is around 0.004π.) The effect of subfilters is mostly on the sidelobes away from the main lobe and the large sidelobes in the neighborhood of the main lobes. So the subfilters have only a minor effect on the transmission power. Therefore the BER performance is also roughly the same.

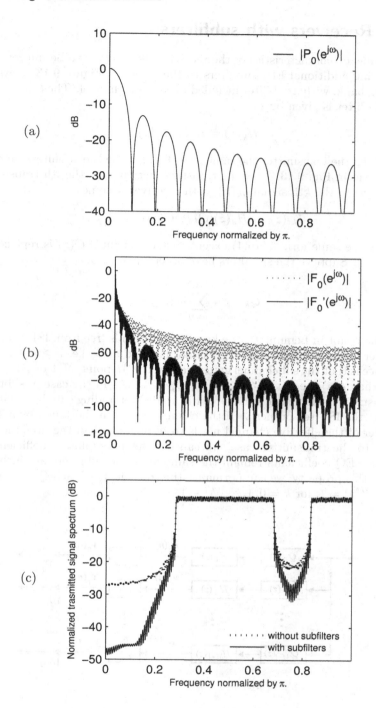

Figure 9.12. Example 9.1. Transmitter subfilters for spectral leakage suppression: (a) the magnitude response of $P_0(e^{j\omega})$; (b) the magnitude response of the prototype filter $F_0'(e^{j\omega})$ normalized with respect to its maximum; (c) the power spectrum of the transmitted signal normalized with respect to its maximum.

9.5   Receivers with subfilters

The frequency characteristics of the receiving filters can also be improved by introducing additional FIR subfilters to the receiver. Figure 9.13 shows the receiving bank, with a subfilter included in each subchannel. The kth effective receiving filter is given by

$$H'_k(z) = H_k(z)Q_k(z).$$

Let us keep the transmitter the same as in Fig. 9.4 (i.e. no subfilters) and use the receiver in Fig. 9.13. Now the transfer function from the ith transmitter input $s_i(n)$ to the kth signal $y_k(n)$ at the receiver becomes

$$T_{ki}(z) = [H_k(z)(Q_k(z)C(z))F_i(z)]_{\downarrow N}.$$

This has the same form as (9.4) except that the channel $C(z)$ is replaced by $Q_k(z)C(z)$. Suppose the subfilters have order equal to β,

$$Q_k(z) = \sum_{n=0}^{\beta} q_k(n)z^{-n}.$$

From the result in Lemma 9.1, we know the system is free from ISI as long as the order of the product $Q_k(z)C(z)$ is not larger than ν, i.e. $\nu \geq \beta + L$. The difference is that the subchannel gains are now the M-point DFT of $q_k(z)*c(n)$ rather than $c(n)$, i.e. $\lambda_k = Q_k(z)C(z)|_{z=e^{j2k\pi/M}}$. As in the case of subfilters for transmitter side, extra cyclic prefix is needed for adding subfilters without affecting the ISI-free property. If the channel has been shortened by a TEQ, the order of the effective channel needs to be smaller than the prefix length. Similar to the transmitter case, we can choose the subfilter coefficients so that the FEQ coefficients remain the same after the subfilters are included. To have this property, we can normalize the coefficients of each $Q_k(z)$ so that $Q_k(e^{j2\pi k/M}) = 1$, for $k = 0, 1, \ldots, M - 1$.

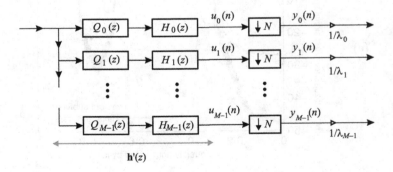

Figure 9.13. Receiving bank with subfilters.

9.5.1 Choosing subfilters as a DFT bank

Let us consider the case that the subfilters are shifted versions of the first subfilter,

$$Q_k(z) = Q_0(zW^k).$$

The new kth receiving filter becomes

$$H_k'(z) = H_k(z)Q_k(z) = W^{-\nu k}H_0(zW^k)Q_0(zW^k) = W^{-\nu k}H_0'(zW^k). \quad (9.23)$$

They are also shifted versions of $H_0'(z)$ (the new prototype filter) except for some scalars. As $H_0(z)$ is a rectangular window of length M, and the new prototype is the product $Q_0(z)H_0(z)$, the new receiving filters are all of length $M + \beta$. Suppose the subfilter $Q_0(z)$ is causal. Let the coefficients of $H_0'(z)$ be $b_i/\sqrt{M}$, and write it as

$$H_0'(z) = \frac{z^{\nu-\beta}}{\sqrt{M}} \sum_{i=0}^{M+\beta-1} b_i z^i.$$

The presence of the advance operator z is due to the original prototype $H_0(z)$ (given in 6.30), which is noncausal. We will call b_i the **receiver window coefficients**. The window comes from the convolution of a rectangular window of length M and a much shorter $q_0(n)$. In practice, the subfilter $q_0(n)$ usually has real coefficients. In this case, b_i will also be real. Similar to the transmitter window, most of the window coefficients are of the same value, except for those at the two ends of the window. A plot of the receiver window is very similar to the plot of a_i in Fig. 9.7. Without loss of generality, we can normalize the DC value of $Q_0(z)$ to unity. Then the $(M - \beta)$ coefficients in the middle of the window are equal to unity. The number of nonunity coefficients at each end is β. When we choose the subfilters as a DFT bank, the new receiving filters again form a DFT bank and thus can be implemented efficiently, as we will see next.

9.5.2 DFT bank implementation

Using the relation $H_k'(z) = W^{-\nu k}H_0'(zW^k)$, we can write the new kth receiving filter as

$$H_k'(z) = \frac{z^{\nu-\beta}}{\sqrt{M}} \sum_{i=0}^{M+\beta-1} b_i W^{k(i-\beta)} z^i.$$

We can express this as

$$H_k'(z) = \frac{z^{\nu-\beta}}{\sqrt{M}} \begin{bmatrix} W^{-\beta k} & W^{-(\beta-1)k} & \cdots W^{k(M-1)} \end{bmatrix} \mathbf{B} \begin{bmatrix} 1 \\ z \\ \vdots \\ z^{M+\beta-1} \end{bmatrix},$$

$$\text{where} \quad \mathbf{B} = \text{diag}\begin{bmatrix} b_0 & b_1 & \cdots & b_{M+\beta-1} \end{bmatrix}.$$

Note that the k-dependent row vector in the above equation can be written as

$$\begin{bmatrix} W^{-\beta k} & W^{-(\beta-1)k} & \cdots W^{k(M-1)} \end{bmatrix} = \begin{bmatrix} 1 & W^k & \cdots W^{k(M-1)} \end{bmatrix} \mathbf{F}_2,$$

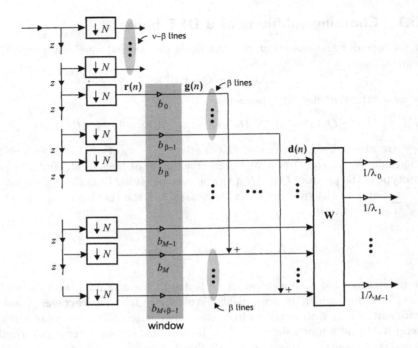

Figure 9.14. Efficient DFT implementation of the receiving bank.

$$\text{where} \quad \mathbf{F}_2 = \begin{bmatrix} \mathbf{0} & \mathbf{I}_{M-\beta} & \mathbf{0} \\ \mathbf{I}_\beta & \mathbf{0} & \mathbf{I}_\beta \end{bmatrix}.$$

Therefore, we have

$$H_k'(z) = \frac{z^{\nu-\beta}}{\sqrt{M}} \begin{bmatrix} 1 & W^k & \cdots & W^{k(M-1)} \end{bmatrix} \mathbf{F}_2 \mathbf{B} \begin{bmatrix} 1 \\ z \\ \vdots \\ z^{M+\beta-1} \end{bmatrix}. \tag{9.24}$$

The new receiving bank $\mathbf{h}'(z)$ as indicated in Fig. 9.13 can be obtained by stacking $H_k'(z)$ together. In (9.24), only the row vector depends on k; stacking these row vectors together gives rise to the DFT matrix $\mathbf{W}$. The receiving bank is thus given by

$$\mathbf{h}'(z) = \begin{bmatrix} H_0'(z) \\ H_1'(z) \\ \vdots \\ H_{M-1}'(z) \end{bmatrix} = z^{\nu-\beta} \mathbf{W} \mathbf{F}_2 \mathbf{B} \begin{bmatrix} 1 \\ z \\ \vdots \\ z^{M+\beta-1} \end{bmatrix}. \tag{9.25}$$

This expression gives us the implementation of the receiver in Fig. 9.14, where we have moved the decimators to the left by using the noble identity for decimators in Section 4.1.2. Note that the first $\nu - \beta$ samples are discarded due to the advance operator $z^{\nu-\beta}$.

In the DFT bank implementation of Fig. 9.14, the received samples are first multiplied by window coefficients b_i. This is illustrated in Fig. 9.15(a).

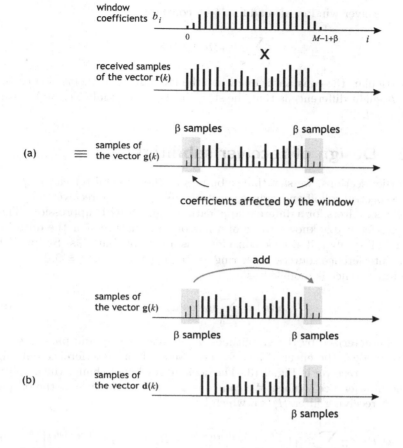

Figure 9.15. Receiver windowing. (a) Multiplying window coefficients; (b) adding the first β samples to the last β samples to produce the samples of $\mathbf{d}(n)$.

When the DC value of $Q_0(z)$ is unity, only the coefficients on the two ends of the window are not equal to one and multiplications are needed only for these coefficients. Then the matrix $\mathbf{F}_2$ in (9.25) performs the operation of adding the first β samples to the last β samples, as shown in Fig. 9.15(b). The resultant vector $\mathbf{d}(n)$, as indicated in Fig. 9.14, is passed over for DFT computation and FEQ. This is the same as the usual receiver windowing described in [143]. When the window has real coefficients, these multiplications and additions will also be real. Compared to the conventional DMT receiver, the new receiver in Fig. 9.14 needs only 2β more multiplications and β more additions per block of outputs. The overhead is very small.

Time domain Nyquist(M) property   As the window is the convolution of a rectangular window and $q_0(n)$, the coefficients b_i can be expressed in a form similarly to those in (9.12). Similar to the transmitter window, the shifts of the receiver window also add up to a constant,

$$\sum_{\ell=-\infty}^{\infty} b_{i+M\ell} = \text{constant}. \tag{9.26}$$

In particular, the constant is one when the DC value of $Q_0(z)$ is normalized to one. A slight difference is that the shifts in (9.26) are multiples of M, instead of N in (9.13).

9.6 Design of receiver subfilters

In earlier sections, we saw that subfilters at the transmitter can help to improve spectral rolloff and reduce spectral leakage. At the receiver side, we will employ subfilters for a different application, namely RFI suppression. The radio interference is known to be of a narrowband nature. For the duration of one DMT symbol, it can be considered as a sum of sinusoids. Suppose there are J interference sources, occurring at frequencies ω_ℓ, for $\ell = 0, 1, \ldots, J-1$. The interference is modeled as

$$v(n) = \sum_{\ell=0}^{J-1} \mu_\ell \cos(\omega_\ell n + \theta_\ell). \tag{9.27}$$

It is characterized by the amplitudes μ_ℓ, frequencies ω_ℓ, and phases θ_ℓ.

   To analyze the effect of interference, we apply an interference-only signal $v(n)$ to the receiver in Fig. 9.13. The decimators do not change the amplitudes of the interference signals. We can consider the interference at the output of the kth receiving filter $H'_k(z)$, which is

$$u_k(n) = \frac{1}{2} \sum_{\ell=0}^{J-1} \mu_\ell \left[H'_k(e^{j\omega_\ell}) e^{j(\omega_\ell n + \theta_\ell)} + H'_k(e^{-j\omega_\ell}) e^{-j(\omega_\ell n + \theta_\ell)} \right]. \tag{9.28}$$

Minimization of the above interference term requires the knowledge of J, μ_ℓ, ω_ℓ, and θ_ℓ. Depending on the availability of this knowledge, two cases will be considered in what follows. In the case when the interference sources are known, we will see that an RFI of the simple form in (9.27) can be completely eliminated if the number of interfering sources $J \le \beta/2$, where β is the order the subfilters.

   Case 1: interference sources are known   If the information of the interference sources is available to the receiver, the subfilters can be individually optimized. When we observe the expression of the kth interference signal $u_k(n)$ in (9.28), we see that its amplitude is a nonlinear function of the kth subfilter coefficients. To simplify the problem, note that the part of $u_k(n)$ due to the ℓth interference source will be small if

$$\mu_\ell^2 \left(|H'_k(e^{j\omega_\ell})|^2 + |H'_k(e^{-j\omega_\ell})|^2 \right)$$

is small. The kth subchannel interference can be mitigated by designing $Q_k(z)$ to minimize

$$\phi'_k = \sum_{\ell=0}^{J-1} \mu_\ell^2 \left(|H'_k(e^{j\omega_\ell})|^2 + |H'_k(e^{-j\omega_\ell})|^2 \right). \tag{9.29}$$

We can write ϕ'_k in a quadratic form similar to that in (9.31) and find the optimal subfilters (Problem 9.9). Such an optimization requires only the amplitudes and frequencies, but not the phases, of the interference sources. When the subfilters are so designed, in general the receiving bank does not have the DFT bank structure in Fig. 9.14 as the subfilters $Q_k(z)$ are not shifted versions of $Q_0(z)$. When the subfilters can be individually optimized it is possible to eliminate completely an interference that is a sum of sinusoids as given in (9.27). In particular, we observe from (9.29) that $\phi'_k = 0$ if both $H'_k(e^{j\omega_\ell})$ and $H'_k(e^{-j\omega_\ell})$ are zero for $j = 0, 1, \ldots, J-1$. This requires that $H'_k(e^\omega)$ have $2J$ zeros on the unit circle. Therefore when

$$J \leq \beta/2,$$

the subchannel interference can be nullified completely. In practice, an interference coming from amplitude-modulation stations or amateur radio is not as simple as a sum of sinusoids. In this case, the RFI cannot be completely removed, but will be significantly suppressed, as to be demonstrated in a simulation example. On the other hand, when the subfilters $Q_k(z)$ are constrained to be shifted versions of $Q_0(z)$, we can design $Q_0(z)$ to minimize the total interference $\sum_{k \in \mathcal{A}} \phi'_k$ [70].

Case 2: interference sources are not known Let us first consider the case when the information of the interference is not available, i.e. the receiver does not know J, μ_ℓ, ω_ℓ, and θ_ℓ. In this case, we can alleviate the effect of interference in the kth subchannel by minimizing

$$\phi_k = \int_{\omega \in \mathcal{O}_k} |H'_k(e^{j\omega})|^2 \, d\omega, \qquad (9.30)$$

where $\mathcal{O}_k$ denotes the stopband of the kth receiving filters. Such an approach also has the advantage that the subfilters need to be designed only once; they need not be redesigned when the interference changes.

Following a procedure similar to that for designing transmitter subfilters, we write $H'_k(z)$ as

$$H'_k(e^{j\omega}) = \boldsymbol{\tau}_k(\omega)\mathbf{q}_k,$$

where $\boldsymbol{\tau}_k(\omega) = H_k(e^{j\omega}) \begin{bmatrix} 1 & e^{-j\omega} & \cdots & e^{-j\beta\omega} \end{bmatrix}$. It follows that the objective function ϕ_k in (9.30) is

$$\phi_k = \mathbf{q}_k^\dagger \mathbf{A}_k \mathbf{q}_k, \quad \text{where} \quad \mathbf{A}_k = \int_{\omega \in \mathcal{O}_k} \boldsymbol{\tau}_k^\dagger(\omega)\boldsymbol{\tau}_k(\omega)d\omega. \qquad (9.31)$$

The matrix $\mathbf{A}_k$ is positive definite because ϕ_k represents the stopband energy of the receiving filters. To minimize ϕ_k, we can choose $\mathbf{q}_k$ as the eigenvector associated with the smallest eigenvalue of $\mathbf{A}_k$. Suppose the stopband $\mathcal{O}_k$ of the kth receiving filter is a shift of $\mathcal{O}_0$ in frequency by $2k\pi/M$. From (9.30), we have

$$\phi_k = \int_{\mathcal{O}_k} |H_k(e^{j\omega})Q_k(e^{j\omega})|^2 \, d\omega = \int_{\mathcal{O}_k} |H_0(e^{j(\omega-k\frac{2\pi}{M})})|^2 |Q_k(e^{j\omega})|^2 \, d\omega.$$

With a change of variables, we observe that the above equation becomes

$$\phi_k = \int_{\mathcal{O}_0} |H_0(e^{j\tau})|^2 |Q_k(e^{j(\tau+2k\pi/M)})|^2 \, d\tau.$$

The above expression means that if $Q_0(z)$ minimizes ϕ_0, then the choice $Q_k(e^{j(\omega+2k\pi/M)}) = Q_0(e^{j\omega})$ will also minimize ϕ_k. In this case

$$\phi_0 = \phi_1 = \cdots = \phi_{M-1}.$$

In other words, $Q_k(z) = Q_0(zW^k)$ minimizes ϕ_k. This implies that there is no loss of generality assuming that $Q_k(z)$ is a frequency shift of $Q_0(z)$. We only need to design $Q_0(z)$ to minimize ϕ_0 and $Q_k(z)$ can be obtained from $Q_0(z)$ by shifting by $k2\pi/M$ in frequency. Therefore the subfilters form a DFT bank and the receiving filters can be implemented efficiently using the DFT bank structure in Fig. 9.14. In the designs of subfilters for the transmitter, we consider two cases: the case when the subfilters are independently optimized and the case when they are constrained to be frequency shifts of a prototype. Here we need not do so because the subfilters will satisfy the frequency-shifting property even though they are designed individually.

Example 9.2 Receiver subfilters for RFI reduction. In this example, we design subfilters to mitigate RFI at the receiver. The DFT size is $M = 512$ and the cyclic prefix length is $\nu = 40$. The channel used in this example is VDSL loop 1 (4500 ft) [8] and the channel noise is AWGN of -140 dBm. The channel is shortened to an order of 30, less than $\nu = 40$, so that subfilters can be included. The order of the subfilters is $\beta = 10$. The transmitter inputs are QAM modulation symbols and the symbol error rate is assumed to be 10^{-7}. Four RFI sources are assumed in the simulations, at 660, 710, 770, and 1050 kHz, respectively, of strength -60, -40, -70, and -55 dBm, respectively. The four RFI sources correspond approximately to tones 153, 164, 178, and 243. The RFI signal is generated as suggested in [8] to simulate actual interference, rather than the simplified model in (9.27). We will consider two different subfilter designs.

- Case 1: Interference sources are known. In this design, the subfilters $Q_k(z)$ are individually optimized by minimizing the objective function ϕ'_k in (9.29). The subfilters are not related by frequency shifts. There are four RFI sources and the order β is 10. Although $\beta \geq 2J$, the interference cannot be completely nullified as the interference signal is not a pure sum of sinusoids.

- Case 2: Interference sources are not known. We choose the subfilters $Q_k(z)$ to be shifted versions of $Q_0(z)$ and only $Q_0(z)$ needs to be designed. The subfilter $Q_0(z)$ is designed without making use of any RFI information. It is the solution to the objective function ϕ_0 in (9.31). In this case the receiving filters form a DFT bank and can be implemented efficiently as in Fig. 9.14. The magnitude response of $Q_0(z)$ is as shown in Fig. 9.16(a). We show the magnitude response of $H'_0(z)$ in Fig. 9.16(b), normalized with respect to its maximum. Also shown in the figure is the magnitude response of the rectangular window ($H_0(z)$). We can see that with subfilter shaping, $H'_0(z)$ has better stopband and hence better frequency separation.

The SINRs (signal to noise interference ratio) of the subchannels are as shown in Fig. 9.16(c). For comparison, we have also shown the subchannel SINRs for the rectangular window. The receivers with subfilters enjoy higher SINRs for the tones that are close to the RFI frequencies. Especially when the information of the RFI sources is known and the subfilters are optimized individually, the RFI is significantly suppressed. As a result, higher transmission rates can be achieved. As an example, Table 9.1 shows the transmission rates using the bit loading formula in (6.38) when the symbol error rate is 10^{-7}. It should be noted that the second subfilter design attains a much higher transmission rate by exploiting the RFI information. When the RFI changes, we need to redesign the subfilters. Both the design and implementation cost are much higher than the original DMT receiver. On the other hand, when the subfilters are independent of the RFI, as in case 1, the implementation cost will be much lower. ∎

Rectangular window	Case 1	Case 2
6.84	7.44	8.54

Table 9.1. Transmission rates given in mega bits per second.

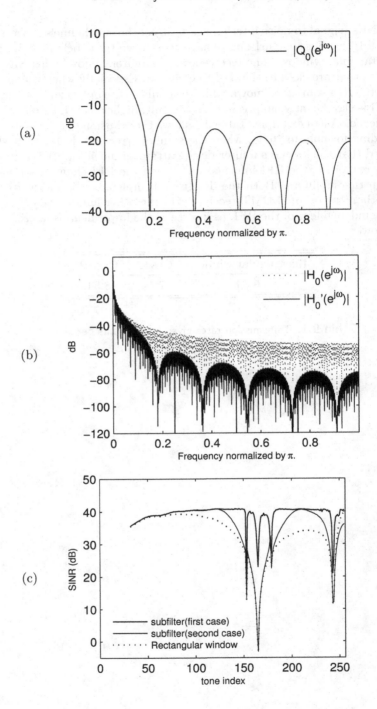

Figure 9.16. **Example 9.2.** Receiver subfilters for RFI reduction: (a) the magnitude response of $Q_0(e^{j\omega})$; (b) the magnitude response of the prototype filter $H_0'(e^{j\omega})$ normalized with respect to its maximum; (c) the subchannel SINRs.

9.7 Zero-padded transceivers

In previous sections, the discussion refers to prefixed transceivers. We can also
apply subfilters to the zero-padded system in Section 6.2. The zero-padded
transceiver also has the filter bank representation in Fig. 9.4. The receiver
for the zero-padded case is not unique. Let us consider the efficient receiver
in (6.15). The transmitting (receiving) filters are frequency-shifted versions
of the first transmitting (receiving) filter except for some scalars. But now
the transmitting prototype is a rectangular window of length M while the
receiving prototype is a length-N rectangular window.

When we inspect the filter bank representation Fig. 9.4 for the zero-padded
transceiver, the ISI-free property continues to hold for any channel $C(z)$ with
order less than or equal to the number of padded zeros ν. Therefore the
results in Lemma 9.1 can be appropriately modified for the zero-padded case.
Subfilters can be employed at the transmitter to shape the transmitted power
spectrum or at the receiver to mitigate the effect of RFI. In particular, we
can have transmitting subfilters as in Fig. 9.6. As long as the order α of the
subfilters $P_k(z)$ and the channel order L satisfy $L + \alpha \leq \nu$, the overall system
still enjoys the ISI-free property. Similarly, we can include receiving subfilters
as in Fig. 9.13 if the order β of the subfilters $Q_k(z)$ satisfies $L + \beta \leq \nu$.
Therefore to add subfilters of order β, the channel needs to be shortened to
an order $L \leq \nu - \beta$.

[68, 89, 102, 113, 141, 155]

9.8 Further reading

Many methods have been developed in the literature to enhance the frequency
separation among the subchannels. For example, to improve the spectral
rolloff of the transmitted signal, a number of continuous-time pulse shaping
filters have been proposed [68, 89, 102, 113, 141, 155]. Usually continuous-time
pulse shapes are designed based on an analog implementation of the transmitter, and in general it is not possible to have a digital implementation [79].
Discrete-time windows have been considered in [15, 86, 138]. The design of
overlapping windows for OFDM with offset QAM over frequency-nonselective
channels was studied fully in [15, 138]. When the channel is distortionless,
orthogonality among the subchannels is preserved by the window [15, 138]
and a better spectral efficiency is achieved. For ISI channels, additional post-
processing can be used to remove interference. More recently, transmitting
windows with the cyclic-prefix property [31, 80] have been considered for
egress control. Windows that are the inverse of a raised cosine function were
optimized in [31] to minimize egress emission. To compensate for the transmitter window, the corresponding receiver requires post-processing equalization
[31, 80]. A joint consideration of spectral rolloff and SNR degradation due
to post-processing was given in [80]. Transmitter windowing for zero-padded
DFT-based transceivers has also been addressed therein. Per-tone windows
were proposed in [20] for shaping the transmitted spectrum. The shaping of
spectra allows more tones to be used for transmission.

Windowing is also often applied at the receiver side. To improve RFI
suppression in DSL applications, receiver windowing has been proposed in

the literature [29, 143]. The clever idea of using extra redundant samples and folding to reduce RFI was given in [143]. For the suppression of sidelobes without using extra redundant samples, it was proposed in [61] to use windows that introduce controlled IBI, removed later using decision feedback. A joint minimization of RFI and channel noise was considered in [123]. The optimal window can be found using the statistics of the RFI and channel noise [123]. Also using the statistics of the channel noise and RFI, a joint design of the TEQ and the receiving window for maximizing bit rates was given in [188]. Minimization of RFI using subfilters was proposed in [83]. A novel combination of raised cosine windowing and per-tone equalization without using extra redundant samples was given in [32]. Per-tone windowing for RFI suppression without using the information of RFI and channel noise statistics has been suggested in [180]. Receiver windowing has also been considered for OFDM applications. To improve the reception of OFDM systems, it was proposed in [98] to use Nyquist windows. Optimal Nyquist windows were considered in [97] to mitigate the effect of additive noise and carrier frequency offsets.

In the context of RFI mitigation, interference cancellation has been suggested in [13]. Using the receiver outputs of unused RFI tones or neighboring tones, the parameters of the RFI signal can be estimated and used to cancel interference on the tones that are affected. Parameter estimation and cancellation for the case when the RFI source is a single sideband signal was developed in [56]. Based on the approach in [13], a more general framework for RFI cancellation was given in [140]. The method is applicable to RFI of all analog modulation schemes, e.g. amplitude modulation, single sideband, double sideband, and so forth.

9.9 Problems

9.1 *Trapezoidal window.* Suppose the transmitter window a_i is the discrete-time Trapezoidal function of length $N + \alpha$ as shown in Fig. P9.1. The coefficients satisfy $a_\alpha = a_{\alpha+1} = \cdots = a_{N-1} = 1$. The two ends of the window drop linearly to zero. The number of nonunity coefficients on each end is α.

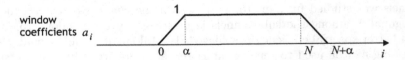

Figure P9.1. Trapezoidal window function.

(a) Does the trapezoidal function satisfy the time domain Nyquist condition in (9.13)?

(b) Suppose we use the trapezoidal window in the DFT implementation in Fig. 9.9. Show that the system in Fig. 9.9 has the equivalent subfilter receiving bank structure in Fig. 9.6.

(c) Find the corresponding subfilter $P_0(z)$ in (b).

9.2 We are given a transmitter window function a_i that has length $N + \alpha$ and satisfies the time domain Nyquist condition in (9.13). In this case does the DFT implementation in Fig. 9.14 always have the equivalent subfilter receiving bank structure in Fig. 9.13? If so, how are the coefficients of $P_0(z)$ related to a_i?

9.3 The transmitting filters $F_k(z)$ in the conventional DMT system form a DFT bank. From Section 9.3 we also know that the new transmitting filters $F_k'(z)$ again form a DFT bank when the subfilters are frequency-shifted versions of the first subfilter. Show that if the subfilters do not form a DFT bank, neither do the new transmitting filters.

9.4 Let $g(n)$ be a sequence satisfying the time domain Nyquist condition $\sum_\ell g(i + N\ell) = 1$. Let $G(e^{j\omega})$ be the Fourier transform of $g(n)$. Show that

$$G(e^{j2\pi k/N}) = 0, \quad \text{for} \quad k = 1, 2, \ldots, N - 1.$$

In other words, the Fourier transform of $g(n)$ has the zero-crossing property.

9.5 Let $h(n)$ be a rectangular window of length N and let $g(n)$ be an arbitrary sequence of length P. Show that the length $P + N - 1$ sequence $h(n) * g(n)$ satisfies the time domain Nyquist(N) property.

9.6 Consider the transmitter in Fig. 9.9. Show that if the subfilter $p_0(n)$ is normalized such that $\sum_{n=0}^{\alpha} p_0(n) = 1$, then the complexity can be reduced to one IDFT plus 2α additions and α multiplications per block.

9.7 We know the receiver in Fig. 9.14 needs only 2β more multiplications and β more additions per block of outputs when it is compared with the conventional DMT receiver. Show that when the window satisfies the Nyquist property, the number of multiplications can be further reduced to β.

9.8 *Transmitter window not satisfying the time domain Nyquist condition.* Suppose the transmitter window coefficients a_i in Fig. 9.9 do not satisfy the time domain Nyquist condition in (9.13). Does the DFT implementation in Fig. 9.9 still have the equivalent receiving bank structure in Fig. 9.6?

9.9 In Section 9.6, it was mentioned that the receiver subfilters can be individually optimized when the information of the interference sources is available. Show that the objective function ϕ_k' in (9.29) can be written as a quadratic form of the kth subfilter vector $\mathbf{q}_k$, i.e. $\mathbf{q}_k^\dagger \boldsymbol{\Phi}_k \mathbf{q}_k$, for some square matrix $\boldsymbol{\Phi}_k$. Find the matrix $\boldsymbol{\Phi}_k$. Is $\boldsymbol{\Phi}_k$ always positive definite?

9.10 *Transmitter windowing without extra redundancy* [80]. The transmitter window discussed in this chapter requires the prefix length $\nu \geq L + \alpha$, where α is the order of the transmitter subfilters and L is the order of the FIR channel. This means we need to have α more redundant samples than the DMT system in Chapter 6. In this problem, we consider the case when there are no extra redundant samples, i.e. $\nu = L$ and $N = M + L$. The transmitter is as shown in Fig. P9.10(a), where $\mathbf{F}_{cp}$ is the $N \times M$ cyclic-prefix insertion matrix defined in (5.14). Suppose the transmitter window has the cyclic-prefix property. That is, the first ν window coefficients are the same as the last ν coefficients,

$$a_i = a_{M+i}, \quad i = 0, 1, \ldots, \nu - 1.$$

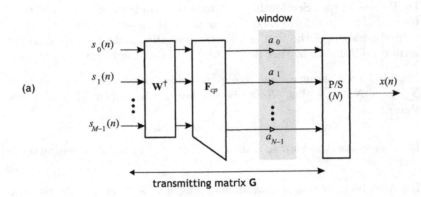

(a)

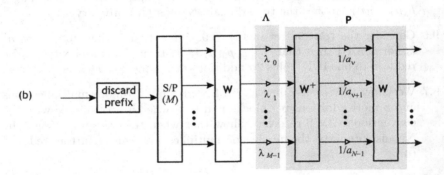

(b)

Figure P9.10. (a) Transmitter window without extra redundancy; (b) ISI-free receiver for the transmitter in (a).

(a) Derive the receiver so that the overall system is ISI-free.

(b) Show that the ISI-free receiver in (a) can be implemented as the structure in Fig. P9.10(b). The coefficients λ_k are the reciprocals of the channel DFT coefficients, $\lambda_k = 1/C_k$, where C_k are the M-point DFT of the channel impulse response. The post-processing matrix $\mathbf{P}$ as indicated in Fig. P9.10(b) depends only on the window but not the channel.

When there are no extra redundant samples, applying windowing at the transmitter results in a receiver that generally depends heavily on the channel. The result in this problem shows that if the window has the cyclic-prefix property, then the channel dependent part of the receiver continues to be the FEQ coefficients only.

9.11 Find a window function of length $M + \nu$ that satisfies both the time domain Nyquist(M) property and the cyclic-prefixed property described in Problem 9.10. Give a sketch of the window.

9.12 In Problem 9.10, the window has the cyclic-prefix property, i.e. $a_i = a_{M+i}$, for $i = 0, 1, \ldots, \nu - 1$. Suppose now that the window does not have this property.

(a) Derive the receiver so that the overall system has the ISI-free property.

(b) Show that the ISI-free receiver in (a) can be implemented as in Fig. P9.12. The post-processing matrix $\mathbf{P}$ is of the form

$$\mathbf{P} = \left[\mathbf{W} \begin{bmatrix} \mathbf{A}_1 & \mathbf{0} \\ \mathbf{0} & \mathbf{A}_2 \end{bmatrix} \mathbf{W}^\dagger + \mathbf{\Lambda W} \begin{bmatrix} \mathbf{0} & \mathbf{C}_0(\mathbf{A}_0 - \mathbf{A}_2) \\ \mathbf{0} & \mathbf{0} \end{bmatrix} \mathbf{W}^\dagger \right]^{-1},$$

where $\mathbf{A}_0$, $\mathbf{A}_1$, and $\mathbf{A}_2$ are diagonal matrices with diagonal elements consisting of, respectively, the first ν window coefficients, the following $M - \nu$ coefficients, and the last ν coefficients. The matrix $\mathbf{\Lambda}$ is as in Problem 9.10 and $\mathbf{C}_0$ is an $L \times L$ lower triangular Toeplitz matrix with the first column given by $\begin{bmatrix} c_0 & c_1 & \cdots & c_{L-1} \end{bmatrix}^T$. Is the post-processing matrix $\mathbf{P}$ still channel-independent as in Problem 9.10?

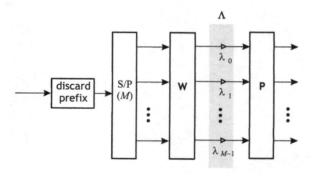

Figure P9.12. Structure of the ISI-free receiver.

9.13 *Effect of subfilters on the subchannel noises.* To mitigate interference, we included subfilters in the receiving bank. These additional subfilters will affect the subchannel output noise. Let the channel noise be AWGN with variance $\mathcal{N}_0$. Let $Q_0(z)$ be the subfilter in the zeroth subchannel with $Q_0(e^{j0}) = 1$.

(a) What is the zeroth subchannel noise variance $\sigma_{e_0}^2$ due to channel noise?

(b) Suppose $Q_0(z)$ has magnitude response as shown in Fig. 9.16(a). Do you expect the noise variance in (a) to be larger or smaller than the case without a subfilter?

(c) Suppose the subfilters are frequency-shifted versions of the first subfilter $Q_0(z)$. What is the kth subchannel noise variance $\sigma_{e_k}^2$ due to channel noise? Is it the same as that in (a)?

9.14 *Zero-padded transceivers.* Consider the zero-padded transceiver in Section 6.2 with the efficient receiver in (6.15).

(a) Apply subfilters to the transmitter as in Fig. 9.6. Suppose the subfilters $P_k(z)$ are shifted versions of the zeroth subfilter $P_0(z)$. Show that the new transmitting filters form a DFT bank and can be implemented efficiently. Draw the DFT implementation.

(b) Suppose the efficient receiver given in (6.15) is used. Apply subfilters to the receiver as in Fig. 9.13. Again assume the subfilters are shifted versions of the first subfilter. Show that the new receiving filters also form a DFT bank and can be implemented efficiently. Draw the DFT implementation.

10

Minimum redundancy FIR transceivers

In earlier chapters we saw that the use of redundancy in block transceivers allows us to remove ISI completely without using IIR filters. When the number of redundant samples per block ν is more than the channel order L, there is no IBI, and we can further achieve zero ISI using a constant receiving matrix. The most notable example is the OFDM system studied in Chapter 6. But the use of redundant samples also decreases the transmission rate. For every M input symbols, the transmitter sends out $N = M + \nu$ samples. The actual transmission rate is decreased by a factor of N/M. There are ν redundant samples in every N samples transmitted. Reducing redundancy leads to a higher transmission rate and hence better bandwidth efficiency. At the same time, we would like the redundancy to be large enough so that the zero-forcing condition can still be satisfied without using IIR filters. A natural question to ask is: for a given channel and N, what is the smallest redundancy such that FIR transceivers exist? In other words, if we are to use an FIR transceiver that achieves zero ISI, what is the largest number of symbols that can be transmitted out of every N samples? This chapter aims to answer the question of minimum redundancy for the existence of **FIR zero-forcing transceivers**.

We will consider general FIR transceivers (Fig. 10.1) in which the filters are not constrained to be DFT filters as in the OFDM system. Moreover the length of the filters can be longer than the block size N. In this case the transmitting and receiving matrices are allowed to have memories, rather than constant matrices as in the OFDM case. We will see that the minimum redundancy depends on the underlying channel $C(z)$, and it can be easily determined from the location of the zeros of the channel $C(z)$ directly by inspection. The topic of minimum redundancy for FIR transceivers was first addressed in [182]. The problem has been studied using different approaches [62, 76, 124, 135]. For consistency with the rest of the book, we will follow the approach given in [76].

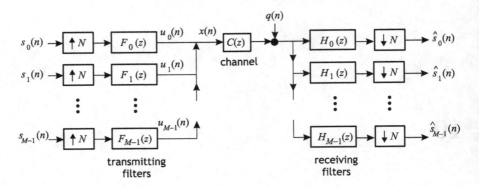

Figure 10.1. M-subchannel filter bank transceiver.

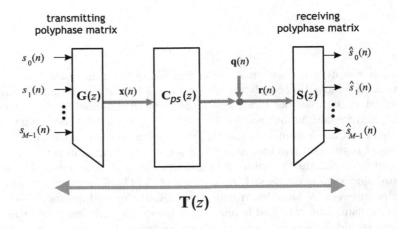

Figure 10.2. Polyphase representation of the transceiver.

10.1 Polyphase representation

In the following discussion, it is often more convenient to work with the polyphase representation (Chapter 5) of the transceiver, which allows us to redraw the filter bank transceiver as in Fig. 10.2. In the figure, the $N \times M$ matrix $\mathbf{G}(z)$ is the polyphase matrix of the transmitting bank and $\mathbf{S}(z)$ is the $M \times N$ polyphase matrix of the receiving bank. The matrices $\mathbf{G}(z)$ and $\mathbf{S}(z)$ are not constrained to be constant matrices as in Chapters 6-8. The $N \times 1$ vector $\mathbf{q}(n)$ is the blocked version of the channel noise $q(n)$. Throughout this chapter, the channel noise will not appear in our discussion as it does not affect the existence of FIR zero-forcing filter bank transceivers. The blocked channel matrix $\mathbf{C}_{ps}(z)$ is a pseudocirculant, which has been introduced earlier in Chapter 5. Some of its properties are exploited in Section 5.4 for eliminating interblock interference. To be more specific, let the channel be an FIR filter of order L:

$$C(z) = c(0) + c(1)z^{-1} + \cdots + c(L)z^{-L}.$$

To avoid degenerate cases, we will consider $C(z)$ with $c(0), c(L) \neq 0$, and $L \geq 1$. The polyphase representation of $C(z)$ is

$$C(z) = C_0(z^N) + C_1(z^N)z^{-1} + \cdots + C_{N-1}(z^N)z^{-N+1}. \tag{10.1}$$

In this chapter, the channel order L can be longer than N. All the polyphase components $C_i(z)$ are in general polynomials in z^{-1}. Then the blocked version $\mathbf{C}_{ps}(z)$ is given by

$$\mathbf{C}_{ps}(z) = \begin{bmatrix} C_0(z) & z^{-1}C_{N-1}(z) & \cdots & z^{-1}C_1(z) \\ C_1(z) & C_0(z) & \cdots & z^{-1}C_2(z) \\ \vdots & \vdots & \ddots & \vdots \\ C_{N-1}(z) & C_{N-2}(z) & \cdots & C_0(z) \end{bmatrix}. \tag{10.2}$$

The study of pseudocirculant matrices turns out to be of vital importance in determining minimum redundancy of FIR transceivers. Some useful properties will be given in Section 10.2. From Fig. 10.2, we know the condition for zero ISI can be given in terms of polyphase representation. There is zero inter-subchannel interference if

$$\mathbf{S}(z)\mathbf{C}_{ps}(z)\mathbf{G}(z) = \mathbf{\Lambda}(z), \tag{10.3}$$

for some diagonal matrix $\mathbf{\Lambda}(z)$. The transceiver is ISI-free (zero inter- and intra-subchannel ISI) if, in addition, the diagonal elements of $\mathbf{\Lambda}(z)$ are merely delays, i.e. $[\mathbf{\Lambda}(z)]_{ii} = \lambda_i z^{-n_i}$, with $\lambda_i \neq 0$. In this case we can always scale and delay the receiver outputs so that

$$\mathbf{S}(z)\mathbf{C}_{ps}(z)\mathbf{G}(z) = \mathbf{I}_M. \tag{10.4}$$

Thus, without loss of generality we can consider this ISI-free condition because delays and scaling do not change the FIR property of the system.

10.2 Properties of pseudocirculants

In this section we introduce some properties of pseudocirculants that will be useful for our subsequent discussion. We will diagonalize pseudocirculants using two types of decomposition. The first one, Smith form decomposition, uses unimodular matrices to diagonalize FIR pseudocirculants. The second one diagonalizes $\mathbf{C}_{ps}(z^N)$, rather than $\mathbf{C}_{ps}(z)$, using DFT matrices and some simple diagonal matrices [157, 159].[1]

---

[1] For a more in-depth treatment of these decompositions, the readers are referred to Chapter 13 of [159] or [58].

10.2.1 Smith form decomposition

A polynomial matrix $\mathbf{A}(z)$ in z^{-1} can be diagonalized by the so-called **unimodular matrices**. We say an $N \times N$ polynomial matrix $\mathbf{U}(z)$ is unimodular if

$$\det \mathbf{U}(z) = c,$$

a nonzero constant [159]. When $\mathbf{U}(z)$ is causal FIR unimodular, so is its inverse $\mathbf{U}^{-1}(z)$, due to the constant determinant property.

An $N \times N$ polynomial matrix $\mathbf{A}(z)$ in z^{-1} can be represented using the **Smith form decomposition**

$$\mathbf{A}(z) = \mathbf{U}(z)\mathbf{\Gamma}_s(z)\mathbf{V}(z), \tag{10.5}$$

where all three matrices in the decomposition are matrix polynomials in the variable z^{-1}. Both $\mathbf{U}(z)$ and $\mathbf{V}(z)$ are unimodular matrices, and $\mathbf{\Gamma}_s(z)$ is a diagonal matrix:

$$\mathbf{\Gamma}_s(z) = \begin{bmatrix} \gamma_0(z) & 0 & \cdots & 0 \\ 0 & \gamma_1(z) & & 0 \\ \vdots & & \ddots & \vdots \\ 0 & 0 & \cdots & \gamma_{N-1}(z) \end{bmatrix}.$$

Moreover the unimodular matrices $\mathbf{U}(z)$ and $\mathbf{V}(z)$ can be so chosen that the polynomials $\gamma_k(z)$ are monic (i.e. the highest power has unity coefficient) and $\gamma_k(z)$ is a factor of $\gamma_{k+1}(z)$ (i.e. $\gamma_k(z)$ divides $\gamma_{k+1}(z)$ for $0 \le k \le N - 2$). In this case the matrix $\mathbf{\Gamma}_s(z)$, called the **Smith form** of $\mathbf{A}(z)$, is unique. The diagonal elements $\gamma_k(z)$ are given by

$$\gamma_k(z) = \Delta_{k+1}(z)/\Delta_k(z),$$

where $\Delta_k(z), k = 1, 2, \ldots, N$, is the greatest common divisor (gcd) of all the $k \times k$ minors of $\mathbf{A}(z)$, and $\Delta_0(z) = 1$. Although $\mathbf{\Gamma}_s(z)$ is unique, the unimodular matrices $\mathbf{U}(z)$ and $\mathbf{V}(z)$ are not. The decomposition can be obtained using a finite number of elementary row and column operations [159].

Return to the blocked version of the transceiver system in Fig. 10.2. When the scalar channel $C(z)$ is causal and FIR, the blocked channel matrix $\mathbf{C}_{ps}(z)$ is also causal and FIR. Let us apply Smith form decomposition to $\mathbf{C}_{ps}(z)$,

$$\mathbf{C}_{ps}(z) = \mathbf{U}(z)\mathbf{\Gamma}_s(z)\mathbf{V}(z), \tag{10.6}$$

where $\mathbf{\Gamma}_s(z)$ is the Smith form of $\mathbf{C}_{ps}(z)$. Let ρ_N denote the number of nonunity terms in the diagonals of the Smith form. This number turns out to be related to the minimum redundancy for the existence of FIR transceivers, as we will see later. With the definition of ρ_N, we can express $\mathbf{\Gamma}_s(z)$ as

$$\mathbf{\Gamma}_s(z) = \mathrm{diag}\begin{bmatrix} 1 & 1 & \cdots & 1 & \gamma_{N-\rho_N}(z) & \cdots & \gamma_{N-1}(z) \end{bmatrix}. \tag{10.7}$$

Example 10.1 In this example, we compute the Smith form of the pseudo-circulant associated with the scalar filter $C(z) = 1 - 2z^{-1} + 2z^{-2} - z^{-3}$. The polyphase components of $C(z)$ with respect to $N = 3$ are, respectively,

$$C_0(z) = 1 - z^{-1}, \quad C_1(z) = -2, \quad \text{and} \quad C_2(z) = 2.$$

The associated pseudocirculant is

$$\mathbf{C}_{ps}(z) = \begin{bmatrix} 1 - z^{-1} & 2z^{-1} & -2z^{-1} \\ -2 & 1 - z^{-1} & 2z^{-1} \\ 2 & -2 & 1 - z^{-1} \end{bmatrix}.$$

This has the following Smith form decomposition:

$$\underbrace{\begin{bmatrix} \frac{1}{2}(1 - z^{-1}) & -1 & 1 \\ -1 & 1 & 0 \\ 1 & 0 & 0 \end{bmatrix}}_{\mathbf{U}(z)} \underbrace{\begin{bmatrix} 1 & 0 & 0 \\ 0 & 1 + z^{-1} & 0 \\ 0 & 0 & (1 + z^{-1})(1 - z^{-1}) \end{bmatrix}}_{\mathbf{\Gamma}_s(z)} \underbrace{\begin{bmatrix} 2 & -2 & 1 - z^{-1} \\ 0 & -1 & 1 \\ 0 & 0 & \frac{1}{2} \end{bmatrix}}_{\mathbf{V}(z)}.$$

Both $\mathbf{U}(z)$ and $\mathbf{V}(z)$ have determinants equal to -1. The Smith form has two nontrivial diagonal terms, $\gamma_1(z) = 1 + z^{-1}$ and $\gamma_2(z) = (1 + z^{-1})(1 - z^{-1})$, where $\gamma_1(z)$ divides $\gamma_2(z)$. Therefore we have $\rho_3 = 2$. ∎

For an arbitrary FIR polynomial matrix, step by step row and column operations are required to compute its Smith form. But for pseudocirculants it is much earlier to obtain its Smith form. We will see in Section 10.5 that, given the zeros of the underlying scalar filter $C(z)$, the Smith form can be determined easily by inspection.

10.2.2 DFT decomposition

It is known that pseudocirculants can also be diagonalized using DFT matrices [159]. An arbitrary pseudocirculant $\mathbf{C}_{ps}(z)$, not necessarily FIR, assumes the following decomposition:

$$\mathbf{C}_{ps}(z^N) = \mathbf{D}(z^{-1})\mathbf{W}\mathbf{\Sigma}(z)\mathbf{W}^\dagger\mathbf{D}(z), \tag{10.8}$$

where $\mathbf{D}(z)$ and $\mathbf{\Sigma}(z)$ are diagonal matrices given, respectively, by

$$\mathbf{D}(z) = \begin{bmatrix} 1 & 0 & \cdots & 0 \\ 0 & z^{-1} & & 0 \\ \vdots & & \ddots & \\ 0 & 0 & \cdots & z^{-N+1} \end{bmatrix},$$

$$\mathbf{\Sigma}(z) = \begin{bmatrix} C(z) & 0 & \cdots & 0 \\ 0 & C(zW) & & 0 \\ \vdots & & \ddots & \\ 0 & 0 & \cdots & C(zW^{N-1}) \end{bmatrix},$$

with $W = e^{-j2\pi/N}$. The matrix $\mathbf{W}$ is the $N \times N$ normalized DFT matrix with $[\mathbf{W}]_{mn} = \frac{1}{\sqrt{N}}e^{-j2mn\pi/N}$, $0 \leq m, n \leq N - 1$. Note that the size of $\mathbf{W}$ is N instead of M as in earlier chapters. The decomposition can be obtained by showing that

$$\mathbf{D}(z)\mathbf{C}_{ps}(z^N)\mathbf{D}(z^{-1}) \tag{10.9}$$

is a circulant matrix (Problem 10.10) and circulants can be diagonalized using DFT matrices.

10.2.3 Properties derived from the two decompositions

Property 10.1. Determinant of $\mathbf{C}_{ps}(z)$ The determinant of $\mathbf{C}_{ps}(z)$ is directly related to that of its Smith form $\boldsymbol{\Gamma}_s(z)$ and to that of the diagonal matrix $\boldsymbol{\Sigma}(z)$ in the DFT decomposition. The two matrices $\mathbf{U}(z)$ and $\mathbf{V}(z)$ in the Smith form decomposition are unimodular and their determinants are constants, so we have

$$\det \mathbf{C}_{ps}(z) = c \det \boldsymbol{\Gamma}_s(z) = c\gamma_0(z)\gamma_1(z)\cdots\gamma_{N-1}(z), \qquad (10.10)$$

where $c = \det \mathbf{U}(z) \det \mathbf{V}(z)$. Now consider the determinants of the two sides of (10.8). We can obtain

$$\det \mathbf{C}_{ps}(z^N) = \det\left(\boldsymbol{\Sigma}(z)\right) = \prod_{k=0}^{N-1} C(zW^k). \qquad (10.11)$$

The above expression implies in particular that $\det \mathbf{C}_{ps}(z)$ is a delay if and only if $C(z)$ is a delay. This expression also leads to the following property that relates the zeros of $\mathbf{C}_{ps}(z)$ to those of the scalar filter $C(z)$.

Property 10.2. Zeros of $\det \mathbf{C}_{ps}(z)$ Suppose the channel $C(z)$ is a causal FIR filter of order L and the zeros are

$$\theta_1, \theta_2, \ldots, \theta_L.$$

Then $\det \mathbf{C}_{ps}(z)$ is also a causal FIR filter of order L and the zeros are

$$\theta_1^N, \theta_2^N, \ldots, \theta_L^N.$$

Proof. Let

$$C(z) = c(0)\left[1 - \theta_1 z^{-1}\right]\left[1 - \theta_2 z^{-1}\right]\cdots\left[1 - \theta_L z^{-1}\right].$$

Property 10.1 tells us that $\det \mathbf{C}_{ps}(z^N) = \prod_{k=0}^{N-1} C(zW^k)$, which is equal to

$$[c(0)]^N \prod_{k=0}^{N-1} \left[1 - \theta_1(zW^k)^{-1}\right]\left[1 - \theta_2(zW^k)^{-1}\right]\cdots\left[1 - \theta_L(zW^k)^{-1}\right].$$

It can be verified that (Problem 10.8)

$$\prod_{k=0}^{N-1} \left[1 - \theta_i(zW^k)^{-1}\right] = 1 - \theta_i^N z^{-N}.$$

It follows from the above expression that the determinant of $\mathbf{C}_{ps}(z)$ can be rearranged as follows:

$$\det \mathbf{C}_{ps}(z) = [c(0)]^N \prod_{i=1}^{L} \left[1 - \theta_i^N z^{-1}\right].$$

This means that $\det \mathbf{C}_{ps}(z)$ is also an FIR filter of order L and the zeros are $\theta_1^N, \theta_2^N, \ldots, \theta_L^N$. ∎

Property 10.3. Zeros of $\gamma_k(z)$ The determinant expression in (10.10) means that the zeros of $\det \mathbf{C}_{ps}(z)$ are distributed among $\gamma_k(z)$. The previous property states that the zeros of $\det \mathbf{C}_{ps}(z)$ are θ_i^N. Therefore the zeros of $\gamma_k(z)$ are also θ_i^N. As $\det \mathbf{C}_{ps}(z)$ has only L zeros, there are at most L nontrivial $\gamma_k(z)$. The number of nonunity terms on the diagonal of the Smith form satisfies $\rho_N \leq \min[L, N]$. When $N > L$, the first $(N-L)$ diagonal terms of the Smith form are equal to one.

Property 10.4. Rank of $\mathbf{C}_{ps}(z)$ For a causal FIR pseudocirculant $\mathbf{C}_{ps}(z)$, a matrix polynomial of z^{-1}, the rank is a function of z. In the Smith form decomposition the unimodular matrices $\mathbf{U}(z)$ and $\mathbf{V}(z)$ are nonsingular for all z. Therefore the rank of $\mathbf{C}_{ps}(z)$ is the same as the rank of its Smith form $\mathbf{\Gamma}_s(z)$. The rank of $\mathbf{\Gamma}_s(z)$ depends on the number of nontrivial terms on the diagonal, i.e. ρ_N. If τ is a zero of $\gamma_{N-\rho_N}(z)$, then, due to the property that $\gamma_i(z)$ divides $\gamma_{i+1}(z)$, we have

$$\gamma_{N-\rho_N}(\tau) = \gamma_{N-\rho_N+1}(\tau) = \cdots = \gamma_{N-1}(\tau) = 0.$$

The rank of $\mathbf{\Gamma}_s(\tau)$ is thus $N - \rho_N$. This is also the smallest rank of $\mathbf{\Gamma}_s(z)$, i.e.

$$\min_z \operatorname{rank}(\mathbf{\Gamma}_s(z)) = N - \rho_N.$$

On the other hand, from the DFT decomposition of $\mathbf{C}_{ps}(z)$ we know that the rank of $\mathbf{C}_{ps}(z^N)$ is the same as the rank of $\mathbf{\Sigma}(z)$ as $\mathbf{W}$ and $\mathbf{D}(z)$ are nonsingular. Therefore, we have

$$\operatorname{rank}(\mathbf{C}_{ps}(z^N)) = \operatorname{rank}(\mathbf{\Sigma}(z)) = \operatorname{rank}(\mathbf{\Gamma}_s(z^N)); \qquad (10.12)$$

the smallest ranks of the three matrices $\mathbf{C}_{ps}(z)$, $\mathbf{\Sigma}(z)$, and $\mathbf{\Gamma}_s(z)$ are the same. We already know that the smallest rank of $\mathbf{\Gamma}_s(z)$ is $N - \rho_N$ from (10.12). Therefore,

$$\min_z \operatorname{rank}(\mathbf{\Gamma}_s(z)) = \min_z \operatorname{rank}(\mathbf{C}_{ps}(z)) = \min_z \operatorname{rank}(\mathbf{\Sigma}(z)) = N - \rho_N,$$

where ρ_N is the number of nonunity terms in the diagonals of the Smith form $\mathbf{\Gamma}_s(z)$. This equation ties the smallest rank of $\mathbf{C}_{ps}(z)$ to the number of nontrivial diagonal terms in its Smith form. For a given FIR channel $C(z)$ and interpolation ratio N, the rank of $\mathbf{C}_{ps}(z)$ can actually be determined more explicitly, as discussed next.

10.2.4 Congruous zeros

The smallest rank of $\mathbf{C}_{ps}(z)$ can also be obtained by inspecting the so-called **congruous zeros** of $C(z)$. Let us first factorize $C(z)$ in terms of its distinct zeros. Suppose $C(z)$ has p distinct zeros, $\alpha_1, \alpha_2, \ldots, \alpha_p$, of multiplicities, respectively, $m_1, m_2, \ldots, m_p$. Then $m_1 + m_2 + \cdots + m_p$ is equal to L, the order of $C(z)$, and

$$C(z) = c(0) \left[1 - \alpha_1 z^{-1}\right]^{m_1} \left[1 - \alpha_2 z^{-1}\right]^{m_2} \cdots \left[1 - \alpha_p z^{-1}\right]^{m_p}.$$

When all the zeros are distinct, we have $p = L$ and $m_1 = m_2 = \cdots = m_p = 1$.

Definition 10.1 *Congruous zeros.* A set of distinct zeros $\{\alpha_1, \alpha_2, \ldots \alpha_q\}$ of $C(z)$ are *congruous with respect to* N if

$$\alpha_1^N = \alpha_2^N = \cdots = \alpha_q^N.$$ ∎

The zeros that are congruous are distinct but their magnitudes are the same and their angles differ by an integer multiple of $2\pi/N$. They can be considered as rotations of each other. We express each as a rotation of α_1:

$$\alpha_j = \alpha_1 W^{-n_j}, \quad 0 \le n_j < N. \tag{10.13}$$

Because the congruous zeros are distinct by definition, the integers n_j are also distinct. In fact, from Property 10.2, we know that $\alpha_1^N, \alpha_2^N, \ldots, \alpha_q^N$ correspond to the zeros of $\det \mathbf{C}_{ps}(z)$. We can say that a set of q congruous zeros are distinct zeros of $C(z)$ that merge as a zero of multiplicity q of $\det \mathbf{C}_{ps}(z)$.

Definition 10.2 The notation μ_N denotes the cardinality of the largest set of congruous zeros of $C(z)$ with respect to N. ∎

The number μ_N represents the largest number of distinct zeros that have the same magnitude and their differences in angles are integer multiples of $2\pi/N$. When $C(z)$ has p distinct zeros, the number μ_N is bounded by $1 \le \mu_N \le p$.

In the earlier example, $C(z) = 1 - 2z^{-1} + 2z^{-2} - z^{-3}$, the zeros are $e^{j\pi/3}$, $e^{-j\pi/3}$, and 1, as shown in Fig. 10.3(a). For $N = 2$, we raise the zeros to the power of 2 and show their locations in Fig. 10.3(b). No two zeros are congruous and $\mu_2 = 1$. The case for $N = 3$ is shown in Fig. 10.3(c). The two zeros, $e^{j\pi/3}$ and $e^{-j\pi/3}$, merge after they are raised to the third power. They are congruous with respect to $N = 3$, and we have $\mu_3 = 2$. When $N = 6$ (Fig. 10.3(d)), the three zeros are congruous and we have $\mu_6 = 3$.

As another example, consider $C(z) = (1 + z^{-1})(1 - z^{-1})^2$. This filter has double zeros at $z = 1$ and one zero at $z = -1$. There are two distinct zeros, -1 and 1, as shown in Fig. 10.4(a). The second and third powers of these zeros are shown respectively in Fig. 10.4(b) and Fig. 10.4(c). For $N = 2$, these two zeros merge and we have $\mu_2 = 2$. For $N = 3$, these two zeros do not merge and we have $\mu_3 = 1$.

The definition of μ_N is based on the congruous zeros of the scalar filter $C(z)$. It turns out that μ_N is actually equal to ρ_N, the number of the non-trivial diagonal terms on the Smith form of the associated pseudocirculant $\mathbf{C}_{ps}(z)$.

Lemma 10.1 For an FIR $N \times N$ pseudocirculant $\mathbf{C}_{ps}(z)$, the number of non-unity terms on the diagonal of its Smith form is the same as the largest number of congruous zeros of $C(z)$, i.e.

$$\rho_N = \mu_N.$$ ∎

Proof. Recall that the diagonal matrix $\mathbf{\Sigma}(z)$ in the DFT decomposition of $\mathbf{C}_{ps}(z^N)$ is FIR if $C(z)$ is FIR. The number of terms on the diagonal of $\mathbf{\Sigma}(z)$ that have common zeros will determine the smallest rank of

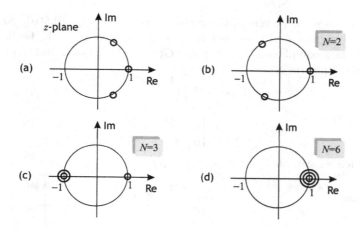

Figure 10.3. (a) The zeros α_i of $C(z)$; (b)–(d) α_i^N for $N = 2, 3, 6$. The channel is
$C(z) = 1 - 2z^{-1} + 2z^{-2} - z^{-3}$.

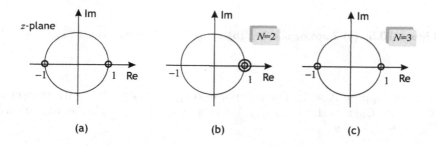

Figure 10.4. (a) The distinct zeros α_i of $C(z)$; (b) α_i^N for $N = 2$; (c) α_i^N for $N = 3$.
The channel is $C(z) = (1 + z^{-1})(1 - z^{-1})^2$.

$\Sigma(z)$. Observe that the zeros of $C(zW^k)$ are those of $C(z)$ rotated by
$2k\pi/N$. If $C(z)$ and $C(zW^k)$ have a common zero α, then

$$C(\alpha) = C(\alpha W^k) = 0.$$

This means that both α and αW^k are zeros of $C(z)$. By Definition 10.1
these zeros are congruous with respect to N. The largest number of
terms on the diagonal of $\Sigma(z)$ that have common zeros is the same as
the largest number of congruous zeros μ_N. Therefore the smallest rank
of $\Sigma(z)$ is $N - \mu_N$. But we know already that the smallest rank of
$C_{ps}(z)$ is $N - \rho_N$. Therefore, we have $\rho_N = \mu_N$. ∎

Example 10.2 Consider the filter $C(z) = 1 + z^{-3}$ with $L = 3$. It has three
distinct zeros at $z = e^{j\pi/3}$, $e^{-j\pi/3}$, and -1 (Fig. 10.5(a)). Consider first

$N = 2$. The second powers of the zeros are shown in Fig. 10.5(b). As no two zeros are congruous with respect to $N = 2$, we have $\mu_2 = 1$. On the other hand, the two polyphase components of $C(z)$ are $C_0(z) = 1$, and $C_1(z) = z^{-1}$. The channel matrix is

$$\mathbf{C}_{ps}(z) = \begin{bmatrix} 1 & z^{-2} \\ z^{-1} & 1 \end{bmatrix}.$$

Its Smith form is

$$\mathbf{\Gamma}_s = \begin{bmatrix} 1 & 0 \\ 0 & 1 - z^{-3} \end{bmatrix},$$

from which we can also obtain $\rho_2 = 1 = \mu_2$.

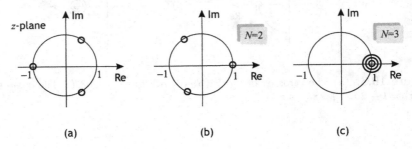

Figure 10.5. (a) Zeros α_i of $C(z)$; (b) α_i^2; (c) α_i^3. The channel is $C(z) = 1 + z^{-3}$.

Now suppose $N = 3$. The polyphase components of $C(z)$ with respect to $N = 3$ are $C_0(z) = 1 + z^{-1}$, $C_1(z) = 0$, and $C_2(z) = 0$. The channel matrix $\mathbf{C}_{ps}(z)$ is given by

$$\mathbf{C}_{ps}(z) = \begin{bmatrix} 1 + z^{-1} & 0 & 0 \\ 0 & 1 + z^{-1} & 0 \\ 0 & 0 & 1 + z^{-1} \end{bmatrix}.$$

The determinant of this matrix has a zero of multiplicity 3, as shown in Fig. 10.5. In this special case, $\mathbf{C}_{ps}(z)$ is the same as its Smith form. The number of nonunity terms ρ_3 on the diagonal is three. This result is also consistent with the fact that the three zeros of $C(z)$ are congruous with respect to $N = 3$, and hence $\mu_3 = 3$. ∎

Example 10.3 Consider the second-order channel $C(z) = 1 + 2z^{-1} + z^{-2}$. The channel has double zeros at $z = -1$. The number of zeros on the unit circle is two but the number of distinct zeros is one. We have $\mu_N = 1$ for all N in this case. For instance, if $N = 2$, the polyphase components of $C(z)$ with respect to $N = 2$ are $C_0(z) = 1 + z^{-1}$ and $C_1(z) = 2$. The channel matrix $\mathbf{C}_{ps}(z)$ is given by

$$\mathbf{C}_{ps}(z) = \begin{bmatrix} 1 + z^{-1} & 2z^{-1} \\ 2 & 1 + z^{-1} \end{bmatrix}.$$

The corresponding Smith form $\mathbf{\Gamma}_s(z)$ is

$$\mathbf{\Gamma}_s(z) = \begin{bmatrix} 1 & 0 \\ 0 & 1 - 2z^{-1} + z^{-2} \end{bmatrix}.$$

The number of nonidentity terms on the diagonal is $\rho_2 = 1$. This number is one for all N in this case. ∎

Lemma 10.1 shows that ρ_N is equal to μ_N, the largest number of congruous zeros, a quantity that can be determined directly by inspecting the zeros of $C(z)$. In Section 10.4 we will see how this number is tied to the minimum redundancy for designing FIR transceivers.

10.3 Transceivers with no redundancy

The filter bank transceiver in Fig. 10.1 is called **minimal** if the interpolation ratio $N = M$ and there is no redundancy. In this section we consider the properties of such systems. We will show that for FIR minimal transceivers, we can achieve only zero inter-subchannel ISI, but not zero ISI. Moreover, for channels that do not have minimum phase, there does not exist an ISI-free minimal transceiver that is *causal and stable*.

10.3.1 FIR minimal transceivers

In the Smith form decomposition of $\mathbf{C}_{ps}(z)$, there are two unimodular FIR matrices $\mathbf{U}(z)$ and $\mathbf{V}(z)$. Their inverses, $\mathbf{U}^{-1}(z)$ and $\mathbf{V}^{-1}(z)$, are also causal FIR unimodular matrices. Suppose we choose the transmitting and receiving matrices as

$$\mathbf{G}(z) = \mathbf{V}^{-1}(z) \quad \text{and} \quad \mathbf{S}(z) = \mathbf{U}^{-1}(z),$$

then both are FIR matrices. The resultant overall transfer matrix $\mathbf{T}(z)$ is

$$\mathbf{T}(z) = \mathbf{S}(z)\mathbf{C}_{ps}(z)\mathbf{G}(z) = \mathbf{\Gamma}_s(z).$$

It is a diagonal matrix, i.e. there is no inter-subchannel ISI. Thus, for FIR minimal filter bank transceivers we can achieve zero inter-subchannel ISI. Although inter-subchannel ISI is canceled, intra-subchannel ISI cannot be removed completely by FIR filtering. In other words, in the no-redundancy case, there is no FIR transceiver that can be ISI-free for frequency-selective channels. To see this, we can consider the determinant of the overall system,

$$\det(\mathbf{T}(z)) = \det(\mathbf{S}(z)\mathbf{G}(z))\det(\mathbf{C}_{ps}(z)).$$

As $C(z)$ is FIR, $\det \mathbf{C}_{ps}(z)$ is also FIR, and it is a delay if and only if $C(z)$ is a delay. Both determinants on the right-hand side of the above equation are FIR. The zero-ISI property requires $\det(\mathbf{T}(z))$ to be a delay. This means that both $\det(\mathbf{S}(z)\mathbf{G}(z))$ and $\det(\mathbf{C}_{ps}(z))$ are merely delays, which in turn implies that $C(z)$ is a delay. So it is not possible to achieve zero ISI using FIR transmitters and receivers for frequency-selective channels.

10.3.2 IIR minimal transceivers

If we are allowed to use IIR filters, one possible ISI-free solution is $\mathbf{G}(z) = \mathbf{V}^{-1}(z)$ and $\mathbf{S}(z) = \mathbf{\Gamma}_s^{-1}(z)\mathbf{U}^{-1}(z)$. Caution must be taken in doing so. The term $\mathbf{\Gamma}_s^{-1}(z)$ may not be stable. In fact, if the channel $C(z)$ does not have minimum phase, there does not exist any ISI-free transceiver that is both *causal* and *stable*.

Lemma 10.2 There exists a causal and stable ISI-free minimal filter bank transceiver if and only if $C(z)$ is a minimum-phase filter. When the condition is satisfied, one such solution is given by

$$\mathbf{G}(z) = \mathbf{V}^{-1}(z), \quad \text{and} \quad \mathbf{S}(z) = \mathbf{\Gamma}_s^{-1}(z)\mathbf{U}^{-1}(z), \qquad (10.14)$$

where $\mathbf{U}(z)$ and $\mathbf{V}(z)$ are the unimodular matrices in (10.6). ∎

Proof. Sufficiency of minimum phase $C(z)$. If $C(z)$ has zeros at θ_ℓ, for $\ell = 1, 2, \ldots, L$, we know that $\det \mathbf{C}_{ps}(z)$ has zeros at θ_ℓ^N and that these are also the zeros of $\gamma_k(z)$. If $C(z)$ has minimum phase, the zeros satisfy $|\theta_\ell| < 1$. The zeros θ_ℓ^N of $\gamma_k(z)$ will also be inside the unit circle. In this case, we can choose $\mathbf{G}(z)$ and $\mathbf{S}(z)$ as in (10.14) for a causal and stable transceiver solution.

Necessity of minimum phase $C(z)$. If $C(z)$ does not have minimum phase, then at least one zero, say θ_1, is outside the unit circle. When the filter bank transceiver is ISI-free, the overall transfer matrix is diagonal and the diagonal elements are merely delays. We have

$$\det\left(\mathbf{S}(z)\mathbf{C}_{ps}(z)\mathbf{G}(z)\right) = cz^{-m_0},$$

for some constant c and m_0. Because $\det \mathbf{C}_{ps}(z)$ contains the factor $(1 - \theta_1^N z^{-1})$, either $\det \mathbf{S}(z)$ or $\det \mathbf{G}(z)$ should contain the factor

$$1/(1 - \theta_1^N z^{-1}).$$

Therefore $\mathbf{S}(z)$ and $\mathbf{G}(z)$ cannot both be stable. In other words, if $C(z)$ does not have minimum phase there does not exist any causal and stable transceiver with the ISI-free property. ∎

Whenever there exists a causal and stable transceiver pair $[\mathbf{G}(z), \mathbf{S}(z)]$, we can use it to generate many other solutions. For example, for any $\mathbf{\Theta}(z)$ that is an $M \times M$ causal and stable transfer matrix with a causal and stable inverse, the pair

$$[\mathbf{G}(z)\mathbf{\Theta}(z), \mathbf{\Theta}^{-1}(z)\mathbf{S}(z)]$$

also forms a causal and stable transceiver. For minimal filter bank transceivers, the existence of a causal and stable solution depends on whether the given channel $C(z)$ has minimum phase. The stability problem also explains why non-minimum-phase channels are difficult to equalize using minimal filter bank transceivers. However, as we have discussed in Chapter 5, this is not the case if a certain redundancy is allowed, i.e. $N > M$. In fact, for most cases, FIR ISI-free filter bank transceivers exist with redundancy as small as 1 (Section 10.4).

For a single-input single-output (SISO) system, it is well known that the inverse of an FIR system is always IIR. The IIR inverse is (i) causal and stable if the original system has minimum phase; and (ii) stable and possibly noncausal if the original system has no zeros on the unit circle. The result in Lemma 10.2 for the minimal transceivers can be viewed as the blocked version of the SISO result.

10.4   Minimum redundancy

In the filter bank transmitter shown in Fig. 10.1, the interpolation ratio is N while the number of subchannels is M. The transmitter sends out N samples for every M inputs. There are $N - M$ redundant samples in every N samples transmitted. The following theorem gives the smallest redundancy that allows the existence of FIR transceivers.

Theorem 10.1 Consider the filter bank transceiver in Fig. 10.1. There exist FIR $\mathbf{G}(z)$ and $\mathbf{S}(z)$ such that the transceiver is ISI-free if and only if the redundancy $K = N - M$ satisfies

$$K \geq \mu_N,$$

where μ_N is the largest number of congruous zeros of $C(z)$ as given in Definition 10.2. When an FIR transceiver does exist, the solution is not unique. One choice of ISI-free FIR transceiver is as follows:

$$\mathbf{G}(z) = \mathbf{V}^{-1}(z) \begin{bmatrix} \mathbf{I}_M \\ \mathbf{0} \end{bmatrix}, \quad \mathbf{S}(z) = \begin{bmatrix} \mathbf{I}_M & \mathbf{0} \end{bmatrix} \mathbf{U}^{-1}(z), \tag{10.15}$$

where $\mathbf{U}(z)$ and $\mathbf{V}(z)$ are the $N \times N$ unimodular matrices in the Smith form decomposition of $\mathbf{C}_{ps}(z)$ in (10.6). The minimum redundancy for FIR transceiver solutions is μ_N, which is also equal to ρ_N. ∎

Proof. We first show that the condition $K \geq \rho_N$ is a sufficient one. That is, if the condition is satisfied the transceiver given in (10.15) is one solution. Then we show that it is also a necessary condition by proving that if $K < \rho_N$, there does not exist any ISI-free FIR transceiver.

Sufficiency. Consider the choice of FIR transmitter $\mathbf{G}(z)$ given in (10.15). Then $\mathbf{C}_{ps}(z)\mathbf{G}(z)$ is equal to

$$\mathbf{U}(z)\mathbf{\Gamma}_s(z)\mathbf{V}(z)\mathbf{V}^{-1}(z) \begin{bmatrix} \mathbf{I}_M \\ \mathbf{0} \end{bmatrix} = \mathbf{U}(z) \begin{bmatrix} \mathbf{I}_M \\ \mathbf{0} \end{bmatrix}.$$

In the above equality, we have used the fact that when $K \geq \rho_N$ the first M elements on the diagonal of $\mathbf{\Gamma}_s(z)$ are equal to unity. Therefore, when $K \geq \rho_N$, the receiver in (10.15) yields an FIR transceiver with zero ISI.

Necessity. Suppose $K < \rho_N$ and there exist FIR $\mathbf{G}(z)$ and $\mathbf{S}(z)$ such that the system is ISI-free. Using Smith form decomposition $\mathbf{C}_{ps}(z) = \mathbf{U}(z)\mathbf{\Gamma}_s(z)\mathbf{V}(z)$, we have

$$[\mathbf{S}(z)\mathbf{U}(z)]\,\mathbf{\Gamma}_s(z)\,[\mathbf{V}(z)\mathbf{G}(z)] = \mathbf{I}_M. \tag{10.16}$$

The smallest rank of $\mathbf{\Gamma}_s(z)$ is $N - \rho_N$, but the rank of the right-hand side of (10.16) is always equal to $M = N - K$, which is greater than $N - \rho_N$ when $K < \rho_N$. So we have a contradiction in this case, and an FIR transceiver with the ISI-free property does not exist. ∎

In the solution given in (10.15) the transmitter pads K zeros to the transmitted inputs while the receiver discards the last K receiver outputs. This solution depends on the unimodular matrices in the Smith form decomposition but not explicitly on the Smith form $\mathbf{\Gamma}(z)$. Note that, as long as the redundancy is large enough, the channel can be equalized and ISI can be eliminated, irrespective of the locations of the channel zeros. Since the unimodular matrices $\mathbf{U}(z)$ and $\mathbf{V}(z)$ in the Smith form decomposition are not unique, $\mathbf{G}(z)$ and $\mathbf{S}(z)$ are not unique. Using the solution given in (10.15), we can derive infinitely many solutions. For example, suppose $[\mathbf{G}(z), \mathbf{S}(z)]$ is an ISI-free solution, and $\mathbf{A}(z)$ is an arbitrary FIR unimodular matrix. Then

$$[\mathbf{G}(z)\mathbf{A}(z), \ \mathbf{A}^{-1}(z)\mathbf{S}(z)]$$

is also an FIR ISI-free solution. More generally, $\mathbf{A}(z)$ can be any FIR matrix with an FIR inverse. It is known that such a matrix can be factorized as a product of an FIR paraunitary matrix and an FIR unimodular matrix. See [158] for more details on the factorization of such matrices.

Example 10.4 Consider the second-order channel in Example 10.3.   For $N = 3$, one set of choices for the unimodular matrices $\mathbf{U}(z)$ and $\mathbf{V}(z)$ is

$$\mathbf{U}(z) = \begin{bmatrix} 1 & 0 & 0 \\ 2 & 1 - 2z^{-1} & 2 \\ 1 & 2 - z^{-1} & 1 \end{bmatrix}, \quad \mathbf{V}(z) = \begin{bmatrix} 1 & z^{-1} & 2z^{-1} \\ 0 & 1 & \frac{1}{3}(2 - z^{-1}) \\ 0 & 0 & -\frac{1}{3} \end{bmatrix}.$$

We know $\rho_3 = 1$ and the minimum redundancy is 1. Let us choose $M = 2$ and $\mathbf{G}(z)$ and $\mathbf{S}(z)$ according to (10.15):

$$\mathbf{G}(z) = \mathbf{V}^{-1}(z) \begin{bmatrix} \mathbf{I}_2 \\ \mathbf{0} \end{bmatrix} = \begin{bmatrix} 1 & -z^{-1} \\ 0 & 1 \\ 0 & 0 \end{bmatrix},$$

$$\mathbf{S}(z) = \begin{bmatrix} \mathbf{I}_2 & \mathbf{0} \end{bmatrix} \mathbf{U}^{-1}(z) = \begin{bmatrix} 1 & 0 & 0 \\ 0 & -1/3 & 2/3 \end{bmatrix}.$$

The overall transfer function is

$$\mathbf{S}(z)\mathbf{C}_{ps}(z)\mathbf{G}(z) = \begin{bmatrix} 1 & 0 & 0 \\ 0 & -1/3 & 2/3 \end{bmatrix} \begin{bmatrix} 1 & z^{-1} & 2z^{-1} \\ 2 & 1 & z^{-1} \\ 1 & 2 & 1 \end{bmatrix} \begin{bmatrix} 1 & -z^{-1} \\ 0 & 1 \\ 0 & 0 \end{bmatrix} = \mathbf{I}_2.$$

In fact, we have seen earlier in Example 10.3 that $\rho_N = 1$ for all N. Therefore the minimum redundancy is 1 for all N.   ∎

Example 10.5 Consider the channel in Example 10.2. For $N = 2$, no two zeros are congruous. We have $\rho_2 = 1$ and thus the minimum redundancy is 1. One set of choices for the unimodular matrices $\mathbf{U}(z)$ and $\mathbf{V}(z)$ is

$$\mathbf{U}(z) = \begin{bmatrix} 1 & 0 \\ z^{-1} & 1 \end{bmatrix} \quad \text{and} \quad \mathbf{V}(z) = \begin{bmatrix} 1 & z^{-2} \\ 0 & 1 \end{bmatrix}.$$

We can choose $\mathbf{G}(z)$ and $\mathbf{S}(z)$ according to (10.15):

$$\mathbf{G}(z) = \mathbf{V}^{-1}(z) \begin{bmatrix} 1 \\ 0 \end{bmatrix} = \begin{bmatrix} 1 \\ 0 \end{bmatrix}, \quad \mathbf{S}(z) = \begin{bmatrix} 1 & 0 \end{bmatrix} \mathbf{U}^{-1}(z) = \begin{bmatrix} 1 & 0 \end{bmatrix}.$$

We only need $K = 1$ for the existence of FIR transceivers for $N = 2$. Now suppose we use $N = 3$. We have computed earlier in Example 10.2 that $\rho_3 = 3$. In this case FIR transceiver solutions do not exist. ∎

From Example 10.5 we observe the following interesting properties.

- When $M = 1$, $K = 1$, and $N = 2$, we have $\rho_2 = 1$ and FIR solutions exist. However, when we increase K to 2, keeping $M = 1$, i.e. $N = 3$, we have $\rho_3 = 3$ and there are no FIR solutions in this case. *This demonstrates that if solutions of FIR transceivers exist for a given K, a solution does not necessarily exist if we increase redundancy from K to $K+1$ and keep M fixed.*

- We also see from this example that the interpolation ratio N can be smaller than the channel order L ($N = 2$ and $L = 3$).

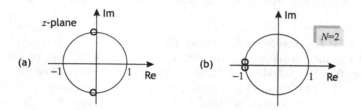

Figure 10.6. (a) Zeros of $C(z)$; (b) the zeros of $\det \mathbf{C}_{ps}(z)$.

Example 10.6 *A channel with almost congruous zeros.* Consider the second-order channel $C(z) = 1 + 2\sin\epsilon z^{-1} + z^{-2}$. The zeros are

$$e^{j(\pi/2+\epsilon)} \quad \text{and} \quad e^{-j(\pi/2+\epsilon)},$$

as shown in Fig. 10.6(a). Let $N = 2$. The channel matrix is given by

$$\mathbf{C}_{ps}(z) = \begin{bmatrix} 1 + z^{-1} & 2z^{-1}\sin\epsilon \\ 2\sin\epsilon & 1 + z^{-1} \end{bmatrix}.$$

When $\epsilon = 0$, the Smith form of $\mathbf{C}_{ps}(z)$ is the same as $\mathbf{C}_{ps}(z)$ and FIR transceivers do not exist. For a small ϵ, the two zeros of $\det\mathbf{C}_{ps}(z)$ are distinct but clustered. That is, the zeros are almost congruous. The zeros of $\det\mathbf{C}_{ps}(z)$ are shown in Fig. 10.6(b). When $\sin\epsilon \neq 0$, it has the following Smith form decomposition:

$$\underbrace{\begin{bmatrix} 1 + z^{-1} & \frac{-1}{2\sin\epsilon} \\ 2\sin\epsilon & 0 \end{bmatrix}}_{\mathbf{U}(z)} \underbrace{\begin{bmatrix} 1 & 0 \\ 0 & 1 + 2z^{-1}\cos(2\epsilon) + z^{-2} \end{bmatrix}}_{\mathbf{\Gamma}_s(z)} \underbrace{\begin{bmatrix} 1 & 1 + \frac{1}{2\sin\epsilon}(1 + z^{-1}) \\ 0 & 1 \end{bmatrix}}_{\mathbf{V}(z)}.$$

If $\epsilon \approx 0$, we have $\mathbf{C}_{ps}(-1) \approx \mathbf{0}$. However, $\text{rank}(\mathbf{\Gamma}_s(-1)) \geq 1$ as long as $\sin\epsilon \neq 0$. Therefore either $\mathbf{U}(-1)$ or $\mathbf{V}(-1)$ is an ill-conditioned matrix, although they are unimodular and have constant determinants. To be more specific, one can compute the condition number. For an $N \times N$ matrix $\mathbf{A}$, the condition number is defined as $||\mathbf{A}|| \, ||\mathbf{A}^{-1}||$, where $|| \cdot ||$ denotes a matrix norm. Let us use the matrix norm that is defined as the maximum of the absolute column sum, i.e. $||\mathbf{A}|| = \max_j \sum_{i=0}^{N-1} |[\mathbf{A}]_{i,j}|$. We can verify that the condition number of $\mathbf{V}(-1)$ is 1, whereas the condition number of $\mathbf{U}(-1)$ is

$$[2 \sin \epsilon]^2 + [1/(2 \sin \epsilon)]^2 \, ,$$

which goes to infinity as ϵ approaches zero. In this case, the receiving matrix in (10.15) is

$$\mathbf{S}(z) = \begin{bmatrix} 1 & 1/(2 \sin \epsilon) \end{bmatrix} .$$

This has a very large coefficient $1/(2 \sin \epsilon)$, which will amplify channel noise although the overall signal gain is still unity ($\mathbf{S}(z)\mathbf{C}_{ps}(z)\mathbf{G}(z) = 1$). To avoid this, we can increase K by 1 and get around the problem of noise amplification.

∎

Example 10.7 Consider the channel $C(z)$ and the power spectrum of the colored channel noise shown in Fig. 10.7. The coefficients of the channel $c(n)$ are $\begin{bmatrix} 0.1659 & 0.3045 & -0.1159 & -0.0733 & -0.0015 \end{bmatrix}$. The channel $C(z)$ has order 4. The channel and the channel noise are drawn from an ADSL environment. The channel is obtained by shortening loop 6 in [7] to five taps and the channel noise is a combination of FEXT crosstalk and AWGN noise. For $N = 5$, the minimum redundancy is 1. We choose $K = 1$ and $M = 4$. The FIR transmitting and receiving matrices are as given in (10.15). The inputs are BPSK symbols, rendering a bit rate of 0.8 bits/sample. The plot of bit error rate versus transmission power is given in Fig. 10.8. For comparison, we also plot the bit error rate performance of a cyclic-prefixed DFT-based transceiver with the same bit rate and relative redundancy, i.e. same K/N or same K/M. The DFT-based transceiver needs at least four redundant samples per block as the channel order is 4. We choose $M = 16$, $K = 4$ for the DFT-based system. The system with minimum redundancy requires a much smaller transmission power for the same bit error rate.

In this example the minimum redundancy is 1 whereas the redundancy is 4 for the DFT-based system. In most cases the minimum redundancy is less than the usual redundancy L. At the same relative redundancy, the system with minimum redundancy has a smaller M, i.e. a shorter block length. We should note that as the system is not DFT-based, the filter bank transceiver has more channel-dependent elements in the design and implementation phases. For example, the transmitter and receiver can only be designed after the channel is known. The transmitting and receiving matrices will both be channel-dependent. We should also note that the design of the filter bank transceiver is not as straightforward as the DFT-based system. The unimodular matrices $\mathbf{V}(z)$ and $\mathbf{U}(z)$ can be ill-conditioned as we have demonstrated in Example 10.6. The computation of the unimodular matrices can also be sensitive to numerical accuracy and channel estimation error. This is even more so for a large N.

∎

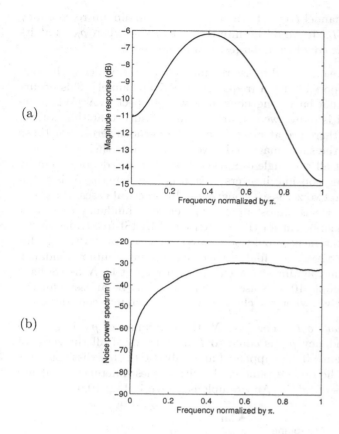

Figure 10.7. (a) Magnitude response of the channel $C(z)$; (b) power spectrum of the additive noise $q(n)$.

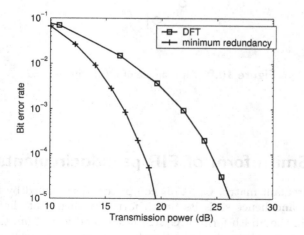

Figure 10.8. Bit error rate for the transceiver with minimum redundancy and the DFT-based system with the same relative redundancy.

For a given FIR channel $C(z)$ of order $L < N$, the minimum redundancy ρ_N is between 1 and L. It would be interesting to know when ρ_N is at its smallest and when it is equal to its largest possible value L.

- When $\rho_N = 1$, we only need to use redundancy $K = 1$, which is the lowest redundancy possible for a frequency-selective channel. This occurs when $C(z)$ does not have congruous zeros with respect to N. When the order of channel is finite, we can always find N such that this is true. This is because there are at most L sets of congruous zeros. For those zeros that have the same magnitude, we can calculate their differences in angle. Collect all the angle differences together and denote them by $2\pi p_\ell / q_\ell$, for some coprime integers p_ℓ and q_ℓ. As long as N is not a multiple of q_i, no two zeros are congruous. In practical cases, the probability that $\rho_N = 1$ is almost unity. Therefore redundancy of $K = 1$ is almost always sufficient for the existence of FIR ISI-free transceivers. However, this statement should be treated with caution. Although for an FIR $C(z)$ we can always find N such that the minimum redundancy is 1, the zeros can be almost congruous with respect to N as we have observed in Example 10.6. Noise amplification may be an issue for the resulting transceiver when the channel noise is taken into consideration.

- *When is ρ_N equal to L?* For $L < N$, the maximum of ρ_N is L. The minimum redundancy ρ_N is equal to L if and only if all the zeros of $C(z)$ are congruous. This happens if and only if $C(z)$ has distinct zeros and these zeros lie on the same circle with angles differences that are integer multiples of $2\pi/N$. An example is shown in Fig. 10.9.

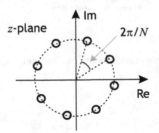

Figure 10.9. An example of $C(z)$ with $\rho_N = L$.

10.5   Smith form of FIR pseudocirculants

A pseudocirculant matrix $\mathbf{C}_{ps}(z)$ is completely determined by its underlying scalar filter and hence so is its Smith form. It turns out that we can very easily obtain the Smith form of $\mathbf{C}_{ps}(z)$ in a closed form from the scalar filter $C(z)$. The diagonal terms in the Smith form can be determined directly by inspection. In the previous section we gave one ISI-free FIR solution, which depends only on the number of nontrivial diagonal terms of the Smith form but not on the Smith form explicitly. Although the Smith form itself is not of importance in the design of FIR transceivers, it is academically satisfying to know that it is closely related to congruent zeros of the scalar filter $C(z)$.

From the properties derived in Section 10.2.3, we already know that the zeros of $\det \mathbf{C}_{ps}(z)$ are distributed among the Smith form's diagonal terms $\gamma_k(z)$. The following lemma provides us a more explicit link between congruous zeros of the scalar filter $C(z)$ and $\gamma_k(z)$.

Lemma 10.3 Given an FIR channel $C(z)$ and its blocked $N \times N$ channel matrix $\mathbf{C}_{ps}(z)$, let $\{\alpha_1, \alpha_2, \ldots, \alpha_q\}$ be a set of congruous zeros of $C(z)$ with respect to N. Suppose no other zeros can be included in the set to form a larger congruous set. Then,

(1) $\operatorname{rank}(\mathbf{C}_{ps}(\alpha_1^N)) = N - q$;

(2) only the last q functions $\gamma_{N-q}(z), \gamma_{N-q+1}(z), \ldots, \gamma_{N-1}(z)$ have the factor $(1 - \alpha_1^N z^{-1})$. ∎

Proof. We express the congruous zeros as $\alpha_j = \alpha_1 W^{-n_j}$. Consider the $(N - n_j)$th diagonal term of $\boldsymbol{\Sigma}(z)$ in (10.8). Observe that

$$C(zW^{N-n_j})|_{z=\alpha_1} = C(\alpha_1 W^{-n_j}) = C(\alpha_j) = 0, \quad \text{for } j = 1, 2, \ldots, q.$$

That is, q terms on the diagonal of $\boldsymbol{\Sigma}(z)$ are equal to zero when $z = \alpha_1$. No other term has a zero at α_1 because this is the largest congruous set containing α_1. Therefore, we have $\operatorname{rank}(\boldsymbol{\Sigma}(\alpha_1)) = N - q$, which implies that

$$\operatorname{rank}(\mathbf{C}_{ps}(\alpha_1^N)) = \operatorname{rank}(\boldsymbol{\Gamma}_s(\alpha_1^N)) = N - q.$$

This, in turn, means that exactly q diagonal terms of the Smith form contain the factor $(1 - \alpha_1^N z^{-1})$. These are the last q terms as $\gamma_k(z)$ divides $\gamma_{k+1}(z)$. ∎

To obtain an exact expression of $\gamma_k(z)$, let us partition all the zeros of $C(z)$ into sets $\mathcal{B}_j$. Each $\mathcal{B}_j$ contains one set of congruous zeros and no other zeros can be included in the set to form a larger congruous set. In addition, when $C(z)$ has multiple zeros, identical zeros are grouped into the same set. Suppose there are a total of s such sets, $\mathcal{B}_1, \mathcal{B}_2, \ldots, \mathcal{B}_s$. Suppose $\mathcal{B}_j$ contains ℓ_j congruous zeros and they are denoted by

$$\alpha_{j,1}, \alpha_{j,2}, \ldots, \alpha_{j,\ell_j}, \tag{10.17}$$

where we have, for convenience, renumbered the zeros of $C(z)$ using double subscripts. (The number of elements in $\mathcal{B}_j$ can be larger than ℓ_j due to identical zeros.) Let the multiplicity of $\alpha_{j,i}$ be $m_{j,i}$ and

$$m_{j,1} \geq m_{j,2} \geq \cdots \geq m_{j,\ell_j}.$$

Then we can write $C(z)$ as

$$C(z) = c(0) \prod_{j=1}^{s} (1 - \alpha_{j,1} z^{-1})^{m_{j,1}} (1 - \alpha_{j,2} z^{-1})^{m_{j,2}} \cdots (1 - \alpha_{j,\ell_j} z^{-1})^{m_{j,\ell_j}}.$$

Without loss of generality, we assume $\ell_1 \geq \ell_2 \cdots \geq \ell_s$. By definition, $\ell_1 = \mu_N$. Lemma 10.3 says that the last ℓ_j terms have the factor $(1 - \alpha_{j,1}^N z^{-1})$. In other words, $\gamma_{N-k}(z)$ contains every factor $(1 - \alpha_{j,1}^N z^{-1})$ that satisfies $\ell_j \geq k$. For $k > \ell_1$, we have $\gamma_{N-k}(z) = 1$. The following theorem gives a closed form expression for the Smith form of $\mathbf{C}_{ps}(z)$. A proof is given in Section 10.6.

Theorem 10.2 The Smith form $\mathbf{\Gamma}_s(z)$ of the pseudocirculant $\mathbf{C}_{ps}(z)$ has diagonal terms given by

$$\gamma_{N-k}(z) = \begin{cases} \displaystyle\prod_{\substack{j=1, \\ \ell_j \geq k}}^{s} (1 - \alpha_{j,1}^N z^{-1})^{m_{j,k}}, & k = 1, 2, \ldots, \ell_1, \\ \\ 1, & \text{otherwise.} \end{cases} \qquad (10.18)$$

$\blacksquare$

For example, suppose $C(z)$ has no multiple zeros and assume the zeros are divided into three sets $\mathcal{B}_1$, $\mathcal{B}_2$, and $\mathcal{B}_3$ with, respectively, $\ell_1 = 4$, $\ell_2 = 3$, and $\ell_3 = 2$ as given in Table 10.1. The zeros in the same column will go to the

Sets	Zeros			
$\mathcal{B}_1$	α_{11}	α_{12}	α_{13}	α_{14}
$\mathcal{B}_2$	α_{21}	α_{22}	α_{23}	
$\mathcal{B}_3$	α_{31}	α_{32}		
	$\downarrow$	$\downarrow$	$\downarrow$	$\downarrow$
	$\gamma_{N-1}(z)$	$\gamma_{N-2}(z)$	$\gamma_{N-3}(z)$	$\gamma_{N-4}(z)$

Table 10.1. An example of zero partition and the respective $\gamma_k(z)$

same function $\gamma_k(z)$. The resultant $\gamma_{N-4}(z)$ has one zero α_{11}^N, $\gamma_{N-3}(z)$ has two zeros $\alpha_{11}^N, \alpha_{21}^N$, and both $\gamma_{N-2}(z)$ and $\gamma_{N-1}(z)$ have three zeros $\alpha_{11}^N, \alpha_{21}^N, \alpha_{31}^N$. All the other diagonal terms are equal to unity if $N > 4$. In summary, we have

$$\gamma_k(z) = 1, \quad \text{for} \quad k \leq N - 4,$$
$$\gamma_{N-4}(z) = 1 - \alpha_{11}^N z^{-1},$$
$$\gamma_{N-3}(z) = (1 - \alpha_{11}^N z^{-1})(1 - \alpha_{21}^N z^{-1}),$$
$$\gamma_{N-2}(z) = \gamma_{N-1}(z) = (1 - \alpha_{11}^N z^{-1})(1 - \alpha_{21}^N z^{-1})(1 - \alpha_{31}^N z^{-1}),$$

Example 10.8 Consider the scalar filter $C(z) = 1 + z^{-3}$ in Example 10.2. It has zeros at $z = e^{j\pi/3}$, $e^{-j\pi/3}$, and -1. When $N = 2$, we have $\mathcal{B}_1 = \{e^{j\pi/3}\}$, $\mathcal{B}_2 = \{e^{-j\pi/3}\}$, and $\mathcal{B}_3 = \{-1\}$. There is only one nontrivial diagonal element $\gamma_1(z)$. The zeros of $\gamma_1(z)$ are those of $C(z)$ raised to the power of 2:

$$\gamma_1(z) = (1 - e^{j2\pi/3} z^{-1})(1 - e^{-j2\pi/3} z^{-1})(1 - (-1)^2 z^{-1}) = 1 - z^{-3}.$$

We immediately get the Smith form as

$$\mathbf{\Gamma}_s(z) = \begin{bmatrix} 1 & 0 \\ 0 & 1 - z^{-3} \end{bmatrix}.$$

Now suppose $N = 3$; the three zeros of $C(z)$ are congruous with respect to 3. We have only one $\mathcal{B}$ set $\{e^{j\pi/3}, e^{-j\pi/3}, -1\}$. Each $\gamma_i(z)$ is of order 1,

$$\gamma_0(z) = \gamma_1(z) = \gamma_2(z) = 1 - z^{-1}.$$

We can conclude that

$$\mathbf{\Gamma}_s(z) = \begin{bmatrix} 1 - z^{-1} & 0 & 0 \\ 0 & 1 - z^{-1} & 0 \\ 0 & 0 & 1 - z^{-1} \end{bmatrix}.$$

∎

One can readily determine the Smith form of the FIR pseudocirculant matrix $\mathbf{C}_{ps}(z)$ by inspection once the zeros of the associated scalar filter $C(z)$ are known. The theorem implies that the number of nontrivial terms in the Smith form is ℓ_1, i.e. the cardinality of the largest congruous set. This number, as shown in the previous section, is exactly the minimum redundancy for the existence of FIR transceivers.

10.6 Proof of Theorem 10.2

When the zeros of $C(z)$ are distinct, Theorem 10.2 follows directly from Lemma 10.3. However, this is not the case when $C(z)$ has multiple zeros. For example, suppose $C(z)$ has congruous zeros α_1 and α_2, with multiplicities m_1 and m_2, respectively. We know that $\det \mathbf{\Gamma}_s(z)$ has $(m_1 + m_2)$ zeros at α_1^N and we know that these zeros will be distributed to the last two diagonal terms. But the exact distribution cannot be determined from Lemma 10.3. It is intuitive to conjecture that m_2 zeros go to $\gamma_{N-2}(z)$ and m_1 zeros go to $\gamma_{N-1}(z)$ if $m_1 > m_2$. And it is indeed how the zeros are distributed. The proof for $C(z)$ with multiple zeros turns out to be more involved. In practical cases, the probability that the zeros $C(z)$ are distinct is almost unity. However, the zeros can be almost congruous (Example 10.6), and the probability for this is not zero. For completeness we present the proof for general channels that can have multiple zeros.

We will see that when we compute the Smith form of $\mathbf{C}_{ps}(z)$, we can consider zeros from each $\mathcal{B}_i$ separately. That is, zeros from different $\mathcal{B}_i$ decouple in the computation of Smith form. As a result we can consider $C(z)$ with only one set of congruous zeros, i.e. $C(z)$ of the form

$$C(z) = (1 - \alpha_1 z^{-1})^{m_1}(1 - \alpha_2 z^{-1})^{m_2} \cdots (1 - \alpha_q z^{-1})^{m_q}, \qquad (10.19)$$

where $\alpha_1, \alpha_2, \ldots, \alpha_q$ are congruous with respect to N with $\alpha_j = \alpha_1 W^{-n_j}$ and $m_1 \geq m_2 \geq \cdots \geq m_q$. We will show that the Smith form of the associated $\mathbf{C}_{ps}(z)$ is given by

$$\mathrm{diag} \begin{bmatrix} 1 & \cdots & 1 & (1 - \alpha_1^N z^{-1})^{m_q} & (1 - \alpha_1^N z^{-1})^{m_{q-1}} & \cdots & (1 - \alpha_1^N z^{-1})^{m_1} \end{bmatrix}.$$
$$(10.20)$$

The proof is organized as follows: We will first show in Section 10.6.1 that the Smith form of $\mathbf{C}_{ps}(z^N)$ is the same as that of $\boldsymbol{\Sigma}(z)$, the diagonal matrix in the DFT decomposition. Therefore we can find the Smith form of $\boldsymbol{\Sigma}(z)$ instead. This is true regardless of whether the zeros of $C(z)$ are distinct or not. In Section 10.6.2, we argue why we only need to consider $C(z)$ with a single set of congruous zeros. Then we give an example for finding the Smith form of $\boldsymbol{\Sigma}(z)$ (Section 10.6.3). This example will motivate the notation and the approach that we use to find the Smith form of $\boldsymbol{\Sigma}(z)$ in Section 10.6.4.

10.6.1 Identical Smith forms

From the DFT decomposition of $\mathbf{C}_{ps}(z)$ in (10.8), we know that

$$\mathbf{D}(z)\mathbf{C}_{ps}(z^N)\mathbf{D}(z^{-1}) = \mathbf{W}\boldsymbol{\Sigma}(z)\mathbf{W}^{\dagger}. \tag{10.21}$$

The left-hand side, although having advances in $\mathbf{D}(z^{-1})$, is a polynomial matrix in z^{-1}, as the right-hand side $\mathbf{W}\boldsymbol{\Sigma}(z)\mathbf{W}^{\dagger}$ is causal. It turns out that the left-hand side has the same Smith form as $\mathbf{C}_{ps}(z^N)$ and also the same Smith form as $\boldsymbol{\Sigma}(z)$. We will prove this with a more general setting.

Let $\mathbf{E}(z)$ be a polynomial matrix in z^{-1} such that

$$\mathbf{B}(z) = \mathbf{D}(z)\mathbf{E}(z)\mathbf{D}(z^{-1})$$

is also a polynomial matrix in z^{-1}. We will show that $\mathbf{B}(z)$ and $\mathbf{E}(z)$ have the same Smith form. Let $\Delta_i(z)$ be the gcd of all the $i \times i$ minors of $\mathbf{E}(z)$ and $\Delta_i'(z)$ be the gcd of all $i \times i$ minors of $\mathbf{B}(z)$. For convenience, we define the set S_i as the collection of all $1 \times i$ index vectors,

$$\begin{bmatrix} k_1 & k_2 & \cdots & k_i \end{bmatrix}, \quad 0 \le k_1 < k_2 < \cdots < k_i \le N - 1. \tag{10.22}$$

For $\mathbf{k}, \mathbf{j} \in S_i$, we define $\mathbf{E}_{\mathbf{k},\mathbf{j}}(z)$ as the $i \times i$ submatrix of $\mathbf{E}(z)$ obtained by keeping the rows in the index vector $\mathbf{k}$ and the columns in the index vector $\mathbf{j}$. For example, if

$$\mathbf{k} = \begin{bmatrix} k_1 & k_2 & \cdots & k_i \end{bmatrix}, \quad \mathbf{j} = \begin{bmatrix} j_1 & j_2 & \cdots & j_i \end{bmatrix},$$

then $\mathbf{E}_{\mathbf{k},\mathbf{j}}(z)$ is the $i \times i$ submatrix of $\mathbf{E}(z)$ obtained by keeping rows $\#k_1$, $\#k_2, \ldots, \#k_i$ and columns $\#j_1, \#j_2, \ldots, \#j_i$. We can write

$$\Delta_i(z) = \underset{\mathbf{k},\mathbf{j}\in S_i}{gcd} \det \mathbf{E}_{\mathbf{k},\mathbf{j}}(z), \quad \Delta_i'(z) = \underset{\mathbf{k},\mathbf{j}\in S_i}{gcd} \det \mathbf{B}_{\mathbf{k},\mathbf{j}}(z).$$

Note that the (k, j)th element of $\mathbf{B}(z)$ is related to $\mathbf{E}(z)$ by $B_{kj}(z) = E_{kj}(z)z^{j-k}$, so we have

$$\Delta_i'(z) = \underset{\mathbf{k},\mathbf{j}\in S_i}{gcd} \left[z^{\sum_{m=1}^{i}(j_m - k_m)} \det \mathbf{E}_{\mathbf{k},\mathbf{j}}(z) \right].$$

We see that $\Delta_i'(z)$ can differ from $\Delta_i(z)$ at most by a delay term. In fact, this term is equal to unity (Problem 10.16), and $\Delta_i'(z) = \Delta_i(z)$. Because the ith diagonal element of the Smith form of $\mathbf{E}(z)$ is $\Delta_i(z)/\Delta_{i-1}(z)$, we can conclude that the Smith forms of $\mathbf{E}(z)$ and $\mathbf{B}(z)$ are the same. Therefore the left-hand side of (10.21) has the same Smith form as $\mathbf{C}_{ps}(z^N)$. The right-hand side of (10.21) is $\boldsymbol{\Sigma}(z)$ sandwiched by unitary matrices $\mathbf{W}$ and $\mathbf{W}^{\dagger}$, which do not change the Smith form. We can conclude that the Smith form of $\mathbf{C}_{ps}(z^N)$ is the same as that of $\boldsymbol{\Sigma}(z)$.

10.6.2   Zeros from different $\mathcal{B}_i$ decouple

For simplicity, suppose $C(z)$ has two sets of congruous zeros. The zeros are partitioned into two sets $\mathcal{B}_1$ and $\mathcal{B}_2$ (the definition of $\mathcal{B}_i$ is given in Section 10.5). We write $C(z)$ as $C_1(z)C_2(z)$, where $C_i(z)$ contains all the zeros in $\mathcal{B}_i$, for $i = 1, 2$. Let $\Delta_i(z)$ be the gcd of all the $i \times i$ minors of $\Sigma(z)$, where $\Sigma(z)$ is the diagonal matrix in the DFT decomposition of $\mathbf{C}_{pz}(z^N)$ given in (10.8). Then we have

$$\Delta_i(z) = \operatorname*{gcd}_{\mathbf{k} \in S_i} \prod_{n=1}^{i} C(zW^{k_n}),$$

where S_i is the collection of all $1 \times i$ index vectors in (10.22). Observe that $C_1(zW^m)$ and $C_2(zW^n)$ do not have any common factor for any m, n because the zeros in $\mathcal{B}_1$ are not congruous with the zeros in $\mathcal{B}_2$. As a result

$$\Delta_i(z) = \left[\operatorname*{gcd}_{\mathbf{k} \in S_i} \prod_{n=1}^{i} C_1(zW^{k_n}) \right] \left[\operatorname*{gcd}_{\mathbf{k} \in S_i} \prod_{n=1}^{i} C_2(zW^{k_n}) \right].$$

This means that in finding the Smith form of $\Sigma(z)$, we can consider zeros from each $\mathcal{B}_i$ separately. We can compute the Smith forms corresponding to $C_1(z)$ and $C_2(z)$. Taking their product, we can then obtain the Smith form corresponding to $C(z)$. Therefore we only need to consider $C(z)$ with a single set of congruous zeros, i.e. $C(z)$ of the form in (10.19). Showing that the Smith form of $\mathbf{C}_{ps}(z)$ is given by (10.20) is equivalent to showing that the Smith form of $\Sigma(z)$ is

$$\operatorname{diag} \begin{bmatrix} 1 & \cdots & 1 & (1 - \alpha_1^N z^{-N})^{m_q} & \cdots & (1 - \alpha_1^N z^{-N})^{m_1} \end{bmatrix}, \qquad (10.23)$$

i.e. the diagonal matrix in (10.20) with z replaced by z^N.

10.6.3   An example of deriving the Smith form of $\Sigma(z)$

We will use one example to bring forward some properties of the Smith form of $\Sigma(z)$ and motivate the construction used in Subsection 10.6.4. Consider the scalar filter $C(z)$ of the form in (10.19) with $q = 3$. Let

$$C(z) = (1 - \alpha_1 z^{-1})^{m_1} (1 - \alpha_2 z^{-1})^{m_2} (1 - \alpha_3 z^{-1})^{m_3},$$

where $\alpha_1, \alpha_2, \alpha_3$ are congruous with respect to N, $\alpha_j = \alpha_1 W^{-n_j}$, and $m_1 \geq m_2 \geq m_3$. Note that the ith diagonal element of $\Sigma(z)$ is $C(zW^i)$. The zeros of $C(zW^i)$ are those of $C(z)$ rotated, i.e. $\alpha_1 W^{-i}, \alpha_2 W^{-i}, \alpha_3 W^{-i}$. They can also be expressed as

$$\alpha_1 W^{-i}, \alpha_1 W^{-(i+n_2)}, \alpha_1 W^{-(i+n_3)},$$

each a different rotation of α_1. As an example, we use

$$N = 6, \ m_1 = 4, \ m_2 = 2, \ m_3 = 1, \ n_2 = 1, \ n_3 = 3.$$

Table 10.2 lists the zeros of $C(zW^i)$ in this case. As each zero of $C(zW^i)$ can be expressed as $\alpha_1 W^{-k_0}$ for some integer k_0, we only show the integer

k_0 in the table. The ith row of the table lists the zeros of $C(zW^i)$. Such a table will be called a **rotation table**. Each column in the table contains the set $\{0, 1, \ldots, N-1\}$, which will be referred to as a **complete rotation**. The table contains a total of $(m_1 + m_2 + m_3)$ complete rotations. More generally, the rotation table contains $(m_1 + m_2 + \cdots + m_q)$ complete rotations.

i	Zeros due to α_1 (k_0 shown) (m_1 columns)				Zeros due to α_2 (k_0 shown) (m_2 columns)		Zeros due to α_3 (k_0 shown) (m_3 columns)
0	0	0	0	0	1	1	3
1	1	1	1	1	2	2	4
2	2	2	2	2	3	3	5
3	3	3	3	3	4	4	0
4	4	4	4	4	5	5	1
5	5	5	5	5	0	0	2

Table 10.2. Rotation table: zeros of $C(zW^i)$ expressed as $\alpha_1 W^{-k_0}$ for some integer k_0 with only the integer k_0 shown in the table.

When we collect all the entries in the rotation table, we obtain a set $\mathcal{A}$ of $N(m_1 + m_2 + m_3)$ elements. Let $\mathcal{A}_{\mathbf{k}}$ be a subset of $\mathcal{A}$ obtained by keeping only those rows of the rotation table in the index vector $\mathbf{k}$. We use the notation $\bar{\mathbf{k}}$ to denote the complement of $\mathbf{k}$. For example, if $N = 6$ and $\mathbf{k} = \begin{bmatrix} 1 & 2 & 5 \end{bmatrix}$, then $\bar{\mathbf{k}} = \begin{bmatrix} 0 & 3 & 4 \end{bmatrix}$. Therefore $\mathcal{A}_{\bar{\mathbf{k}}}$ is the subset of $\mathcal{A}$ obtained by deleting the rows in the rotation table corresponding to the index vector $\mathbf{k}$. As an example, Table 10.3 shows the elements of $\mathcal{A}_{\bar{\mathbf{k}}}$ for $\mathbf{k} = \begin{bmatrix} 1 & 2 & 5 \end{bmatrix}$. It is obtained by deleting rows #1, #2, and #5 from the rotation table. We also divide $\mathcal{A}_{\bar{\mathbf{k}}}$ into three groups according to the three distinct zeros α_1, α_2, and α_3. The ℓth group contains the entries of $\mathcal{A}_{\bar{\mathbf{k}}}$ due to α_ℓ as shown in Table 10.3. There are $(N - i)m_\ell$ elements in the ℓth group when $\mathbf{k}$ is of dimension $1 \times i$. In the following we show how the rotation table is related to the computation of $\Delta_i(z)$.

Now let us compute the gcd of all the $i \times i$ minors of $\Sigma(z)$ to obtain its Smith form. Denote the gcd by $\Delta_i(z)$. The $N \times N$ minor of $\Sigma(z)$ is $\det \Sigma(z)$, and we have $\Delta_6(z) = \prod_{i=0}^{N-1} C(zW^i)$. We may note that such a product can also be obtained using the rotation table. Observe that the zeros of the product are exactly those listed in the rotation table! Having one complete rotation gives rise to the factor

$$(1 - \alpha_1 z^{-1})(1 - \alpha_1 z^{-1}W^{-1}) \cdots (1 - \alpha_1 z^{-1}W^{-(N-1)}) = (1 - \alpha_1^N z^{-N}).$$

The table contains $(m_1 + m_2 + m_3)$ complete rotations. It follows that $\Delta_6(z)$

i	Zeros due to α_1 (m_1 columns)				Zeros due to α_2 (m_2 columns)		Zeros due to α_3 (m_3 columns)
0	0	0	0	0	1	1	3
~~1~~	~~1~~	~~1~~	~~1~~	~~1~~	~~2~~	~~2~~	~~4~~
~~2~~	~~2~~	~~2~~	~~2~~	~~2~~	~~3~~	~~3~~	~~5~~
3	3	3	3	3	4	4	0
4	4	4	4	4	5	5	1
~~5~~	~~5~~	~~5~~	~~5~~	~~5~~	~~0~~	~~0~~	~~2~~
	(group 1)				(group 2)		(group 3)

Table 10.3. Elements of $A_{\bar{k}}$ with $\mathbf{k} = [1\ 2\ 5]$, obtained by deleting rows #1, #2, and #5 from the rotation table in Table 10.2.

is

$$\Delta_6(z) = \prod_{i=0}^{N-1} C(zW^i) = (1 - \alpha_1^N z^{-N})^{m_1 + m_2 + m_3} = (1 - \alpha_1^6 z^{-6})^7.$$

The term $\Delta_5(z)$ is the gcd of all possible combinations of five terms on the diagonal of $\mathbf{\Sigma}(z)$. We can obtain $\Delta_5(z)$ by examining the rotation table. Considering the zeros of five arbitrarily chosen diagonal terms of $\mathbf{\Sigma}(z)$ is the same as considering five arbitrarily chosen rows of the rotation table, or equivalently the rotation table with one row deleted. Therefore we can consider the subset of $\mathcal{A}$ with one row in the rotation table deleted, i.e. $\mathcal{A}_{\bar{k}}$, where $0 \le k \le 5$. We can obtain $\Delta_5(z)$ by finding the intersection

$$\bigcap_{k=0}^{5} \mathcal{A}_{\bar{k}}.$$

Note that we only need to consider complete rotations because the non-identity diagonal terms of the Smith form of $\mathbf{\Sigma}(z)$ has only factors of the form $(1 - \alpha_1^N z^{-N})$. Let us look into each $\mathcal{A}_{\bar{k}}$. Observe that the set $\mathcal{A}_{\bar{0}}$ is obtained by deleting row #0 from the rotation table. We see from Table 10.2 that to have complete rotations, the numbers missing in group 2 are two 1s and group 3 lacks the number 3. However, we can find these missing numbers in group 1. Therefore $\mathcal{A}_{\bar{0}}$ contains three complete rotations. In a similar manner, we can verify that each $\mathcal{A}_{\bar{k}}$, for $0 \le k \le 5$ contains three complete rotations. In fact, the intersection $\bigcap_{k=0}^{5} \mathcal{A}_{\bar{k}}$ is exactly $\{0, 1, \ldots, 5, 0, 1, \ldots, 5, 0, 1, \ldots, 5\}$, i.e. three complete rotations and no other elements. Suppose this is not true and there is one more entry, say ℓ_0, in the intersection in addition to the three complete rotations. For each $\mathcal{A}_{\bar{k}}$, all the elements in groups 2 and 3 are already contained in the three complete rotations. This additional ℓ_0 must come from group 1. However, the set $\mathcal{A}_{\bar{\ell}_0}$, obtained by deleting the ℓ_0th row, does not have ℓ_0 in group 1. Therefore, the intersection $\bigcap_{k=0}^{5} \mathcal{A}_{\bar{k}}$ contains no other elements than the three complete rotations. We have

$$\Delta_5(z) = (1 - \alpha_1^6 z^{-6})^3.$$

The term $\Delta_4(z)$ is the gcd of all possible combinations of 4 terms on the diagonal of $\Sigma(z)$. To obtain $\Delta_4(z)$, again we can use the rotation table. We construct subsets $\mathcal{A}_{\bar{k}}$ by deleting two rows from the rotation table in Table 10.2 and find their intersection, i.e.

$$\bigcap_{\mathbf{k} \in S_2} \mathcal{A}_{\bar{k}}.$$

For example, when $\mathbf{k} = \begin{bmatrix} 0 & 1 \end{bmatrix}$, rows #0 and #1 are deleted from the rotation table. Group 3 lacks the numbers 3 and 4 to complete a rotation. We note that these two numbers can be found in group 2, hence $\mathcal{A}_{\bar{k}}$ in this case contains a complete rotation. In a similar way, we can verify that each $\mathcal{A}_{\bar{k}}$ for $\mathbf{k} \in S_2$ contains a complete rotation. That is, the intersection $\bigcap_{\mathbf{k} \in S_2} \mathcal{A}_{\bar{k}}$ is $\{0, 1, \ldots, 5\}$. Thus, $\Delta_4(z) = 1 - \alpha_1^6 z^{-6}$. Using a similar approach, we can find $\bigcap_{\mathbf{k} \in S_3} \mathcal{A}_{\bar{k}} = \phi$, an empty set, and thus $\Delta_3(z) = 1$. It follows that $\Delta_2(z) = 1$ and $\Delta_1(z) = 1$. The above discussion demonstrates that we can obtain $\Delta_i(z)$ by examining the rotation table. Although working with the rotation table may seem roundabout at first, we will see it is actually much easier when we consider a more general case of $C(z)$ later.

Denote the ith diagonal element of the Smith form of $\Sigma(z)$ by $\beta_i(z)$. Then $\beta_i(z) = \Delta_{i+1}(z)/\Delta_i(z)$ are given by

$$\beta_5(z) = (1 - \alpha_1^6 z^{-6})^4, \quad \beta_4(z) = (1 - \alpha_1^6 z^{-6})^2, \quad \beta_3(z) = 1 - \alpha_1^6 z^{-6},$$

$$\beta_0(z) = \beta_1(z) = \beta_2(z) = 1.$$

We see that the nonunity terms are of the form $(1 - \alpha_1^N z^{-N})^{\ell_i}$, and the exponents ℓ_i happen to be the same as m_1, m_2, and m_3. In particular,

$$\beta_{N-i}(z) = (1 - \alpha_1^N z^{-N})^{m_i}, \quad i = 1, 2, 3.$$

This turns out to be true in general, as we will see next.

10.6.4   Smith form of $\Sigma(z)$

To show that the Smith form of $\Sigma(z)$ is of the form in (10.23), we need to show that

$$\beta_{N-i}(z) = (1 - \alpha_1^N z^{-N})^{m_i}, \quad i = 1, 2, \ldots, q.$$

This is equivalent to showing that

$$\Delta_{N-i}(z) = (1 - \alpha_1^N z^{-N})^{\sum_{\ell=i+1}^{q} m_\ell}, \quad i = 0, 1, \ldots, q - 1. \qquad (10.24)$$

As in Subsection 10.6.3, we form the rotation table by listing the zeros of $C(zW^i)$ in a row, and obtain a table of $N(m_1 + m_2 + \cdots + m_q)$ entries. The set $\mathcal{A}$ is the collection of elements in the rotation table and it contains

$(m_1 + m_2 + \cdots + m_q)$ complete rotations. We will first look at $\Delta_N(z)$ and then the general term $\Delta_{N-i}(z)$.

To consider $\Delta_N(z)$, we can examine the set $\mathcal{A}$, which consists of $(m_1 + m_2 + \cdots + m_q)$ complete rotations, i.e. the set $\{0, 1, \ldots, N-1\}$ repeating $(m_1 + m_2 + \cdots + m_q)$ times. Each complete rotation contributes the factor $(1 - \alpha_1^N z^{-N})$ to $\Delta_N(z)$. Therefore,

$$\Delta_N(z) = (1 - \alpha_1^N z^{-N})^{(m_1+m_2+\cdots+m_q)}.$$

Now let us consider $\Delta_{N-i}(z)$. To obtain the zeros of $\Delta_{N-i}(z)$, we can inspect the intersection

$$\bigcap_{k \in S_i} \mathcal{A}_{\bar{k}},$$

where $\mathcal{A}_{\bar{k}}$ is a set of $(N-i)(m_1+m_2+\cdots+m_q)$ elements obtained by deleting i rows from the rotation table. The deleted rows correspond to the i indices in the index vector $\mathbf{k}$. We divide $\mathcal{A}_{\bar{k}}$ into q groups according to the distinct zeros $\alpha_1, \alpha_2, \ldots, \alpha_q$ as in Section 10.6.3. The ℓth group contains the entries of $\mathcal{A}_{\bar{k}}$ due to α_ℓ and it contains $(N-i)m_\ell$ elements. We will show that $\Delta_{N-i}(z)$ is as given in (10.24) by proving the following two claims.

Claim 1. The set $\bigcap_{k \in S_i} \mathcal{A}_{\bar{k}}$ contains $\sum_{\ell=i+1}^{q} m_\ell$ complete rotations.

Claim 2. It contains no other elements.

Proof of Claim 1   Each $\mathcal{A}_{\bar{k}}$ is obtained by deleting i rows from the rotation table. None of the groups of $\mathcal{A}_{\bar{k}}$ contains a complete rotation as i numbers are missing in each column. To form a complete rotation, numbers in each group need to be combined with entries from other groups. We claim that for each of the last $q-i$ groups of $\mathcal{A}_{\bar{k}}$, the missing numbers can be found in the first i groups and therefore there will be $(m_{i+1} + m_{i+2} + \cdots + m_q)$ complete rotations.

Let us consider the qth group of $\mathcal{A}_{\bar{k}}$ and suppose the missing i numbers are $\ell_1, \ell_2, \ldots, \ell_i$. We need to show that $\ell_1, \ell_2, \ldots, \ell_i$ can be found in the first i groups. We will prove it by contradiction. Suppose this is not true and the number ℓ_1 cannot be found in any of the first i groups. Then ℓ_1 is missing in a total of $i+1$ groups (the first i groups plus the qth group). Note that each column of the rotation table contains a complete rotation. That a number is missing in a group means that it is in the rows that are deleted when $\mathcal{A}_{\bar{k}}$ is constructed. We may also observe that, for every row in the rotation table, a number appears in at most one group. This is because the groups correspond to distinct zeros and in each row the same number cannot appear in more than one group. In forming $\mathcal{A}_{\bar{k}}$, we have i rows removed. In these i rows, ℓ_1 can appear in at most i rows or appear in at most i groups. It cannot be missing from $i+1$ groups. Therefore we must be able to find ℓ_1 in one of the first i groups. Suppose ℓ_1 appears in the k_0th group of $\mathcal{A}_{\bar{k}}$, one of the first i groups. Because $m_{k_0} \geq m_q$, we can find ℓ_1 at least m_q times. Similarly, we can find $\ell_2, \ldots, \ell_i$ in the first i groups of $\mathcal{A}_{\bar{k}}$. Therefore we have m_q complete rotations due to the zeros in the qth group.

Using the same approach we can always find the missing numbers for groups $\#(i+1), \#(i+2), \ldots, \#(q-1)$. We will have problems using the same approach only if the same number is missing in more than one of the last $(q - i)$ groups. Suppose ℓ_1 is missing in both groups $\#q$ and $\#(q - 1)$. We know ℓ_1 appears m_q times in one of the first i groups as shown earlier. Can we find the same number m_{q-1} more times for group $\#(q - 1)$? The answer is in the affirmative. The reason is as follows. If ℓ_1 is missing in both groups $\#q$ and $\#(q - 1)$, this means that ℓ_1 appears in two deleted rows already. As only i rows are deleted, ℓ_1 can appear in at most $(i - 2)$ other deleted rows; we can find ℓ_1 in at least two of the first i groups of $\mathcal{A}_{\bar{\mathbf{k}}}$. In conclusion, we have shown that, in each $\mathcal{A}_{\bar{\mathbf{k}}}$, there are $(m_q + m_{q-1} + \cdots + m_{i+1})$ complete rotations, which proves the result in Claim 1.

Proof of Claim 2   From Claim 1, we know that all the numbers in the last $(q - i)$ groups of $\mathcal{A}_{\bar{\mathbf{k}}}$ are already in the intersection $\bigcap_{\mathbf{k} \in S_i} \mathcal{A}_{\bar{\mathbf{k}}}$. If the intersection contains another number, say n_0, this number must come from the first i groups. Consider only the first i groups of the whole rotation table. The number n_0 will appear in exactly i rows, say rows $\#k_1, \#k_2, \ldots, \#k_i$. Then for $\mathbf{k} = \begin{bmatrix} k_1 & k_2 & \cdots & k_i \end{bmatrix}$, $\mathcal{A}_{\bar{\mathbf{k}}}$ does not have n_0 in any of the first i groups, which implies that $\bigcap_{\mathbf{k} \in S_i} \mathcal{A}_{\bar{\mathbf{k}}}$ cannot contain n_0. Therefore we can conclude that $\bigcap_{\mathbf{k} \in S_i} \mathcal{A}_{\bar{\mathbf{k}}}$ contains no entries other than $(m_{i+1} + m_{i+2} + \cdots + m_q)$ complete rotations.

10.7 Further reading

The use of filter banks to introduce redundancy in the transmitted signal was proposed by Xia in 1996 [181, 182]. The scheme was termed filter bank precoding. Using filter bank precoding, redundancy was added and ISI cancellation for FIR channels achieved with FIR transmitters and receivers. It has been shown that FIR transceivers exist under very general conditions even if the redundancy K is as small as unity [182]. The filters in filter bank transceivers are LTI filters. In [135], time-varying systems were employed for designing FIR transceivers. Suppose the channel is FIR with distinct roots and the number of subchannels M is larger than the channel order. It was shown in [135] that, we can always find a channel-independent time-varying transmitter such that FIR time-varying receivers exist. By extending the results in [135], necessary and sufficient conditions that guarantee the existence of FIR zero-forcing equalizers were presented in [124]. A reduction in the length of the FIR equalizer can thus be achieved. Precoding was combined with OFDM to reduce redundancy and improve robustness against channel spectral nulls in [184].

Filter bank precoding has also been extended to MIMO (multi-input multi-output) systems. Minimum redundancy for the existence of FIR transceivers in the case of MIMO channels was considered in [66]. Similar to the case of single-input single-output channels, the minimum-redundancy condition can be given in terms of the congruous zeros of the channel matrix.

10.8 Problems

10.1 Suppose we have a scalar filter $C(z) = (1 + z^{-2})(1 - z^{-1})$; assume the block size is $N = 2$.

    (a) Determine the corresponding pseudocirculant matrix $\mathbf{C}_{ps}(z)$.

    (b) Find a Smith form decomposition of $\mathbf{C}_{ps}(z)$.

    (c) What is the minimum redundancy?

10.2 Repeat Problem 10.1 for $N = 3$.

10.3 Give one example of $C(z)$ such that the corresponding 3×3 $\mathbf{C}_{ps}(z)$ has $\rho_3 = 3$.

10.4 Find the general expression of an Lth-order channel $C(z)$ such that $\mu_N = N$.

10.5 Let $C(z) = (1 + z^{-2})(1 - z^{-1})$, assume the block size $N = 4$, and let the corresponding pseudocirculant matrix be denoted as $\mathbf{C}_{ps}(z)$.

    (a) What are the zeros of $\det \mathbf{C}_{ps}(z)$?

    (b) Find the Smith form of $\mathbf{C}_{ps}(z)$.

    (c) Determine the diagonal matrix $\mathbf{\Sigma}(z)$ in the DFT decomposition of $\mathbf{C}_{ps}(z)$.

10.6 Find an ISI-free FIR transceiver $(\mathbf{G}(z), \mathbf{S}(z))$ for Problem 10.2. Can we find another FIR transceiver solution?

10.7 Suppose the channel is the third-order FIR filter $C(z) = (1 + z^{-2})(1 - z^{-1})$.

    (a) What is the smallest block size N such that $\rho_N = 1$?

    (b) Find all interpolation ratios N for which FIR transceivers exist with redundancy $K = 1$.

10.8 Show that

$$\prod_{k=0}^{N-1} [1 - \alpha(zW^k)^{-1}] = 1 - \alpha^N z^{-N},$$

where $W = e^{-j2\pi/N}$.

10.9 Suppose a channel $C(z)$ is FIR with distinct zeros and the blocked $N \times N$ channel matrix is $\mathbf{C}_{ps}(z)$. It is known that $\det \mathbf{C}_{ps}(z)$ has q distinct zeros $\beta_1, \beta_2, \ldots, \beta_q$ with multiplicities respectively $\ell_1, \ell_2, \ldots, \ell_q$, with $\ell_1 \geq \ell_2 \geq \cdots \geq \ell_q$. Determine the Smith form $\mathbf{C}_{ps}(z)$.

10.10 Show (10.9) using the following two steps.

    (a) Show that the matrix $\mathbf{D}(z)\mathbf{C}_{ps}(z^N)\mathbf{D}(z^{-1})$ is circulant.

    (b) Use (a) to prove the DFT decomposition in (10.8).

10.11 Suppose $C(z)$ is an FIR channel and the corresponding $N \times N$ channel matrix is $\mathbf{C}_{ps}(z)$. The Smith form of $\mathbf{C}_{ps}(z)$ has ρ_N nonunity diagonal terms. Let redundancy $K \geq \rho_N$. Suppose the transmitter does nothing other than zero padding, i.e.

$$\mathbf{G}(z) = \begin{bmatrix} \mathbf{I}_M \\ \mathbf{0} \end{bmatrix}.$$

Is it always possible to find an FIR zero-forcing receiver for the zero-padding transmitting matrix given above? If your answer is *yes*, find a zero-forcing receiver. If not, justify your answer.

10.12 Consider the channel in Example 10.6. Suppose the channel noise is AWGN with variance $\mathcal{N}_0$.

    (a) Find a Smith form decomposition for $\mathbf{C}_{ps}(z)$ when $N = 3$. Use the decomposition to find an FIR transceiver solution with minimum redundancy.

    (b) Determine the total noise power for the outputs of the receiver in (a).

10.13 Consider an FIR channel $C(z) = \sum_{n=0}^{L} c(n)z^{-n}$. Show that FIR transceivers with no redundancy can achieve the ISI-free property only if $C(z)$ is a delay.

10.14 Suppose we block a scalar channel by using a delay chain of z^{-p} rather than z^{-1} as in Fig. P5.3 of Problem 5.4 (where p and N are coprime). The resulting blocked channel matrix will be different from the pseudo-circulant given in (10.2).

(a) In the discussion of minimal transceivers in Section 10.3, we saw that solutions of causal stable zero-forcing transceivers exist if the scalar channel has minimum phase. Does the new blocking scheme affect the existence of causal stable zero-forcing transceivers?

(b) For a given interpolation ratio N, is the minimum redundancy affected by the new blocking scheme?

10.15 Suppose the diagonal matrix $\mathbf{\Sigma}(z)$ in the DFT decomposition of $\mathbf{C}_{ps}(z)$ in (10.8) is real. What does that tell us about the scalar filter $C(z)$?

10.16 In Section 10.6.1 we considered two matrices related by

$$\mathbf{B}(z) = \mathbf{D}(z)\mathbf{E}(z)\mathbf{D}(z^{-1}),$$

where $\mathbf{E}(z)$ is a polynomial matrix in z^{-1} such that $\mathbf{B}(z)$ is also a polynomial matrix in z^{-1}, and $\mathbf{D}(z)$ is as given in (10.8). Let $\Delta_i(z)$ be the gcd of all the $i \times i$ minors of $\mathbf{E}(z)$ and $\Delta_i'(z)$ the gcd of all $i \times i$ minors of $\mathbf{B}(z)$. Show that $\Delta_i(z) = \Delta_i'(z)$.

Appendix A

Mathematical tools

In this appendix, we summarize some mathematical inequalities and theorems that are used in the book. For more detailed descriptions of these materials, please see [52].

Arithmetic-mean and geometric-mean (AM-GM) inequality

Given a set of non-negative numbers a_i for $0 \leq i \leq M - 1$, the arithmetic mean (AM) and geometric mean (GM) are respectively defined as

$$AM = \frac{1}{M} \sum_{i=0}^{M-1} a_i,$$

$$GM = \left(\prod_{i=0}^{M-1} a_i \right)^{1/M}.$$

The AM–GM inequality says that

$$AM \geq GM,$$

with equality if and only if $a_0 = a_1 = \cdots = a_{M-1}$.

Matrix inversion lemma

Let $\mathbf{P}$ and $\mathbf{R}$ be, respectively, $M \times M$ and $N \times N$ invertible matrices. Then

$$(\mathbf{P} + \mathbf{QRS})^{-1} = \mathbf{P}^{-1} - \mathbf{P}^{-1}\mathbf{Q} \left(\mathbf{SP}^{-1}\mathbf{Q} + \mathbf{R}^{-1} \right)^{-1} \mathbf{SP}^{-1}.$$

- When $M = N$ and $\mathbf{Q} = \mathbf{S} = \mathbf{I}$, the lemma reduces to

$$(\mathbf{P} + \mathbf{R})^{-1} = \mathbf{P}^{-1} - \mathbf{P}^{-1} \left(\mathbf{P}^{-1} + \mathbf{R}^{-1} \right)^{-1} \mathbf{P}^{-1}.$$

- When $\mathbf{P}^{-1}$ has a simple form and the rank of $\mathbf{R}$ is much smaller than M, then the lemma can be employed for an efficient computation of $(\mathbf{P} + \mathbf{R})^{-1}$. For example, when $\mathbf{R}$ has rank 1, $\mathbf{R} = \mathbf{qs}^T$ for some $M \times 1$ vectors $\mathbf{q}$ and $\mathbf{s}$. The inverse becomes

$$(\mathbf{P} + \mathbf{R})^{-1} = \mathbf{P}^{-1} - \frac{(\mathbf{P}^{-1}\mathbf{q})(\mathbf{s}^T\mathbf{P}^{-1})}{1 + \mathbf{s}^T\mathbf{P}^{-1}\mathbf{q}}.$$

Singular value decomposition

Given an $M \times N$ matrix $\mathbf{A}$ of rank r, its singular value decomposition (SVD) is given by

$$\mathbf{A} = \mathbf{U}\mathbf{D}\mathbf{V}^{\dagger},$$

where $\mathbf{U}$ and $\mathbf{V}$ are, respectively, $M \times M$ and $N \times N$ unitary matrices. The matrix $\mathbf{D}$ is a diagonal matrix and its r nonzero diagonal entries satisfy $d_{00} \geq d_{11} \geq \cdots \geq d_{r-1,r-1} > 0$.

- The quantities d_{ii} are the singular values of $\mathbf{A}$. If we let λ_i be the nonzero eigenvalues of $\mathbf{A}^{\dagger}\mathbf{A}$ or $\mathbf{A}\mathbf{A}^{\dagger}$ arranged in a nonincreasing order, then $d_{ii} = \sqrt{\lambda_i}$ for $0 \leq i \leq r - 1$.

- The column vectors of $\mathbf{U}$ and $\mathbf{V}$ are known as the left and right singular vectors of $\mathbf{A}$ due to their locations in the decomposition. These vectors are the eigenvectors of $\mathbf{A}\mathbf{A}^{\dagger}$ and $\mathbf{A}^{\dagger}\mathbf{A}$, respectively.

- The singular values of a matrix are unique, whereas the singular vectors are not.

Hadamard inequality

For any $M \times M$ positive semidefinite matrix $\mathbf{A}$, the Hadamard inequality says that

$$\prod_{k=0}^{M-1} [\mathbf{A}]_{kk} \geq \det[\mathbf{A}].$$

Furthermore, if $\mathbf{A}$ is positive definite, then equality holds if and only if $\mathbf{A}$ is diagonal.

Fischer inequality

Let $\mathbf{A}$ be a positive definite matrix. Partition $\mathbf{A}$ as

$$\mathbf{A} = \begin{bmatrix} \mathbf{B} & \mathbf{C} \\ \mathbf{C}^{\dagger} & \mathbf{D} \end{bmatrix},$$

where $\mathbf{B}$ and $\mathbf{D}$ are square. Then the Fischer inequality says that

$$\det[\mathbf{A}] \leq \det[\mathbf{B}] \det[\mathbf{D}].$$

- The inequality becomes an equality if and only if $\mathbf{C} = \mathbf{0}$.

- The Hadamard inequality can be proven by repeatedly applying the Fischer inequality. Thus the latter can be viewed as a generalization of the former.

Rayleigh–Ritz theorem

Let λ_{min} and λ_{max} be, respectively, the largest and smallest eigenvalues of a Hermitian matrix $\mathbf{A}$. Then

$$\min_{\mathbf{v} \neq 0} \frac{\mathbf{v}^\dagger \mathbf{A} \mathbf{v}}{\mathbf{v}^\dagger \mathbf{v}} = \lambda_{min},$$

$$\max_{\mathbf{v} \neq 0} \frac{\mathbf{v}^\dagger \mathbf{A} \mathbf{v}}{\mathbf{v}^\dagger \mathbf{v}} = \lambda_{max}.$$

The minimum and maximum values of λ_{min} and λ_{max} are achieved when $\mathbf{v}$ are the eigenvectors of $\mathbf{A}$ associated with λ_{min} and λ_{max}, respectively.

Convex and concave functions

A function $f(t)$ is said to be **convex** over an interval (T_0, T_1) if, for every $t_0, t_1 \in (T_0, T_1)$ and for all $0 < \lambda < 1$,

$$\lambda f(t_0) + (1 - \lambda) f(t_1) \geq f(\lambda t_0 + (1 - \lambda) t_1).$$

The function $f(t)$ is said to be **strictly convex** when equality holds if only if $t_0 = t_1$. If the above inequality is reversed, $f(t)$ is said to be **concave** (or **strictly concave** when equality holds if and only if $t_0 = t_1$).

- A function $f(t)$ is (strictly) convex if and only if $-f(t)$ is (strictly) concave.

- If the second derivative of $f(t)$ exists and it satisfies $f''(t) \geq 0$ for all $t \in (T_0, T_1)$, then $f(t)$ is convex (it is strictly convex if $f''(t) > 0$). On the other hand, if $f''(t) \leq 0$ for all $t \in (T_0, T_1)$, then $f(t)$ is concave (it is strictly concave if $f''(t) < 0$).

- Given a set of M numbers, $t_0, t_1, \ldots, t_{M-1} \in (T_0, T_1)$, the convexity of $f(t)$ implies that

$$\sum_{i=0}^{M-1} \lambda_i f(t_i) \geq f\left(\sum_{i=0}^{M-1} \lambda_i t_i\right),$$

where $0 < \lambda_i < 1$, and $\sum_{i=0}^{M-1} \lambda_i = 1$. If $f(t)$ is strictly convex, then equality holds if and only if $t_0 = t_1 = \cdots = t_{M-1}$. On the other hand, the concavity of $f(t)$ would imply

$$\sum_{i=0}^{M-1} \lambda_i f(t_i) \leq f\left(\sum_{i=0}^{M-1} \lambda_i t_i\right).$$

When $f(t)$ is strictly concave, equality holds if and only if $t_0 = t_1 = \cdots = t_{M-1}$.

Appendix B

Review of random processes

This appendix gives a brief overview of random processes. A more detailed treatment of this topic can be found in [111, 179].

B.1 Random variables

A real random variable $x(\eta)$ is a real value that is assigned to every outcome η of a random experiment. The argument η is usually dropped for convenience. The probability of $x \le a$ can be described by the *cumulative density function* (cdf)

$$F_x(a) \triangleq P(x \le a).$$

We can also compute it using the *probability density function* (pdf) $f_x(x)$,

$$P(x \le a) = \int_{-\infty}^{a} f_x(x)dx,$$

where the pdf $f_x(x)$ is the derivative of $F_x(x)$. In communications, a frequently used random variable is the Gaussian. For a real Gaussian random variable, the pdf is

$$f_x(x) = \frac{1}{\sqrt{2\pi}\sigma_x}e^{-(x-m_x)^2/2\sigma_x^2}, \quad -\infty < x < \infty,$$

where m_x and σ_x^2 are, respectively, the mean and variance of x.

Expected values Let $g(x)$ be a function of a random variable x. The expected value of $g(x)$ is defined as

$$E[g(x)] = \int_{-\infty}^{\infty} g(x)f_x(x)dx.$$

Some commonly used expected values are the mean, mean squared value, and variance, given respectively by

$$\text{mean} \quad m_x = E[x],$$
$$\text{mean squared value} = E[x^2],$$
$$\text{variance} \quad \sigma_x^2 = E[(x - m_x)^2].$$

When $m_x = 0$, we say the random variable is zero-mean. In this book, all random variables have zero-mean unless otherwise mentioned. For a zero-mean random variable, the variance is equal to its mean squared value. As the transmitted signals and noise are often modeled as random variables in digital communication systems, we will call the mean squared value the *power* of the random variable.

Two or more random variables Two random variables x and y can be described by the *joint probability density function* $f_{xy}(x, y)$. Let $g(x, y)$ be a function of x and y. The expected value of $g(x, y)$ is defined as

$$E[g(x, y)] = \int_{-\infty}^{\infty} \int_{-\infty}^{\infty} g(x, y) f(x, y) dx dy.$$

Two random variables x and y are *independent* if

$$f_{xy}(x, y) = f_x(x) f_y(y).$$

A useful quantity that often comes into the discussion of two random variables is the *cross correlation*, given by $R_{xy} = E[xy]$. We say x and y are *uncorrelated* if

$$R_{xy} = E[x]E[y],$$

and we say x and y are *orthogonal* if $R_{xy} = 0$. Note that the independence implies uncorrelatedness but the converse is usually not true. However, for Gaussian random variables, the property that two random variables are uncorrelated also implies that they are independent. For two random variables x and y that are zero-mean and uncorrelated, the sum $w = x + y$ has variance given by $\sigma_w^2 = \sigma_x^2 + \sigma_y^2$.

Similarly, a set of M random variables $x_0, x_1, \ldots, x_{M-1}$, can be described by a joint pdf. The expected value, independence, and uncorrelatedness can be defined in a similar manner.

Complex random variables A complex random variable $x = a + jb$ is a complex quantity whose real part a and imaginary part b are both random variables. For a complex random variable x, the mean, mean squared value, and variance are defined respectively as

$$m_x = E[x], \quad E[|x|^2], \quad E[|x - m_x|^2].$$

For two complex random variables x and y, the *cross correlation* is

$$R_{xy} = E[xy^*].$$

Similar to the real case, we say x and y are *uncorrelated* if $R_{xy} = E[xy^*] = E[x]E[y^*]$, and they are *orthogonal* if $R_{xy} = 0$. We say a random variable is *circularly symmetric complex Gaussian* if its real and imaginary parts are independent Gaussian variables with the same variance. In passband communications, the channel noise is often modeled as a zero-mean circularly symmetric complex Gaussian random variable. Its pdf is given in (2.5) and it is plotted in Fig. 2.7. In the following discussion, the random variables are assumed to be complex unless otherwise mentioned.

Random vectors   A random vector $\mathbf{x}$ of dimensions $M \times 1$ is a collection of M random variables, $\mathbf{x} = \begin{bmatrix} x_0 & x_1 & \cdots & x_{M-1} \end{bmatrix}^T$. Its mean

$$E[\mathbf{x}] = \begin{bmatrix} E[x_0] & E[x_1] & \cdots & E[x_{M-1}] \end{bmatrix}^T$$

is also a vector. The *cross correlation matrix* of two random vectors $\mathbf{x}$ and $\mathbf{y}$ is defined as $E[\mathbf{xy}^\dagger]$. We call the $M \times M$ matrix $\mathbf{R}_x = E[\mathbf{xx}^\dagger]$ the *autocorrelation matrix* of $\mathbf{x}$. For example, let $M = 3$; then we have

$$\mathbf{R}_x = \begin{bmatrix} E[|x_0|^2] & R_{x_0 x_1} & R_{x_0 x_2} \\ R^*_{x_0 x_1} & E[|x_1|^2] & R_{x_1 x_2} \\ R^*_{x_0 x_2} & R^*_{x_1 x_2} & E[|x_2|^2] \end{bmatrix},$$

where $R_{x_i x_j}$ denotes the cross correlation between x_i and x_j. The diagonal of $\mathbf{R}_x$ consists of the mean squared values $E[|x_i|^2]$ while the off-diagonal terms are the cross correlations among the random variables. In the zero-mean case $\mathbf{R}_x$ is also said to be the *covariance matrix*. Observe that an autocorrelation matrix is always Hermitian, i.e. $\mathbf{R}_x^\dagger = \mathbf{R}_x$. Moreover, it can be shown that $\mathbf{R}_x$ is positive semidefinite. In the special case that the elements of $\mathbf{x}$ are orthogonal to one another, the autocorrelation matrix is diagonal. If, in addition, the random variables have the same variance σ_x^2, the autocorrelation matrix becomes $\sigma_x^2 \mathbf{I}$.

B.2   Random processes

A random process $x(n, \eta)$ is a function that is assigned to every outcome η of an experiment. We usually write it as $x(n)$ and drop the argument η for convenience. For a particular outcome, $x(n)$ is a fixed time function, which we call a *realization* or a *sample function*. We can view $x(n)$ for every n as a random variable. Often times, the first- and second-order statistics are very useful in applications involving random processes, in particular the mean and autocorrelation functions. In general the mean function $E[x(n)]$ depends on the time index n. The *autocorrelation function* $R_x(m, n)$ is defined as the cross correlation between $x(m)$ and $x(n)$,

$$R_x(m, n) = E[x(m)x^*(n)].$$

This is generally a function of both time indices m and n. The random processes that arise in the application of communication systems can often be assumed to have certain stationary properties, e.g. wide sense stationary and cyclo wide sense stationary properties.

Wide sense stationary processes

A random process $x(n)$ is said to be wide sense stationary (WSS) if $x(n)$ satisfies the following two conditions:

(1) $E[x(n)] = m_x$ (a constant independent of n);
(2) $E[x(n)x^*(n-k)] = R_x(k)$ (a function of k).

(B.1)

The mean function $E[x(n)]$ is independent of n and the autocorrelation function $E[x(n)x^*(n-k)]$ depends only on the *time lag* k, but not n.

Power spectrum   For a WSS process, the power spectrum (or *power spectral density*) is defined as the Fourier transform of $R_x(k)$,

$$S_x(e^{j\omega}) = \sum_{k=-\infty}^{\infty} R_x(k)e^{-j\omega k}.$$

Conversely $R_x(k)$ is the inverse Fourier transform of $S_x(e^{j\omega})$. In particular, the power of $x(n)$ can be obtained from its power spectrum using

$$R_x(0) = \int_0^{2\pi} S_x(e^{j\omega})d\omega/2\pi.$$

It can be shown that $S_x(e^{j\omega}) \geq 0$ for all ω; the power spectrum is a nonnegative function.   The area under $S_x(e^{j\omega})$ from ω_0 to ω_1 can be viewed as the power of $x(n)$ in the frequency range (ω_0, ω_1).

Jointly WSS processes   Two random processes $x(n)$ and $y(n)$ are said to be *jointly WSS* if each of them is WSS and their cross correlation $E[x(n)y^*(n-k)]$ is a function of time lag k only, independent of n. In other words, their cross-correlation function is

$$R_{xy}(k) = E[x(n)y^*(n-k)].$$

We say that $x(n)$ and $y(n)$ are *uncorrelated* if $R_{xy}(k) = m_x m_y^*$, for all k. If one of the processes has zero-mean, then uncorrelatedness implies $R_{xy}(k) = 0$ for all k. When we have two jointly WSS random processes $x(n)$ and $y(n)$, zero-mean and uncorrelated with each other, their sum $w(n) = x(n) + y(n)$ is also a WSS random process with zero-mean, and $\sigma_w^2 = \sigma_x^2 + \sigma_y^2$.

Computation of experimental statistics   A WSS process is said to be *ergodic* if the statistical averages, e.g. $E[x(n)]$, $R_x(k)$, are equal to the corresponding time averages for any single realization. The ergodicity assumption allows us to estimate the expected values by using time averages obtained from a single realization. In practice, the number of available samples of a realization $x_0(n)$ is often finite. For example, given N samples of a realization $x_0(n)$, we can estimate the mean by

$$\widehat{m}_x = \frac{1}{N} \sum_{n=0}^{N-1} x_0(n).$$

The autocorrelation function can be estimated by

$$\widehat{R}_x(k) = \frac{1}{N-k} \sum_{n=k}^{N-1} x_0(n)x_0^*(n-k), \quad k = 0, 1, \ldots, N-1.$$

The accuracy of the estimate improves as the number of available samples N increases.

White processes

A random process is said to be white if $x(m)$ and $x(n)$ are uncorrelated for any $m \neq n$, i.e.

$$E[x(m)x^*(n)] = E[x(m)]E[x^*(n)], \quad m \neq n.$$

When a white process is also WSS, we have

$$R_x(k) = \begin{cases} m_x^2 + \sigma_x^2, & k = 0, \\ m_x^2, & \text{otherwise.} \end{cases}$$

If, in addition, $x(n)$ has zero-mean, then

$$R_x(k) = \delta(k)\sigma_x^2.$$

In this case the power spectrum is a constant given by

$$S_x(e^{j\omega}) = \sigma_x^2, \forall \omega.$$

Such a process is usually referred to as *white noise*. A WSS process is said to be *colored* if it is not white. The channel noise $q(n)$ in wireless environments can often be assumed to be an additive white Gaussian noise (AWGN) with zero-mean. For wired applications, the noise is often colored additive Gaussian noise (AGN). For AWGN and AGN, wide sense stationarity and zero-mean are usually implicitly assumed.

Vector random processes

An $M \times 1$ vector random process $\mathbf{x}(n) = \begin{bmatrix} x_0(n) & x_1(n) & \cdots & x_{M-1}(n) \end{bmatrix}^T$ is a vector of M random processes, where each entry $x_i(n)$ is a random process. For each n, $\mathbf{x}(n)$ is a random vector. We say $\mathbf{x}(n)$ is WSS if both $E[\mathbf{x}(n)]$ and $E[\mathbf{x}(n)\mathbf{x}^\dagger(n-k)]$ are independent of n. Then the autocorrelation is

$$\mathbf{R}(k) = E[\mathbf{x}(n)\mathbf{x}^\dagger(n-k)],$$

which is an $M \times M$ matrix for every time lag k. When $\mathbf{x}(n)$ is WSS, any pair of components $x_m(n)$ and $x_i(n)$ are jointly WSS.

Cyclo wide sense stationary processes

A random process $x(n)$ is said to be cyclo wide sense stationary with period M (abbreviated as CWSS(M)) if it satisfies the following two conditions:

$$
\begin{align}
&(1) \quad E[x(n+M)] = E[x(n)], \\
&(2) \quad E[x(n)x^*(n-k)] = E[x(n+M)x^*(n+M-k)].
\end{align}
\tag{B.2}
$$

These two properties imply that the mean function $E[x(n)]$ and the mean squared function $E[|x(n)|^2]$ are both periodic with period M. Also the auto correlation $E[x(n)x^*(n-k)]$ is a periodic function of n. We define the average autocorrelation function as

$$R_x(k) = \frac{1}{M} \sum_{n=0}^{M-1} E[x(n)x^*(n-k)],$$

which is a function independent of n. The average power of $x(n)$ or the average mean squared value is

$$R_x(0) = \frac{1}{M} \sum_{n=0}^{M-1} E[|x(n)|^2].$$

Similar to the WSS case, we define the average power spectrum of a CWSS process as the Fourier transform of the average autocorrelation function,

$$S_x(e^{j\omega}) = \sum_{\ell=-\infty}^{\infty} R_x(\ell)e^{-j\omega\ell}.$$

We will omit the word "average" and call $R_x(k)$ the autocorrelation function and $S_x(e^{j\omega})$ the power spectrum of $x(n)$ when it is clear from the context that $x(n)$ is a CWSS process.

B.3   Processing of random variables and random processes

Passage through LTI systems

When we pass a WSS process $x(n)$ through a stable LTI system $H(z)$, the output $y(n)$ (Fig. B.1) is also WSS. In particular, $y(n)$ has mean and auto-correlation functions given respectively by

$$\begin{aligned} E[y(n)] = m_y &= m_x \sum_n h(n), \\ R_y(k) &= R_x(k) * h(k) * h^*(-k). \end{aligned} \tag{B.3}$$

The power spectrum of $y(n)$ is related to that of $x(n)$ by

$$S_y(e^{j\omega}) = S_x(e^{j\omega})|H(e^{j\omega})|^2.$$

When $x(n)$ is AWGN (WSS with zero-mean), $S_y(e^{j\omega}) = \sigma_x^2|H(e^{j\omega})|^2$; the power spectrum of $y(n)$ assumes the shape of $|H(e^{j\omega})|^2$. In this case, we can compute the output power σ_y^2 using

$$\sigma_y^2 = \sigma_x^2 \int_0^{2\pi} |H(e^{j\omega})|^2 \frac{d\omega}{2\pi}.$$

The input power is amplified by a factor of $\int_0^{2\pi} |H(e^{j\omega})|^2 d\omega/2\pi$, which is equal to the energy of the filter.

When we pass a CWSS(M) process $x(n)$ through a stable LTI filter $H(z)$, as shown in Fig. B.1, the output is also a CWSS(M) process. The (average) autocorrelation function of the output $y(n)$ is given by

$$R_y(k) = R_x(k) * h(k) * h^*(-k). \tag{B.4}$$

We can obtain the (average) power spectrum of the output by taking the Fourier transform of the above expression, $S_y(e^{j\omega}) = S_x(e^{j\omega})|H(e^{j\omega})|^2$.

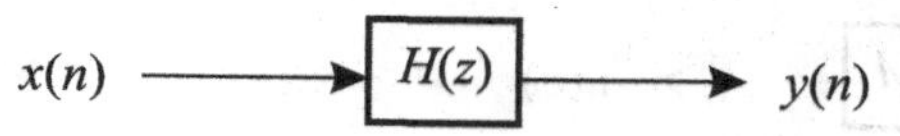

Figure B.1. Passage of a WSS process through an LTI system.

Passage through multirate blocks

Suppose we pass a WSS process $x(n)$ through a decimator (Fig. B.2(a)), then the output is given by $u(n) = x(Mn)$. It can be shown that $u(n)$ is also a WSS process with mean and autocorrelation function given, respectively, by

$$m_u = m_x, \quad R_u(k) = R_x(Mk).$$

On the other hand, when we pass a WSS process $x(n)$ through an expander (Fig. B.2(b)), the output $v(n)$ has mean function given by

$$E[v(n)] = \begin{cases} m_x, & n \text{ is a multiple of } M, \\ 0, & \text{otherwise.} \end{cases}$$

Furthermore,

$$E[v(n)v^*(n-k)] = \begin{cases} R_x(k/M), & n \text{ and } k \text{ are integer multiples of } M, \\ 0, & \text{otherwise.} \end{cases}$$

The expander output $v(n)$ is no longer WSS. It follows from the definition of CWSS random processes that $v(n)$ is CWSS(M). It turns out that the (average) autocorrelation function of $v(n)$ is simply an expanded and scaled version of $R_x(k)$; that is,

$$R_v(k) = \frac{1}{M}(R_x(k))_{\uparrow M} = \begin{cases} \frac{1}{M}R_x(k/M), & k \text{ is an integer multiple of } M, \\ 0, & \text{otherwise,} \end{cases}$$

(B.5)

where $(R_x(k))_{\uparrow M}$ denotes the M-fold expanded version of $R_x(k)$. We can verify this by first examining the expression

$$R_v(k) = \frac{1}{M} \sum_{n=0}^{M-1} E[v(n)v^*(n-k)].$$

Only the term $E[v(0)v^*(-k)]$ can be nonzero, and it is nonzero only when k is an integer multiple of M. In this case, the term $E[v(0)v^*(-k)]$ is equal to $R_x(k/M)$. When k is not a multiple of M,

$$R_v(k) = \frac{1}{M}E[v(0)v^*(-k)] = 0.$$

From (B.5), we can obtain the power spectrum of $v(n)$:

$$S_v(e^{j\omega}) = \frac{1}{M}S_x(e^{jM\omega}).$$

It is a scaled and "squeezed" version of the input power spectrum.

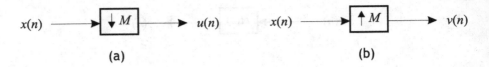

$x(n) \longrightarrow \boxed{\downarrow M} \longrightarrow u(n)$ $x(n) \longrightarrow \boxed{\uparrow M} \longrightarrow v(n)$

(a)   (b)

Figure B.2. Passage of a WSS process through multirate building blocks.

Passage through matrices

In Fig. B.3, we pass an $M \times 1$ random vector $\mathbf{x}$ through an $N \times M$ matrix $\mathbf{P}$. The output $\mathbf{y} = \mathbf{Px}$ has autocorrelation matrix $\mathbf{R}_y = E[\mathbf{yy}^\dagger]$ given by

$$\mathbf{R}_y = \mathbf{P}E[\mathbf{xx}^\dagger]\mathbf{P}^\dagger = \mathbf{PR}_x\mathbf{P}^\dagger.$$

In many cases the DFT matrix has been found to be very useful in the processing of random vectors. Consider the case that $\mathbf{P}$ is the $M \times M$ normalized DFT matrix $\mathbf{W}$. When the input vector $\mathbf{x}$ has uncorrelated elements, its autocorrelation matrix is diagonal, and in this case $\mathbf{R}_y$ is an $M \times M$ matrix sandwiched between $\mathbf{W}$ and $\mathbf{W}^\dagger$. From Theorem 5.2, we know matrices of such a form are circulant. Furthermore, the variances $\sigma_{y_i}^2$ are of the same value and it is equal to the average of the variances $\sigma_{x_i}^2$.

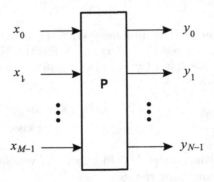

$x_0 \longrightarrow$ $\quad$ $\longrightarrow y_0$

$x_1 \longrightarrow$ $\quad \mathbf{P} \quad$ $\longrightarrow y_1$

$\vdots$ $\qquad\qquad$ $\vdots$

$x_{M-1} \longrightarrow$ $\qquad$ $\longrightarrow y_{N-1}$

Figure B.3. Passage of a random vector through a matrix.

Blocking

In many applications, we block a scalar random process to obtain a vector random process, as shown in Fig. B.4. The vector random process is

$$\mathbf{x}(n) = \begin{bmatrix} x(Mn) & x(Mn+1) & \cdots & x(Mn+M-1) \end{bmatrix}^T.$$

When $x(n)$ is WSS, so is the blocked version $\mathbf{x}(n)$. In some applications, we are only concerned with $\mathbf{R}(0)$ but not $\mathbf{R}(k)$ for $k \neq 0$, although the underlying

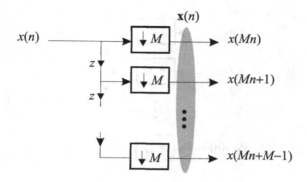

Figure B.4. Blocking of a scalar random process $x(n)$.

process is a vector random process. For example, when we consider one-shot block transmission in Chapters 6 and 7, only $\mathbf{R}(0)$ is relevant. The autocorrelation matrix $\mathbf{R}(0)$ of $\mathbf{x}(n)$ is a Hermitian positive semidefinite Toeplitz matrix. Suppose the autocorrelation function of $x(n)$ is r_k, then, for $M = 3$,

$$\mathbf{R}(0) = \begin{bmatrix} r_0 & r_1 & r_2 \\ r_1^* & r_0 & r_1 \\ r_2^* & r_1^* & r_0 \end{bmatrix}.$$

We say $\mathbf{R}(0)$ is the $M \times M$ autocorrelation matrix associated with the scalar process $x(n)$. In the special case that the scalar WSS process $x(n)$ is white, $\mathbf{R}(0)$ reduces to the simple matrix $\sigma_x^2 \mathbf{I}$. For the blocking system in Fig. B.4, we can verify that the output vector random process $\mathbf{x}(n)$ is WSS when the input $x(n)$ is CWSS(M).

Unblocking

Consider the interleaving (unblocking) system of expanders followed by a delay chain, as shown in Fig. B.5. The system interleaves the M inputs to produce a single output $x(n)$. If the inputs $y_k(n)$ are jointly WSS, then the interleaved output $x(n)$ is CWSS(M). To see this, we first verify that the mean function of $x(n)$ is periodic. Let us write $n = n_0 + n_1 M$, where $n_0 = ((n))_M$ and the notation $((\cdot))_M$ denotes modulo M operation. We see that $E[x(n)] = E[y_{n_0}(n_1)]$. Also note that $E[x(n + M)] = E[y_{n_0}(1 + n_1)]$, which is equal to $E[y_{n_0}(n_1)]$ because $y_{n_0}(\ell)$ is WSS. To see if it satisfies the second property in (B.2), let $n - k = \ell_0 + \ell_1 M$, where $\ell_0 = ((n - k))_M$. Then

$$E[x(n)x^*(n - k)] = E[y_{n_0}(n_1)y_{\ell_0}^*(\ell_1)] = R_{y_{n_0}y_{\ell_0}}(n_1 - \ell_1).$$

Similarly, we can verify that

$$E[x(n + M)x^*(n + M - k)] = E[y_{n_0}(n_1 + 1)y_{\ell_0}^*(\ell_1 + 1)] = R_{y_{n_0}y_{\ell_0}}(n_1 - \ell_1)$$

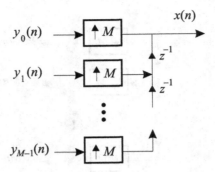

Figure B.5. Interleaving (unblocking) of M random processes.

and the second property in (B.2) holds. Conversely, it can be shown that any CWSS(M) random process can be viewed as the output of the unblocking system in Fig. B.5 with jointly WSS inputs.

B.4   Continuous-time random processes

In this book, we mostly deal with discrete-time random processes. Continuous-time random processes also arise in some of the discussions. A continuous-time random process $x_a(t, \eta)$ is a continuous-time function that is assigned to every outcome η of an experiment. Like the discrete-time case, we usually write it as $x_a(t)$ and drop the argument η for convenience. For a particular outcome, $x_a(t)$ is a fixed continuous-time function.

WSS processes   For continuous-time random processes, the WSS property can be defined in a similar manner. We say that $x_a(t)$ is WSS if both $E[x_a(t)]$ and $E[x_a(t)x_a^*(t-\tau)]$ are independent of the variable t. The autocorrelation of $x_a(t)$ is the continuous-time function $R_{x_a}(\tau) = E[x_a(t)x_a^*(t-\tau)]$. The power spectrum (or *power spectral density*) is defined as the Fourier transform of $R_{x_a}(\tau)$,

$$S_{x_a}(j\Omega) = \int_{-\infty}^{\infty} R_{x_a}(\tau)e^{-j\Omega\tau}\,d\tau.$$

Conversely, $R_{x_a}(\tau)$ is the inverse Fourier transform of $S_{x_a}(j\Omega)$. When $x_a(t)$ is white, WSS, and zero-mean, its autocorrelation function is an impulse $R_{x_a}(\tau) = \sigma_{x_a}^2 \delta(\tau)$ and the power spectrum is a constant $S_{x_a}(j\Omega) = \sigma_{x_a}^2$, for all Ω.

When we pass a continuous-time WSS process $x_a(t)$ through a stable LTI filter $H(j\Omega)$, as shown in Fig. B.6, the output is also a WSS process. The autocorrelation function of the output $y_a(t)$ is given by

$$R_{y_a}(\tau) = R_{x_a}(\tau) * h(\tau) * h^*(-\tau). \tag{B.6}$$

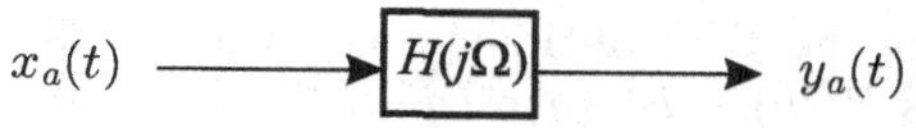

$$x_a(t) \longrightarrow \boxed{H(j\Omega)} \longrightarrow y_a(t)$$

Figure B.6. Passage of a WSS process through a continuous-time LTI system.

The power spectrum of $y_a(t)$ is related to that of $x_a(t)$ by

$$S_{y_a}(j\Omega) = S_{x_a}(j\Omega)|H(j\Omega)|^2.$$

For input $x_a(t)$ that is white, WSS, and zero mean, the output power spectrum $S_{y_a}(j\Omega) = \sigma_{x_a}^2 |H(j\Omega)|^2$ assumes the shape of $|H(j\Omega)|^2$.

CWSS processes We say that a continuous-time random process $x_a(t)$ is cyclo wide sense stationary with period T (denoted by CWSS(T)) if

$$
\begin{align}
&(1) \quad E[x_a(t+T)] = E[x_a(t)], \\
&(2) \quad E[x_a(t)x_a^*(t-\tau)] = E[x_a(t+T)x_a^*(t+T-\tau)].
\end{align}
\tag{B.7}
$$

Both the mean function $E[x_a(t)]$ and the mean squared function $E[|x_a(t)|^2]$ are periodic with period T. The (average) autocorrelation function is defined as

$$R_{x_a}(\tau) = \frac{1}{T}\int_0^T E[x_a(t)x_a^*(t-\tau)]dt,$$

which is a function independent of t. We define the (average) power spectrum as the Fourier transform of the (average) autocorrelation function,

$$S_{x_a}(j\Omega) = \int_{-\infty}^{\infty} R_{x_a}(\tau)e^{-j\Omega\tau}\,d\tau.$$

Like the discrete-time case, the word "average" will be omitted when it is clear from the context that $x_a(t)$ is a CWSS process.

When we pass a continuous-time CWSS(T) process $x_a(t)$ through a stable LTI filter $H(j\Omega)$, as shown in Fig. B.6, the output is also a CWSS(T) process. The (average) autocorrelation function of the output $y_a(t)$ is given by

$$R_{y_a}(\tau) = R_{x_a}(\tau) * h(\tau) * h^*(-\tau).$$
$$\tag{B.8}$$

Taking the Fourier transform of $R_{y_a}(\tau)$, we can obtain the (average) power spectrum of the output as $S_{y_a}(j\Omega) = S_{x_a}(j\Omega)|H(j\Omega)|^2$.

Passage through C/D and D/C converters

When we pass a continuous-time random process $r_a(t)$ through the system of $p_2(t)$ followed by a C/D converter, as shown in Fig. B.7, we get a discrete-time random process $r(n)$. The process $r(n)$ is given by

$$r(n) = w_a(nT) = (r_a * p_2)(t)\Big|_{t=nT}.$$

When $r_a(t)$ is WSS, the discrete-time process $r(n)$ is also WSS. The autocorrelation function is $R_r(k) = E[r(n)r^*(n-k)] = E[w_a(nT)w_a^*(nT-kT)]$. We can also write it as

$$R_r(k) = R_{w_a}(kT) = R_{r_a}(\tau) * p_2(\tau) * p_2^*(-\tau)\Big|_{\tau=kT},$$

which is a sampled version of $R_{w_a}(\tau)$. Note that $R_r(0) = R_{w_a}(0)$, that is $r(n)$ and $w_a(t)$ have the same power. Consider the case where $r_a(t)$ is white, WSS, and zero mean. Suppose the receiving pulse $p_2(t)$ is an ideal lowpass filter with unity gain in the passband and cutoff frequency π/T. Then $w_a(t)$ has a lowpass power spectrum and $S_{w_a}(j\Omega) = \sigma_{r_a}^2$ in the passband. It follows that $R_{w_a}(0) = \sigma_{r_a}^2/T$ and $r(n)$ is a discrete-time white noise (WSS and zero-mean) with $S_r(e^{j\omega}) = \sigma_{r_a}^2/T$.

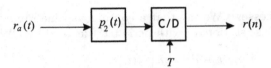

Figure B.7. Passage of a continuous-time random process through $p_2(t)$ followed by a C/D converter.

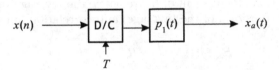

Figure B.8. Passage of a discrete-time random process through a D/C converter followed by $p_1(t)$.

Now consider the D/C converter followed by a transmitting pulse $p_1(t)$ shown in Fig. B.8. Assume the discrete-time input $x(n)$ is WSS. When the transmitting pulse $p_1(t)$ is an ideal lowpass filter, the output $x_a(t)$ is WSS [131]. When $p_1(t)$ is not ideal, the output $x_a(t)$ is in general a continuous-time CWSS(T) process. Furthermore, when the discrete-time input $x(n)$ is CWSS(M), the output $x_a(t)$ is CWSS(MT). In the following we will consider the case that $x(n)$ is CWSS(M), as a WSS $x(n)$ can be viewed as a CWSS(M) process with $M = 1$.

For the system shown in Fig. B.8, we know the output can be expressed as

$$x_a(t) = \sum_n x(n)p_1(t - nT).$$

We can verify $x_a(t)$ is CWSS(MT) by directly substituting the above expression into (B.7). It turns out that the (average) autocorrelation function of $x_a(t)$ is related to that of $x(n)$ in a simple manner. To derive $R_{x_a}(\tau)$, we first note that

$$E\left[x_a(t)x_a^*(t-\tau)\right] = E\left[\sum_n \sum_m x(n)x^*(m)p_1(t-nT)p_1^*(t-\tau-mT)\right]$$

$$= \sum_n \sum_m E\left[x(n)x^*(m)\right]p_1(t-nT)p_1^*(t-\tau-mT)$$

$$= \sum_n \sum_k E\left[x(n)x^*(n-k)\right]p_1(t-nT)p_1^*(t-\tau-(n-k)T)$$

$$= \sum_k \sum_\ell \sum_{i=0}^{M-1} E\left[x(M\ell+i)x^*(M\ell+i-k)\right]$$
$$\times \ p_1(t-(M\ell+i)T)p_1^*(t-\tau-(M\ell+i-k)T),$$

where we have used a change of variables $k = n-m$ to obtain the third equality and a change of variables $n = M\ell+i$ to obtain the fourth equality. As $x(n)$ is CWSS(M), we have $E\left[x(M\ell+i)x^*(M\ell+i-k)\right] = E\left[x(i)x^*(i-k)\right]$. It follows that

$$E\left[x_a(t)x_a^*(t-\tau)\right] = \sum_k \sum_\ell \sum_{i=0}^{M-1} E\left[x(i)x^*(i-k)\right]$$
$$\times \ p_1(t-(M\ell+i)T)p_1^*(t-\tau-(M\ell+i-k)T).$$

Therefore, we have

$$R_{x_a}(\tau) = \frac{1}{MT}\int_0^{MT} E[x_a(t)x_a^*(t-\tau)]dt,$$

$$= \frac{1}{MT}\sum_k \sum_{i=0}^{M-1} E\left[x(i)x^*(i-k)\right]$$
$$\times \sum_\ell \int_0^{MT} p_1(t-(M\ell+i)T)p_1^*(t-\tau-(M\ell+i-k)T)dt$$

$$= \frac{1}{MT}\sum_k \sum_{i=0}^{M-1} E\left[x(i)x^*(i-k)\right]$$
$$\times \int_{-\infty}^{\infty} p_1(t-iT)p_1^*(t-\tau-(i-k)T)dt. \tag{B.9}$$

Let us define

$$h(\tau) = p_1(\tau) * p_1^*(-\tau),$$

then $R_{x_a}(\tau)$ can be written as

$$R_{x_a}(\tau) = \frac{1}{T}\sum_k h(\tau-kT)\frac{1}{M}\sum_{i=0}^{M-1} E\left[x(i)x^*(i-k)\right]$$

$$= \frac{1}{T}\sum_k R_x(k)h(\tau-kT). \tag{B.10}$$

Taking the Fourier transform of the above expression, we obtain the (average) power spectrum of $x_a(t)$ as

$$S_{x_a}(j\Omega) = \frac{1}{T}S_x(e^{j\Omega T})|P_1(j\Omega)|^2, \tag{B.11}$$

where $P_1(j\Omega)$ is the Fourier transform of the transmitting pulse $p_1(t)$.

References

[1] K. Van Acker, G. Leus, M. Moonen, O. van de Wiel, and T. Pollet, "Per tone equalization for DMT-based systems," *IEEE Trans. Commun.*, vol. 49, pp. 109–119, Jan. 2001.

[2] N. Ahmed and K. R. Rao, *Orthogonal transforms for digital signal processing*. Secaucus, NJ: Springer Verlag, 1975.

[3] A. N. Akansu, P. Duhamel, X. Lin, and M. de Courville, "Orthogonal transmultiplexers in communication: a review," *IEEE Trans. Signal Process.*, vol. 46, pp. 979–995, April 1998.

[4] A. N. Akansu and P. R. Haddad, *Multiresolution signal decomposition: transforms, subbands, wavelets*. Boston: Kluwer Academic Publishers, 2001.

[5] N. Al-Dhahir and J. M. Cioffi, "Block transmission over dispersive channels: transmit filter optimization and realization, and MMSE-DFE receiver performance," *IEEE Trans. Inf. Theory*, vol. 42, pp. 56–64, Jan. 1996.

[6] N. Al-Dhahir and J. M .Cioffi, "Optimum finite-length equalization for multicarrier transceivers," *IEEE Trans. Commun.*, vol. 44, pp. 56–64, Jan. 1996.

[7] American National Standards Institute,"Network and Customer Installation Interfaces–Asymmetric Digital Subscriber Line (ADSL) Metallic Interface," ANSI T1.413, 1998.

[8] American National Standards Institute, "Very-high speed digital subscriber Lines (VDSL)-Metallic Interface," ANSI T1.424, 2004.

[9] J. Armstrong, "Analysis of new and existing methods of reducing intercarrier interference due to carrier frequency offset in OFDM," *IEEE Trans. Commun.*, vol. 47, pp. 365–369, March 1999.

[10] G. Arslan, B. L. Evans, and S. Kiaei, "Equalization for discrete multitone transceivers to maximize bit rate," *IEEE Trans. Signal Process.*, vol. 49, pp. 3123–3135, Dec. 2001.

[11] M. Bellanger, "On computational complexity in digital transmultiplexer filters," *IEEE Trans. Commun.*, vol. 30, pp. 1461–1465, July 1982.

[12] J. A. C. Bingham, "Multicarrier modulation for data transmission: an idea whose time has come," *IEEE Commun. Mag.*, vol. 26, pp. 5–14, May 1990.

[13] J. A. C. Bingham, "RFI suppression in multicarrier transmission systems," in *Proc. IEEE Global Telecommunications Conf.*, 1996, vol. 2, pp. 1026–1030.

[14] J. A. C. Bingham, *ADSL, VDSL, and Multicarrier Modulation*. Wiley-Interscience, 2000.

[15] H. Bölcskei, P. Duhamel, and R. Hleiss, "Design of pulse shaping OFDM/OQAM systems for high data-rate transmission over wireless channels," in *Proc. IEEE Int. Conf. Communications*, 1999, pp. 559–564.

[16] S. Boyd and L. Vandenberghe, *Convex Optimization*. Cambridge: Cambridge University Press, 2004.

[17] J. K. Cavers, "An analysis of pilot symbol assisted modulation for Rayleigh fading channels," *IEEE Trans. Veh. Technol.*, vol. 40, pp. 686–693, 1991.

[18] R. W. Chang, "Synthesis of band-limited orthogonal signals for multichannel data transmission," *Bell Syst. Tech. J.*, no. 45, pp. 1775–1796, Dec. 1966.

[19] R. W. Chang and R. A. Gibby, "A theoretical study of performance of an orthogonal multiplexing data transmission scheme," *IEEE Trans. Commun. Technol.*, pp. 529–540, Aug. 1968.

[20] C.-Y. Chen and S.-M. Phoong, "Per tone shaping filters for DMT transmitters," in *Proc. IEEE Int. Conf. Acoustic, Speech, and Signal Processing*, May 2004, pp. 1061–1064.

[21] C.-Y. Chi, C.-C. Feng, C.-H. Chen, and C.-Y. Chen, *Blind Equalization and System Identification: Batch Processing Algorithms, Performance and Applications*. Springer-Verlag, 2006.

[22] T.-D. Chiueh and P.-Y. Tsai, *OFDM Baseband Receiver Design for Wireless Communications*. Wiley, 2007

[23] P. S. Chow, J. M. Cioffi, and J. A. C. Bingham, "A practical discrete multitone transceiver loading algorithm for data transmission over spectrally shaped channels," *IEEE Trans. Commun.*, pp 773–775, April 1995.

[24] J. S. Chow, J. C. Tu, and J. M. Cioffi, "A discrete multitone transceiver system for HDSL applications," *IEEE J. Sel. Areas Commun.*, vol. 9, pp. 895–908, Aug. 1991.

[25] P. S. Chow, J. C. Tu, and J. M. Cioffi, "Performance evaluation of a multichannel transceiver system for ADSL and VHDSL services," *IEEE J. Sel. Areas Commun.*, vol. 9, pp. 909–919, Aug. 1991.

[26] Y.-H. Chung, and S.-M. Phoong, "Low complexity zero-padding zero-jamming DMT systems," in *Proc. European Signal Processing Conf.*, 2006.

[27] L. J. Cimini, "Analysis and simulation of a digital mobile channel using orthogonal frequency division multiplexing," *IEEE Trans. Commun.*, vol. 30, pp. 665–675, July 1985.

[28] J. M. Cioffi, *Equalization,* Chapter 3, Lecture notes, Stanford University, CA. http://www.stanford.edu/group/cioffi/ee379a/

[29] L. De Clercq, M. Peeters, S. Schelstraete, and T. Pollet, "Mitigation of radio interference in xDSL transmission," *IEEE Commun. Mag.*, vol. 38, pp. 168–173, March 2000.

[30] R. E. Crochiere and L. R. Rabiner, *Multirate Digital Signal Processing,* Englewood Cliffs, NJ: Prentice Hall PTR, 1983.

[31] G. Cuypers, K. Vanbleu, G. Ysebaert, M. Moonen "Egress reduction by intra-symbol windowing in DMT-based transmissions," in *Proc. IEEE Int. Conf. Acoustics, Speech and Signal Processing*, 2003, pp. 532–535.

[32] G. Cuypers, K. Vanbleu, G. Ysebaert, M. Moonen, and P. Vandaele, "Combining raised cosine windowing and per tone equalization for RFI mitigation in DMT receivers," in *Proc. IEEE Int. Conf. Commun.*, 2003, pp. 2852–2856.

[33] Y. Ding, T. N. Davidson, J.-K. Zhang, Z.-Q. Luo, and K. M. Wong, "Minimum BER block precoders for zero-forcing equalization," in *Proc. IEEE Int. Conf. Acoustics, Speech, and Signal Processing*, pp. 2261–2264, 2002.

[34] Y. Ding, T. N. Davidson, Z.-Q. Luo, and K. M. Wong, "Minimum BER block precoders for zero-forcing equalization," *IEEE Trans. Signal Process.*, vol. 51, pp. 2410-2423, Sept. 2003.

[35] M. Dong and L. Tong "Optimal design and placement of pilot symbols for channel estimation," *IEEE Trans. Signal Process.*, vol. 50, pp. 3055–3069, Dec. 2002.

[36] X. Dong, W.-S. Lu, and C. K. Soong, "Linear interpolation in pilot symbol assisted channel estimation for OFDM," *IEEE Trans. Wireless Commun.*, vol. 6, pp. 1910–1920, May 2007.

[37] I. Djokovic, "MMSE equalizers for DMT systems with and without crosstalk," *Conf. Record 31st Asilomar Conf. Signals, Systems, and Computers*, Nov. 1997, pp. 545–549.

[38] G. D. Dudevoir, J. S. Chow, S. Kasturia, and J. M. Cioffi, Vector coding for T1 transmission in the ISDN digital subscriber loop, in *Proc. IEEE Int. Conf. Communications*, pp. 536–540, June 1989.

[39] European Telecommunication Standard Institute, "Radio broadcasting systems; digital audio broadcasting (DAB) to mobile, portable and fixed receivers," ETS 300 401, 1997.

[40] European Telecommunication Standard Institute, "Digital video broadcasting; framing, structure, channel coding and modulation for digital terrestrial television (DVB-T)," ETS 300 744, 1997.

[41] D. Falconer, S. L. Ariyavisitakul, A. Benyamin-Seeyar, and B. Eidson, "Frequency domain equalization for single-carrier broadband wireless systems," *IEEE Commun. Mag.*, vol. 40, pp. 58–66, April 2002.

[42] B. Farhang-Boroujeny and M. Ding, "Design methods for time-domain equalizers in DMT transceivers," *IEEE Trans. Commun.*, vol. 49, pp. 554–562, March 2001.

[43] N. J. Fliege, *Multirate Digital Signal Processing: Multirate Systems–filter banks–wavelets.* Chichester: Wiley, 2000.

[44] G. B. Giannakis, "Filterbanks for blind channel identification and equalization," *IEEE Signal Process. Lett.*, vol. 4, pp. 184–187, June 1997.

[45] G. B. Giannakis, Y. Hua, P. Stoica, and L. Tong, *Signal Processing Advances in Wireless and Mobile Communications, Volume 1: Trends in Channel Estimation and Equalization.* Upper Saddle River, NJ: Prentice Hall PTR, 2000.

[46] G. B. Giannakis, Z. Wang, A. Scaglione, and S. Barbarossa, "AMOUR-generalized multicarrier transceivers for blind CDMA regardless of multipath," *IEEE Trans. Commun.*, vol. 48, pp. 2064–2076, Dec. 2000.

[47] G. Gu and E. F. Badran, "Optimal design for channel equalization via the filterbank approach," *IEEE Trans. Signal Process.*, vol. 52, pp. 536–545, Feb. 2004.

[48] S. H. Han and J. H. Lee, "An overview of peak-to-average power ratio reduction techniques for multicarrier transmission," *IEEE Wireless Commun.*, vol. 12, pp. 56–65, April 2005.

[49] F. J. Harris, *Multirate Signal Processing for Communication Systems.* Upper Saddle River, NJ: Prentice Hall PTR, 2004.

[50] S. Haykin, *Communication Systems.* Wiley, 2000.

[51] A. Hjørungnes, P. S. R. Diniz, and M. L. R. de Campos, "Jointly minimum BER transmitter and receiver FIR MIMO filters for binary signal vectors," *IEEE Trans. Signal Process.*, vol. 52, pp. 1021–1036, April 2004.

[52] R. A. Horn and C. R. Johnson, *Matrix Analysis.* Cambridge: Cambridge University Press, 1985.

[53] D. Huang and K. B. Letaief, "Carrier frequency offset estimation for OFDM systems using null subcarriers," *IEEE Trans. Commun.*, vol. 54, pp. 813–823, May 2006.

[54] IEEE, "Wireless LAN Medium Access Control (MAC) and Physical Layer (PHY) specifications: High-speed Physical Layer in the 5 GHz Band," IEEE Standard 802.11a, 1999.

[55] IEEE, "Air Interface for Fixed Broadband Wireless Access Systems", IEEE Standard 802.16, 2004.

[56] B.-J. Jeong and K.-H. Yoo, "Digital RFI canceller for DMT based VDSL," *Electron. Lett.*, vol. 34, pp. 1640–1641, Aug. 1998.

[57] A. E. Jones, T. A. Wilkinson, and S. K. Barton, "Block coding scheme for reduction of peak to mean envelope power ratio of multicarrier transmission scheme," *Electron. Lett.*, vol. 30, pp. 2098–2099, Dec. 1994.

[58] T. Kailath, *Linear Systems*, Upper Saddle River, NJ: Prentice Hall, Inc., 1980.

[59] I. Kalet, "The multitone channel," *IEEE Trans. Commun.*, vol. 37, pp. 119–124, Feb. 1989.

[60] I. Kalet, "Multitone modulation," in *Subband and Wavelet Transforms: Design and Applications*, A. N. Akansu and M. J. T. Smith, Eds., Boston, MA: Kluwer, 1995.

[61] S. Kapoor, and S. Nedic, "Interference suppression in DMT receivers using windowing," in *Proc. IEEE Int. Conf. Commun.*, 2000, pp. 778–782.

[62] T. Karp and G. Schuller, "Joint transmitter/receiver design for multicarrier data transmission with low latency time, in *Proc. IEEE Int. Conf. Acoustic, Speech, and Signal Processing*, May 2001, pp. 2401–2404.

[63] S. Kasturia, J. T. Aslanis, and J. M. Cioffi, "Vector coding for partial response channels," *IEEE Trans. Inform. Theory*, vol. 36, pp. 741–762, July 1990.

[64] S. M. Kay, *Fundamentals of Statistical Signal Processing, Volume I: Estimation Theory*, Englewood Cliffs, New Jersey: Prentice Hall PTR, 1993.

[65] S. M. Kay, *Fundamentals of Statistical Signal Processing, Volume II: Detection Theory*, Upper Saddle River, NJ: Prentice Hall PTR, 1998.

[66] A. V. Krishna and K. V. S. Hari, "Filter bank precoding for FIR equalization in high-rate MIMO communications," *IEEE Trans. Signal Process.*, vol. 54, pp. 1645–1652, June 2006.

[67] B. P. Lathi, *Modern Digital and Analog Communication Systems*, New York: Oxford University Press, 1998.

[68] N. Laurenti and L. Vangelista, "Filter design for the conjugate OFDM-OQAM system," in *1st Int. Workshop Image and Signal Processing and Analysis*, June 2000, pp. 267–272.

[69] J. W. Lechleider, "High bit rate digital subscriber lines: a review of HDSL progress," *IEEE J, Sel. Areas Commun.*, vol. 9, pp. 769–784, Aug. 1991.

[70] C.-C. Li and Y.-P. Lin "Receiver window designs for radio frequency interference suppression," in *Proc. European Signal Processing Conf.*, 2006.

[71] J. Li, G. Liu, and G. B. Giannakis, "Carrier frequency offset estimation for OFDM-based WLANs," *IEEE Signal Process. Lett.*, vol. 8, pp. 80–82, March 2001.

[72] T. Li and Z. Ding, "Joint transmitter-receiver optimization for partial response channels based on nonmaximally decimated filterbank precoding technique," *IEEE Trans. Signal Process.*, vol. 47, pp. 2407–2414, Sept. 1999.

[73] Y.-P. Lin and S.-M. Phoong, "Perfect discrete wavelet multitone modulation for fading channels," in *Proc. 6th IEEE Int. Workshop Intelligent Signal Processing and Communication Systems*, Nov. 1998.

[74] Y.-P. Lin and S.-M. Phoong, "ISI free FIR filterbank transceivers for frequency selective channels," *IEEE Trans. Signal Process.*, vol. 49, pp. 2648–2658, Nov. 2001.

[75] Y.-P. Lin and S.-M. Phoong, "Optimal ISI free DMT transceivers for distorted channels with colored noise," *IEEE Trans. Signal Process.*, vol. 49, pp. 2702–2712, Nov. 2001.

[76] Y.-P. Lin and S.-M. Phoong, "Minimum redundancy ISI free FIR filterbank transceivers, *IEEE Trans. Signal Process.*, vol. 50, pp. 842–853, April 2002.

[77] Y.-P. Lin and S.-M. Phoong, "Analytic BER comparison of OFDM and single carrier systems," in *Proc. Int. Workshop Spectral Methods and Multirate Signal Processing*, 2002.

[78] Y.-P. Lin and S.-M. Phoong, "BER minimized OFDM systems with channel independent precoders," *IEEE Trans. Signal Process.*, vol. 51, pp. 2369–2380, Sept. 2003.

[79] Y.-P. Lin and S.-M. Phoong, "OFDM transmitters: analog representation and DFT based implementation," *IEEE Trans. Signal Process.*, vol. 51, pp. 2450–2453, Sept. 2003.

[80] Y.-P. Lin and S.-M. Phoong, "Window designs for DFT based multicarrier system," *IEEE Trans. Signal Process.*, vol. 53, pp. 1015–1024, March 2005.

[81] Y.-P. Lin, Y.-P. Lin, and S.-M. Phoong, "A frequency domain based TEQ design for DSL systems," in *Proc. IEEE Int. Symp. Circuits and Systems*, May 2006, pp. 1575–1578.

[82] Y.-P. Lin, L.-H. Liang, P.-J. Chung, and S.-M. Phoong, "An eigen-based TEQ design for VDSL systems," *IEEE Trans. Signal Process.*, vol. 55, pp. 290–298, Jan. 2007.

[83] Y.-P. Lin, C.-C. Li, and S.-M. Phoong, "A filterbank approach to window designs for multicarrier systems," *IEEE Circuits Syst. Mag.*, pp. 19–30, first quarter, 2007.

[84] C. H. Liu, S.-M. Phoong, and Y.-P. Lin, "ISI-free block transceivers for unknown frequency selective channels," *IEEE Trans. Signal Process.*, vol. 55, pp. 1564–1567, April 2007.

[85] H. Liu and U. Tureli, "A high-efficiency carrier estimator for OFDM communications," *IEEE Commun. Lett.*, vol. 2, pp. 104–106, April 1998.

[86] R. W. Lowdermilk, "Design and performance of fading insensitive orthogonal frequency division multiplexing (OFDM) using polyphase filtering techniques," in *Conf. Record 30th Asilomar Conf. Signals, Systems and Computers*, Nov. 1996, pp. 674–678.

[87] R. K. Martin, J. Balakrishnan, W. A. Sethares, and C. R. Johnson Jr., "A blind, adaptive TEQ for multicarrier systems," *IEEE Signal Process. Lett.*, vol. 9, pp. 341–343, Nov. 2002.

[88] R. K. Martin, K. Vanbleu, M. Ding, G. Ysebaert, M. Milosevic, B. L. Evans, M. Moonen, and C. R. Johnson Jr., "Unification and evaluation of equalization structures and design algorithms for discrete multitone modulation systems," *IEEE Trans. Signal Process.*, vol. 53, pp. 3880–3894, Oct. 2005.

[89] K. Matheus and K.-D. Kammeyer, "Optimal design of a multicarrier systems with soft impulse shaping including equalization in time or frequency direction," in *Proc. IEEE Global Telecommunications Conf.*, Nov. 1997, vol. 1, pp. 310-314.

[90] P. J. W. Melsa, R. C. Younce, and C. E. Rohrs, "Impulse response shortening for discrete multitone transceivers," *IEEE Trans. Commun.*, vol. 44, pp. 1662–1672, Dec. 1996.

[91] U. Mengali and A. N. D'Andrea, *Synchronization Techniques for Digital Receivers*. New York: Plenum, 1997.

[92] A. Mertins, "MMSE design of redundant FIR precoders for arbitrary channel lengths," *IEEE Trans. Signal Process.*, vol. 51, pp. 2402–2409, Sept. 2003.

[93] H. Meyr, M. Moeneclaey, and S. A. Fechtel, *Digital Communication Receivers: Synchronization, Channel Estimation, and Signal Processing*. Wiley-Interscience, 1997.

[94] M. Milosevic, L. F. C. Pessoa, B. L. Evans, and R. Baldick, "DMT bit rate maximization with optimal time domain equalizer filter bank architecture," *Conf. Record 36th Asilomar Conf. Signals, Systems, and Computers*, Nov. 2002, vol. 1, pp. 377–382.

[95] P. H. Moose, "A technique for orthogonal frequency division multiplexing frequency offset correction," *IEEE Trans. Commun.*, vol. 42, pp. 2908–2914, Oct. 1994.

[96] B. Muquet, Z. Wang, G. B. Giannakis, M. de Courville, and P. Duhamel, "Cyclic prefixing or zero padding for wireless multicarrier transmissions?" *IEEE Trans. Commun.*, vol. 50, pp. 2136–2148, Dec. 2002.

[97] S. H. Müller-Weinfurtner, "Optimal Nyquist windowing in OFDM receivers," *IEEE Trans. Commun.*, vol. 49, pp. 417–420, March 2001.

[98] C. Muschallik, "Improving an OFDM reception using an adaptive Nyquist windowing," *IEEE Trans. Consum. Electron.*, vol. 42, pp. 259–269, Aug. 1996.

[99] M. Narasimha and A. Peterson, "Design of a 24-channel transmultiplexer," *IEEE Trans. Acoust., Speech, Signal Process.*, vol. 27, pp. 752–762, Dec. 1979.

[100] R. Negi and J. M. Cioffi, "Blind OFDM symbol synchronization in ISI channels," *IEEE Trans. Commun.*, vol. 50, pp. 1525–1534, Sept. 2002.

[101] R. Negi and J. Cioffi, "Pilot tone selection for channel estimation in a mobile OFDM System," *IEEE Trans. Consum. Electron.*, vol. 44, pp. 1122–1128, 1998.

[102] H. Nikookar and R. Prasad, "Optimal waveform design for multicarrier transmission through a multipath channel," in *Proc. IEEE Vehicular Technology Conf.*, May 1997, vol. 3, pp. 1812–1816.

[103] S. Ohno and G. B. Giannakis, "Optimal training and redundant precoding for block transmissions with application to wireless OFDM," *IEEE Trans. Commun.*, vol. 50, pp. 2113–2123, Oct. 2002.

[104] S. Ohno, "Performance of single-carrier block transmissions over multipath fading channels with linear equalization," *IEEE Trans. Signal Process.*, vol. 54, pp. 3678–3687, Oct. 2006.

[105] A. V. Oppenheim and R. W. Schafer, *Discrete-Time Signal Processing.* Upper Saddle River, NJ: Prentice Hall, 1999.

[106] D. P. Palomar, J. M. Cioffi, and M. A. Lagunas, "Joint Tx-Rx beamforming design for multicarrier MIMO channels: A unified framework for convex optimization," *IEEE Trans. Signal Process.*, vol. 51, pp. 2381–2401, Sept. 2003.

[107] D. P. Palomar, M. A. Lagunas, and J. M. Cioffi, "Optimum linear joint transmit-receive processing for MIMO channels with QoS constraints," *IEEE Trans. Signal Process.*, vol. 52, pp. 1179–1197, May 2004.

[108] D. P. Palomar, M. Bengtsson, and B. Ottersten, "Minimum BER linear transceivers for MIMO channels via primal decomposition," *IEEE Trans. Signal Process.*, vol. 53, pp. 2866–2882, Aug. 2005.

[109] A. Pandharipande and S. Dasgupta, "Optimum DMT-based transceivers for multiuser communications," *IEEE Trans. Commun.*, vol. 51, pp. 2038–2046, Dec. 2003.

[110] A. Pandharipande, and S. Dasgupta, "Optimum multiflow biorthogonal DMT with unequal subchannel assignment," *IEEE Trans. Signal Process.*, vol. 53, pp. 3572–3582, Sept. 2005.

[111] A. Papoulis and S. U. Pillai, *Probability, Random Variables and Stochastic Processes.* 4th ed. New York: McGraw-Hill Publishing Co., 2002.

[112] M. Park, H. Jun, J. Cho, N. Cho, D. Hong, and C. Kang, "PAPR reduction in OFDM transmission using Hadamard transform," in *Proc. IEEE Int. Conf. Communications*, June 2000, vol. 1, pp. 430–433.

[113] M. Pauli and P. Kuchenbecker, "On the reduction of the out-of-band radiation of OFDM-signals," in *Proc. IEEE Int. Conf. Communications*, June 1998, vol. 3, pp. 1304–1308.

[114] S. C. Pei and C. C. Tseng, "A new eigenfilter based on total least squares error criterion," *IEEE Trans. Circuits Syst. I*, vol. 48, pp. 699–709, June 2001.

[115] S. C. Pei and C. C. Tseng, "An efficient design of a variable fractional delay filter using a first-order differentiator," *IEEE Signal Process. Lett.*, vol. 10, pp. 307–310, Oct. 2003.

[116] A. Peled and A. Ruiz, "Frequency domain data transmission using reduced computational complexity algorithms," in *Proc. IEEE Int. Conf. Acoustics, Speech, and Signal Processing*, 1980, pp. 964–967.

[117] S.-M. Phoong, Y. Chang, and C.-Y. Chen, "DFT modulated filter bank transceivers for multipath fading channels," *IEEE Trans. Signal Process.*, vol. 53, pp. 182–192, Jan. 2005.

[118] T. Pollet, M. Van Bladel, and M. Moeneclaey, "BER sensitivity of OFDM systems to carrier frequency offset and Wiener phase noise," *IEEE Trans. Commun.*, vol. 43, pp. 191-193, 1995.

[119] H. V. Poor and S. Verdu, "Probability of error in MMSE multiuser detection," *IEEE Trans. Inf. Theory*, vol. 43, May 1997.

[120] J. G. Proakis, *Digital Communications.* New York: McGraw-Hill, 2001.

[121] R. Ramesh, "A multirate signal processing approach to block decision feedback equalization," in *IEEE Global Telecommunications Conf.*, Dec. 1990, vol. 3, pp. 1685–1690.

[122] D. J. Rauschmayer, *ADSL/VDSL Principles: A Practical and Precise Study of Asymmetric Digital Subscriber Lines and Very High Speed Digital Subscriber Lines*, Macmillan Technical Publishing, 1999.

[123] A. J. Redfern, "Receiver window design for multicarrier communication systems," *IEEE J. Sel. Areas Commun.*, vol. 20, pp. 1029–1036, June 2002.

[124] C. B. Ribeiro, M. L. R. de Campos, and P. S. R. Diniz, "FIR equalizers with minimum redundancy," in *Proc. IEEE Int. Conf. Acoustics, Speech, and Signal Processing*, 2002, vol. III, pp. 2673–2676.

[125] J. Rinne and M. Renfors, "Pilot spacing in orthogonal frequency division multiplexing systems," *IEEE Trans. Consumer Electron.*, vol. 42, pp. 959–962, 1996.

[126] Y. Rong, S. A. Vorobyov, and A. B. Gershman, "Linear block precoding for OFDM systems based on maximization of mean cutoff rate," *IEEE Transactions on Signal Process.*, vol. 53, pp. 4691–4696, Dec. 2005.

[127] B. R. Saltzberg, "Performance of an efficient parallel data transmission system," *IEEE Trans. Commun. Tech.*, vol. 15, pp. 805–811, Dec. 1967.

[128] H. Sari, G. Karam, and I. Jeanclaude, "An analysis of orthogonal frequency-division multiplexing for mobile radio applications," in *IEEE Vehicular Technology Conf.*, June, 1994, pp. 1635–1639.

[129] H. Sari, G. Karam, and I. Jeanclaude, "Frequency-domain equalization of mobile radio and terrestrial broadcast channels," in *Proc. IEEE Global Telecommunications Conf.*, 1994, vol. 1, pp. 1–5.

[130] H. Sari, G. Karam, and I. Jeanclaude, "Transmission techniques for digital terrestrial TV broadcasting," *IEEE Commun. Mag.*, vol. 3, pp. 100–109, Feb. 1995.

[131] V. P. Sathe and P. P. Vaidyanathan, "Effects of multirate systems on the statistical properties of random signals," *IEEE Trans. Signal Process.*, vol. 41, pp. 131–146, Jan. 1993.

[132] A. Scaglione and G. B. Giannakis, "Design of user codes in QS-CDMA systems for MUI elimination in unknown multipath," *IEEE Commun. Lett.*, vol. 3, pp. 25–27, Feb. 1999.

[133] A. Scaglione, S. Barbarossa, and G. B. Giannakis, "Fading-resistant and MUI-free codes for CDMA systems," in *Proc. IEEE Int. Conf. Acoustic, Speech, and Signal Processing*, March 1999, pp. 2687–2690.

[134] A. Scaglione, S. Barbarossa, and G. B. Giannakis, "Filterbank transceivers optimizing information rate in block transmissions over dispersive channels," *IEEE Trans. Inf. Theory*, vol. 45, pp. 1019–1032, April 1999.

[135] A. Scaglione, G. B. Giannakis, and S. Barbarossa "Redundant filterbank precoders and equalizers part I: unification and optimal designs," *IEEE Trans. Signal Process.*, vol. 47, pp. 1988–2006, July 1999.

[136] T. M. Schmidl and D. C. Cox, "Robust frequency and timing synchronization for OFDM," *IEEE Trans. Commun.*, vol. 45, pp. 1613–1621, Dec. 1997.

[137] M. K. Simon, S. M. Hinedi, W. C. Lindsey, *Digital Communication Techniques: Signal Design and Detection*. Upper Saddle River, NJ: Prentice Hall, 1994.

[138] P. Siohan, C. Siclet, and N. Lacaille, "Analysis and design of OFDM/OQAM systems based on filterbank theory," IEEE *Trans. Signal Process.*, vol. 50, pp. 1170–1183, May 2002.

[139] F. Sjöberg, M. Isaksson, P. Deutgen, R. Nilsson, P. Ödling, and P. O. Börjesson, "Performance evaluation of the Zipper duplex method," in *Proc. IEEE Int. Conf. Communications*, 1998, pp. 1035–1039.

[140] F. Sjöberg, R. Nilsson, P. O. Börjesson, P. Ödling, B. Wiese, and J. A. C. Bingham, "Digital RFI suppression in DMT-based VDSL systems," *IEEE Trans. Circuits Syst. I, Reg. Papers*, vol. 51, pp. 2300–2312, Nov. 2004.

[141] S. B. Slimane, "Performance of OFDM systems with time-limited waveforms over multipath radio channels," in *Proc. IEEE Global Telecommunications Conf.*, 1998, pp. 962–967.

[142] X. Song and S. Dasgupta, "Optimum ISI-free DMT systems with integer bit loading and arbitrary data rates: when does orthonormality suffice?" in *Proc. IEEE Int. Conf. Acoustic, Speech, and Signal Processing*, May 2006, vol. 4, pp. 117–120.

[143] P. Spruyt, P. Reusens, and S. Braet, "Performance of improved DMT transceiver for VDSL," ANSI T1E1.4, Doc. 96-104, Apr. 1996.

[144] T. Starr, J. M. Cioffi, and P. J. Silvermann, *Understanding Digital Subscriber Line Technology*. Upper Saddle River, NJ: Prentice Hall, 1999.

[145] T. Starr, M. Sorbara, J. M. Cioffi, and P. J. Silvermann, *DSL Advances*. Upper Saddle River, NJ: Prentice Hall, 2003.

[146] G. Strang and T. Nguyen, *Wavelets and Filter Banks*. Wellesley, MA: Wellesley-Cambridge Press, 1996.

[147] B. Su and P. P. Vaidyanathan, "Subspace-based blind channel identification for cyclic prefix systems using few received blocks," *IEEE Trans. Signal Process.*, vol. 55, pp. 4979–4993, Oct. 2007.

[148] J. Tellado and J. M. Cioffi "Efficient algorithms for reducing PAR in multicarrier systems," in *Proc. IEEE Int. Symp. Information Theory*, 1998, pp. 191.

[149] J. Tellado, *Multicarrier Modulation with Low PAR: Applications to DSL and Wireless*. Kluwer Academic Press, 2000.

[150] C. Tepedelenlioglu and R. Challagulla, "Low-complexity multipath diversity through fractional sampling in OFDM," *IEEE Trans. Signal Process.*, vol. 52, pp. 3104–3116, Nov. 2004.

[151] A. Tkacenko, P. P. Vaidyanathan, and T. Q. Nguyen, "On the eigenfilter design method and its applications: a tutorial," *IEEE Trans. Circuits Syst. II, Analog Digit. Signal Process.*, vol. 50, pp. 497–517, Sept. 2003.

[152] A. Tkacenko and P. P. Vaidyanathan, "A low-complexity eigenfilter design method for channel shortening equalizers for DMT systems," *IEEE Trans. Commun.*, vol. 51, pp. 1069–1072, July 2003.

[153] L. Tong, B. M. Sadler, and M. Dong, "Pilot assisted wireless transmissions: general model, design criteria, and signal processing," *IEEE Signal Process. Mag.*, vol. 21, pp. 12–25, Nov. 2004.

[154] J. C. Tu, J. S. Chow, G. P. Dudevoir, and J. M. Cioffi, "Crosstalk-limited performance of a computationally efficient multichannel transceiver for high rate digital subscriber lines," in *Proc. IEEE Global Telecommunications Conf.*, 1989, vol. 3, pp. 1940–1944.

[155] A. Vahlin and N. Holte, "Optimal finite duration pulses for OFDM," *IEEE Trans. Commun.*, vol. 44, pp. 10–14, Jan. 1996.

[156] P. P. Vaidyanathan and T. Q. Nguyen, "Eigenfilters: a new approach to least squares FIR filter design and applications including Nyquist filters," *IEEE Trans. Circuits Syst.*, vol. 34, pp. 11–23, Jan. 1987.

[157] P. P. Vaidyanathan and S. K. Mitra, "Polyphase networks, block digital filtering, LPTV systems, and alias-free QMF banks: a unified approach based on pseudocirculants, " *IEEE Trans. Acoust., Speech, Signal Processs.*, vol. 36, pp. 381–391, March 1988.

[158] P. P. Vaidyanathan, "How to capture all FIR perfect reconstruction QMF banks with unimodular matrices?" in *Proc. IEEE Int. Symp. Circuits and Systems*, May 1990, pp. 2030–2033.

[159] P. P. Vaidyanathan, *Multirate Systems and Filter Banks*. Englewood Cliffs, NJ: Prentice-Hall, 1993.

[160] P. P. Vaidyanathan and B. Vrcelj, "Biorthogonal partners and applications," *IEEE Trans. Signal Process.*, vol. 49, pp. 1013–1028, May 2001.

[161] P. P. Vaidyanathan, "Filter banks in digital communications," *IEEE Circuits Syst. Mag.*, vol. 1, pp. 4–25, 2001.

[162] P. P. Vaidyanathan and B. Vrcelj, "Transmultiplexers as precoders in modern digital communication: a tutorial review," in *Proc. IEEE Int. Symp. Circuits and Systems*, May 2004, vol. 5, pp. 405–412.

[163] P. P. Vaidyanathan, "On equalization of channels with ZP precoders," in *IEEE Int. Symp. on Circuits and Systems*, 2007, pp. 1329–1332.

[164] P. P. Vaidyanathan, *The Theory of Linear Prediction*. Morgan and Claypool Publishers, 2008.

[165] K. Vanbleu, G. Ysebaert, G. Cuypers, M. Moonen, and K. Van Acker, "Bitrate maximizing time-domain equalizer design for DMT-based systems," *IEEE Trans. Commun.*, vol. 52, pp. 871–876, June 2004.

[166] J.J. Van de Beek, M. Sandell, and P. O. Börjesson, "ML estimation of time and frequency offset in OFDM systems," *IEEE Trans. Signal Process.*, vol. 45, pp. 1800–1805, July 1997.

[167] M. Vemulapalli, S. Dasgupta, and A. Pandharipande, "A new algorithm for optimum bit loading with a general cost," in *Proc. IEEE Int. Symp. on Circuits and Systems*, May 2006, pp. 21–24.

[168] M. Vetterli, "Perfect transmultiplexers," in *Proc. Int. Conf. on Acoustic, Speech, and Signal Processing*, April 1986, pp. 2567–2570.

[169] M. Vetterli and J. Kovacevic, *Wavelets and Subband Coding*. Englewood Cliffs, NJ: Prentice Hall PTR, 1995.

[170] A. Viholainen, T. H. Stitz, J. Alhava, T. Ihalainen, and M. Renfors, "Complex modulated critically sampled filter banks based on cosine and sine modulation," in *Proc. IEEE Int. Symp. Circuits and Systems*, 2002, pp. 833–836.

[171] B. Vrcelj and P. P. Vaidyanathan, "MIMO biorthogonal partners and applications," *IEEE Trans. Signal Process.*, vol. 50, pp. 528–543, March 2002.

[172] B. Vrcelj and P. P. Vaidyanathan, "Fractional biorthogonal partners in channel equalization and signal interpolation," *IEEE Trans. Signal Process.*, vol. 51, pp. 1928–1940, July 2003.

[173] T. Walzman and M. Schwartz, "Automatic equalization using the discrete frequency domain," *IEEE Trans. Inf. Theory*, vol. 19, pp. 59–68, Jan. 1973.

[174] C.-L. Wang, C.-H. Chang, J. L. Fan, and J. M. Cioffi, "Discrete hartley transform based multicarrier modulation," in *Proc. IEEE Int. Conf. Acoustics, Speech, and Signal Processing*, 2000, pp. 2513–1516.

[175] Z. Wang, X. Ma, and G. B. Giannakis, "OFDM or single-carrier block transmissions?" *IEEE Trans. Commun.*, vol. 52, pp. 380–394, March 2004.

[176] C.-C. Weng and P. P. Vaidyanathan, "Joint optimization of transceivers with fractionally spaced equalizers," in *Proc. IEEE Int. Conf. Acoustics, Speech and Signal Processing*, March 2008, pp. 2913–2916.

[177] C.-C. Weng and P. P. Vaidyanathan, "Joint optimization of transceivers with decision feedback and bit loading," in *Proc. 42nd IEEE Asilomar Conf. on Signals, Systems, and Computers*, Oct. 2008, pp. 1310–1314.

[178] S. Weinstein and P. Ebert, "Data transmission by frequency-division multiplexing using the discrete Fourier transform," *IEEE Trans. Commun. Tech.*, vol. 19, pp. 628–634, Oct. 1971.

[179] J. W. Woods and H. Stark , *Probability and Random Processes with Applications to Signal Processing*, 3rd ed. Upper Saddle River, NJ: Prentice Hall, 2001.

[180] Y.-W. Wu, Y.-P. Lin, C.-C. Li, and S.-M. Phoong "Design of time domain equalizers incorporating radio frequency interference suppression," in *Proc. IEEE Asia Pacific Conf. Circuits and Systems*, Dec. 2006, pp. 446–449.

[181] X.-G. Xia, "Applications of nonmaximally decimated multirate filterbanks in partial response channel ISI cancellation," in *Proc. IEEE-SP Int. Symp. Time-Frequency and Time-Scale Analysis*, June 1996, pp. 65–68.

[182] X.-G. Xia, "New precoding for intersymbol interference cancellation us-
      ing nonmaximally decimated multirate filterbanks with ideal FIR equal-
      izers," *IEEE Trans. Signal Process.*, vol. 45, pp. 2431–2441, 1997.

[183] X.-G. Xia, W. Su, and H. Liu, "Filterbank precoders for blind equaliza-
      tion: polynomial ambiguity resistant precoders (PARP)," *IEEE Trans.
      Signal Process.*, vol. 48, pp. 193–209, Feb. 2001.

[184] X.-G. Xia, "Precoded and vector OFDM robust to channel spectral
      nulls and with reduced cyclic prefix length in single transmit antenna
      systems," *IEEE Trans. Commun.*, vol. 49, pp. 1363–1374, Aug. 2001.

[185] X.-G. Xia, *Modulated Coding for Intersymbol Interference Channels.*
      New York: Marcel Dekker, Inc., 2001.

[186] A. Yasotharan, "Multirate zero-forcing Tx-Rx design for MIMO channel
      under BER constraints," *IEEE Trans. Signal Process.*, vol. 54, pp. 2280–
      2301, June 2006.

[187] P. Yip and K. R. Rao, "Fast Discrete Transforms," in *Handbook of
      Digital Signal Processing*, D. F. Elliott, Ed. San Diego, CA: Academic
      Press, 1987.

[188] G. Ysebaert, K. Vanbleu, G. Cuypers, and M. Moonen, "Joint window
      and time domain equalizer design for bit rate maximization in DMT-
      receivers," *IEEE J. Sel. Areas Commun.*, vol. 20, pp. 1029–1036, June
      2002.

[189] J. Zhang, E. K. P. Chong, and D. N. C. Tse, "Output MAI distribution
      of linear MMSE multiuser receivers in DS-CDMA system," *IEEE Trans.
      Inf. Theory*, vol. 47, pp. 1128–1144, March 2001.

[190] J. Zhang and E. K. P. Chong, "Linear MMSE multiuser receivers: MAI
      conditional weak convergence and network capacity," *IEEE Trans. Inf.
      Theory*, vol. 48, pp. 2114–2122, July 2002.

Index

Printed in the United States
By Bookmasters